NEURAL SYSTEMS: ANALYSIS AND MODELING

NEURAL SYSTEMS: ANALYSIS AND MODELING

Edited by

Frank H. Eeckman
Lawrence Livermore National Laboratory

SPRINGER SCIENCE+BUSINESS MEDIA, LLC

Library of Congress Cataloging-in-Publication Data

Neural systems: analysis and modeling / edited by Frank H. Eeckman.
p. cm.
"The collected papers of the 1991 Conference on Analysis and Modeling of Neural Systems (AMNS) and the papers presented at the satellite symposium on compartmental modeling, held July 23-26, in San Francisco, California" -- Introd.
Includes bibliographical references and index.
ISBN 978-1-4613-6581-5 ISBN 978-1-4615-3560-7 (eBook)
DOI 10.1007/978-1-4615-3560-7
1. Neural circuitry--Computer simulation--Congresses. 2. Neural networks (Computer science)--Congresses. I. Eeckman, Frank H.
QP363.3.N447 1992
591.1'88--dc20 92-18660
CIP

Originally published by Kluwer Academic Publishers in 1993
Softcover reprint of the hardcover 1st edition 1993

Printed on acid-free paper.

TABLE OF CONTENTS

Part II - Sensory Systems

INTRODUCTION

This volume contains the collected papers of the 1991 Conference on Analysis and Modeling of Neural Systems (AMNS), and the papers presented at the satellite symposium on compartmental modeling, held July 23-26, in San Francisco, California.

Analysis and Modeling 1991 is the second and, for the time being, the last conference of its kind in California. Next year, 1992, it will be a bigger meeting, the format will be different, and it will be called Computation and Neural Systems 1992 (CNS*92).

Analysis and Modeling of Neural Systems was conceived by Nora Smiriga, Muriel Ross, Edwin Lewis and John Miller. D. Curtis Deno and I joined the organizing committee a few weeks later. In 1991, Edward Keller replaced Curtis Deno.

In 1990, Analysis and Modeling listed 21 invited talks and 18 in 1991. The oral presentations were by invitation, submissions were accepted for posters only. In 1990, there were 36 poster presentations and in 1991 there were 30. The decrease in the number of presentations was intentional in order to make the schedule less demanding. However, AMNS 1991 featured an extra day of tutorials and a satellite symposium on software development for compartmental modeling.

I had hoped to get these proceedings out by the end of January, 1992, but various unforseen circumstances got in the way and made it impossible to meet that deadline. I thank Thomas Stone and Stephanie Faulkner at Kluwer for their help and understanding.

This book is organized in the same way as the first Analysis and Modeling. There are three main parts: Techniques, Sensory Systems, and Motor Systems, each of these comprised of two or more sections. Each section has its own introduction, listing all the papers it contains; cross references to other papers on related topics in other sections are also provided. Key areas of focus in 1991 were development, signal processing techniques and modeling tools.

Chris Ghinazzi and Debora Williams did a great job in coordinating the 1991 meeting and the special event banquet.

I would like to thank Tim Axelrod, Joachim Buhmann, Jim Barbieri, D. Curtis Deno, Edward Keller, Barbara Liepe, Edwin Lewis, W. Geoffrey Owen, John Miller, Muriel Ross, Josef Skrzypek, and Nora Smiriga for their help and assistance.

PART I

ANALYSIS AND MODELING TOOLS AND TECHNIQUES

SECTION 1

SECTION 2

SECTION 1

ANALYSIS

The analysis section concentrates on data and mathematical analysis. In this section, the authors have successfully applied techniques for signal processing theory, information theory, and dynamical systems theory to analyze biological neural systems.

William Bialek gives an overview of the evidence for optimal real-time processing in the nervous system. The topics range from fly vision to bat echolocation.

Fred Rieke et al. discuss information capacity and coding efficiency in sensory neurons from the point of view of the organism.

Michael Lyons and Lionel Harrison analyze a class of models for the formation of ocular dominance stripes using an approximate technique for solving the nonlinear equations. The results indicate that stripe formation in the visual cortex is the result of the bilateral symmetry of the influence of the two eyes. See also **Ken Miller** in the Vision section.

Bill Baird and Frank Eeckman illustrate that a hierarchical sensory-motor network can be constructed out of oscillatory modules, whose interconnections contain higher order synapses. The model is presented in the framework of dynamical systems theory.

Bruce Parnas and Edwin Lewis present an integrate-and-fire spike model with an intrinsic noise source that can mimic the spike-response characteristics of many classes of neurons.

Frederic Theunissen et al. describe an approach to linearization of spiking equations that uses both noise and shunting currents.

Optimal Real-Time Signal Processing in the Nervous System

William Bialek

NEC Research Institute
4 Independence Way
Princeton, New Jersey 08540

Introduction

Most of us share the qualitative impression that the nervous system is a remarkable computational device. Part of the evidence for this belief is the tremendous difficulty in imitating neural functions using man-made hardware — we are a long way from having machines which see or hear. Yet it is easy to find tasks where even rather modest modern computers far outperform relatively sophisticated animals. What then is so special about neural computation?

Sensory signal processing provides a favorable case in which to quantify the performance of the nervous system and thus put some precision into our qualitative impressions. Any sensory system consists of measuring devices, encoders, and several layers of processing circuitry. At each stage of sensory processing there are fundamental physical and information theoretic limits to performance. If we understand the task facing the organism, we can thus construct a theory of the optimal or ideal sensory system that accomplishes the biologically relevant task with a precision or efficiency which reaches the physical limit. It is a remarkable fact that many real biological sensory systems reach these fundamental limits to their performance [14,15].

In this short review I will give an overview of the theory for the optimal design of sensory systems, at each stage giving examples of how the theory connects with the observed properties of real neural systems. I hope to convince you that we have the outlines of truly predictive theory for neural computation in the first few layers of the sensory systems. These ideas are drawn from work done in collaboration with L. Kruglyak, M. Landolfa, E. R. Lewis, J. P. Miller, K. Moortgat, W. G. Owen, F. Rieke, D. Ruderman, R. R. de Ruyter van Steveninck, D. Warland, W. Yamada, and A. Zee.

Receptor sampling

In designing a sensory system our first problem is to arrange a limited number of receptor cells to provide an effective sampling of the world. The best studied example of this problem is provided by the insect compound eye, which is even discussed Feynman's undergraduate lectures on physics [24]. If the individual facets of the compound eye have a size d, then by geometrical optics the bee divides the world into pixels of angular dimension $\delta\phi \sim d/R$ radians, where R is the eye radius. Making the facets smaller improves the angular resolution until geometrical optics breaks down and diffraction takes over. Then if we have a physical aperture of linear dimension d the effective aperture is of angular subtense $\delta\phi \sim \lambda/d$, where λ is a typical wavelength of the light being used to form the image. To optimize the angular resolution (minimize $\delta\phi$) one must seek a compromise between these different regimes, which predicts that $d \sim \sqrt{\lambda R}$. Feynman notes that this is prediction is correct within ten percent for the species of bee he was looking at.

Feynman's discussion of the compound eye has antecedents which can traced back into the last century [35]. By 1952 Barlow had surveyed some twenty-seven different insect eyes, finding that the square-root relation between facet size and eye size holds, to a good approximation, over a factor of 25 variation in R [4]. But there are examples of insects which apparently undersample the world by as much as a factor of three, and compound eyes of creatures dwelling in murky waters can have facets as much as 100 times as large as expected from the diffraction limit. Perhaps physicists are just too simple-minded, and biology is governed by the historical accidents of evolution rather than by any clean theoretical principles. But perhaps the physics of the problem is more subtle. Specifically, is the maximization of angular resolution really the relevant design criterion for an imaging system?

When it's dark an insect which divides the world into tiny pixels suffers the problem that each pixel captures only a few photons, leading to a very limited number of "grey levels" which can be distinguished at each sample point. If the aperture of each photoreceptor is enlarged the angular resolution is degraded but the intensity resolution within each pixel is improved. Similarly, if the receptor cells have limited time resolution and the insect flies at high speed, large numbers of tiny pixels would carry redundant information because of motion blur. Thus there should be some tradeoff between angular resolution and intensity resolution which is governed by the overall light intensity (photon arrival rate), the time resolution and noise characteristics of the receptors, the dynamics of the insect's flight, and the statistical structure of the images one is attempting to capture.

Snyder, Stavenga, and Laughlin [55-57] suggested that the tradeoff of angular resolution and signal-to-noise could be quantified by optimizing the eye's information capacity. This approach results in formal versions of the intuitive prediction that insects flying slowly in bright light should reach the diffraction limit, while quick night-flyers should open up their apertures. There is a critical light intensity below which diffraction-limited sampling ceases to be the optimal solution, and this critical intensity is near the level of moderate daylight. In the application of these results to the well-studied case of *Musca domestica,* the relatively high angular velocities of the bug play a crucial role in the correct prediction of undersampling.

I think one should view these results as semi-quantitative. A quantitative treatment requires more careful discussion of receptor time resolution in relation to the measurable signal and noise spectra, as well as some appreciation of the statistics of natural images. Nonetheless the work of Snyder *et al.* is crucial, for it teaches us that a single theoretical principle can predict a diversity of function in organisms adapted to different problems.

The auditory system also faces an interesting sampling problem, as discussed by Zweig *et al* [61]. Given a collection of transducer (hair) cells, the ear divides the range of interesting frequencies among these cells, in some cases using the mechanical tuning properties of accessory structures to insure that each cell only receives a limited frequency band. In mammals this external mechanical tuning is evident in the basilar membrane, which supports a "traveling wave" along its length; tuning is achieved by the fact that the local mechanical properties of the membrane vary with distance along the membrane [60]. If the mechanical properties vary too slowly, the system cannot focus different frequency bands onto different hair cells; if mechanical properties vary too rapidly one will build a mirror for the traveling wave. The natural compromise is to operate almost at the threshold for wave reflection. Analysis of experiments on basilar membrane motion suggests that the mammalian ear indeed operates near this point. On the other hand, de Boer and MacKay have suggested that one can build completely reflectionless systems which achieve arbitrarily precise focusing [20]. There are, however, subtle mathematical points in comparing these different points of view; I think there is an important open problem here regarding the relevance of the "cochlear compromise" to the design of the inner ear.

Analog pre-processing

Receptor cells are inevitably noisy. In case of vision, noise must be a serious problem for any organism which is active in the darkest times of night. In many organisms the dark-adapted retina is capable of providing reliable responses to individual photons [10,12]. Entire organisms, including people, are capable of giving behavioral responses to small numbers of photons, and the statistics of these responses reflect the statistics of photon arrivals at the retina [28,49]. Any photodetector has a level of dark noise which determines the reliability of photon counting, and many years ago Barlow suggested that the dark noise of the photoreceptors would limit the reliability of perception in the dark-adapted state [6]; clearly the dark noise of human observers [49] cannot be less than the receptor dark noise. Baylor and co-workers have measured the dark noise levels in photoreceptors from several different species, including the macaque monkey whose visual system is so close to our own [11,12], and the dark noise level of the entire human observer is in fact equal to the dark noise level of the receptors.

Barlow's suggestion has been confirmed beautifully in measurements on the behavioral and neural responses of a frog, where it possible to vary the temperature over a range corresponding to more than order of magnitude in receptor cell dark noise levels. Aho *et al.* [1] demonstrate that the "output" (behavior or retinal ganglion cell) noise level follows the temperature dependence measured for the "input" (receptor cell) noise; if one extrapolates to 37°C the data match those for monkeys and humans. These observations strongly suggest that under dark-adapted conditions — where the signal-to-noise ratio for vision is by definition low — organisms make optimal use of the photoreceptor signals, extracting all the information relevant to their behavioral tasks and adding little if any noise in the process [9,23]. What is the structure of such an optimal processor for photoreceptor signals?

A collection of photoreceptors receives photons at a time-varying rate $R(t)$. This is the rate of a random Poisson process, with the photons themselves arriving at times t_μ. These arrivals are signaled by current pulses of stereotyped waveform $I_0(t - t_\mu)$, but these pulses are embedded in a background of noise $\delta I(t)$, with spectral density $N(\Omega)$. The receptor current

$$I(t) = \sum_\mu I_0(t - t_\mu) + \delta I(t). \tag{1}$$

The simplest problem of computational vision is just to estimate the light intensity — the time-varying photon arrival rate $R(t)$ — from the current trace. We are especially interested in the limit where photons arrive only rarely (it's dark outside), so the signal-to-noise ratio is low.

Rieke, Owen and I [16,42] have approached this problem of optimal estimation in two ways. First, in the spirit of Wiener's original discussion [59], we try to find a filter $f(\tau)$ such that if we pass the current through this filter the result is the best possible estimator of $R(t)$ in the least-square sense. The result, at low SNR, is that the optimal filter

$$\tilde{f}(\Omega) = \int d\tau e^{+i\Omega\tau} f(\tau) = S_R(\omega)\frac{\tilde{I}_0^*(\Omega)}{N(\Omega)}. \tag{2}$$

This optimal filter for the reconstruction of the light intensity has two pieces. First there is a term $\tilde{I}_0^*(\Omega)/N(\Omega)$ which is determined entirely by the signal and noise characteristics of the receptor cell. It is a "matched filter" which serves to insure that the output amplitude at each frequency reflects the signal-to-noise ratio at that frequency. The second term, $S_R(\Omega)$, is the power spectrum of the fluctuations in photon arrival rate, and incorporates our *a priori* knowledge about the signals which will be impinging upon the receptor cell in a given natural or experimental situation.

The decomposition of the reconstruction filter into a matched filter and an ensemble-specific filter can be seen more generally using a second, probabilistic point of view. Everything that we know about the signal by virtue of observing the receptor current is contained in the conditional probability distribution $P[R(t)|I(t)]$. Any estimate of what is happening in the outside world is controlled by the structure of this distribution. The possible signals $R(t)$ live in a high dimensional space, for example the space of all the

Fourier coefficients needed to specify the function. Before we make any measurements on the receptor current, we know only that $R(t)$ is chosen from some *a priori* distribution $P[R(t)]$ which we imagine as a cloud of points in this high dimensional space. Once we observe the current $I(t)$ this cloud is shifted and possibly narrowed around our best estimate of the real signal. The problem is analogous to what happens to a collection of molecules as they diffuse. At first we have a cloud in random positions, but if we apply an external force, e.g. in electrophoresis or centrifugation, the cloud takes a definite mean trajectory and narrows around the position of minimum energy. To understand the problem of estimating the signal in a noisy background we have to understand this effective "force" on the distribution of possible waveforms and its relation to our observations on the receptor current.

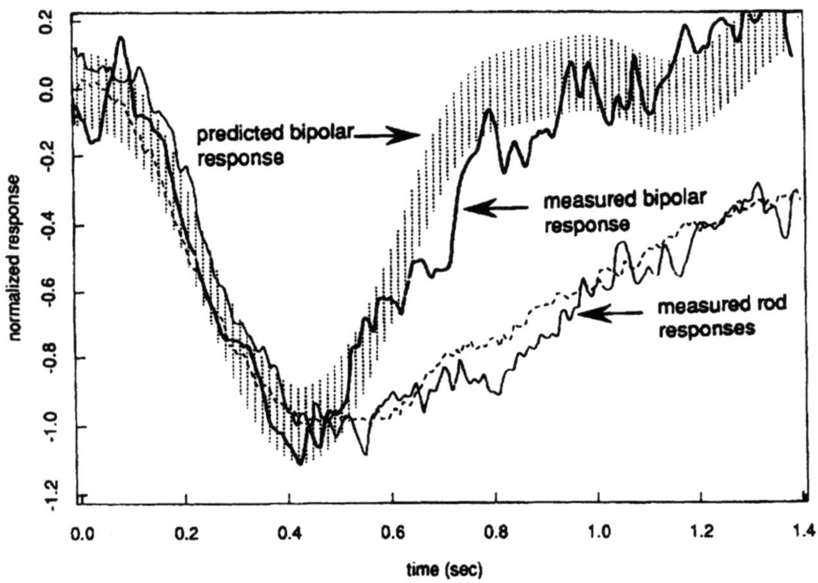

Figure 1: Bipolar cell impulse response and the first stage of optimal processing. Hashed lines show parameter-free prediction of the bipolar cell response to a dim flash at $t = 0$, assuming that bipolar cell voltage is proportional to the ouput of the matched filter as described in the text. Solid line shows the actual impulse response of a bipolar cell in the tiger salamander retina under dark-adapted conditions. For comparison, two records of rod cell voltage responses are also shown. Details are given in Refs. [16,42].

In the low SNR limit we have found that the effective force which determines our estimate of the stimulus waveform is just proportional to the output of the matched filter. Thus *the optimal estimate of any signal in the low photon flux limit is determined by the receptor currents passed through the matched filter.* The matched filter is the *universal* first stage in optimal signal processing at low photon flux. We know that in the vertebrate retina all photoreceptor signals pass through the bipolar cells, and this is anatomically the first stage of signal processing. If the first mathematical stage of optimal signal processing can be identified with the first anatomical stage, then we have a parameter-free theory for the dynamics of the bipolar cells! This is not a "model" for the bipolar cell; there is nothing in this prediction which incorporates the mechanisms in real bipolar cells. Optimal processing provides us with a theory of what the nervous system *should* compute rather than a highly parameterized model of how it computes.

In Fig. 1 we see the comparison between theory and experiment, with data from the tiger salamander retina. The agreement is excellent, and as far as we know this is the first case in which it has been possible to give a truly predictive theory for the computational dynamics of a neuron.

In the analysis of pre-processing at low signal-to-noise ratio we have isolated a stage of processing which consists of a (matched) linear filter. But no cell is linear in response to all inputs, and signals in the natural world do not become arbitrarily large. Is there an optimal strategy for mapping the limited range of natural signals onto the limited dynamic range of the cell? Laughlin [31] suggested that on information theoretic grounds one would like to map equally probable ranges of the input signal into equal ranges of the output variable. If this is correct then a cell's normalized input/output relation should be proportional to the cumulative probability distribution for the natural input signals.

Laughlin was interested in the second order cells of the fly retina, the large monopolar cells. To characterize the statistics of natural stimuli he synthesized an aperture for a photodetector with dimensions corresponding to the angular subtense of the cell's receptive field, then measured the distribution of contrast as seen through this aperture in a reasonably natural scene. The saturation behavior of the cell was characterized by measuring the peak voltage response to contrast pulses against a fixed mean background illumination. The results showed good agreement with the predictions from a hypothesis of "matched coding." Clearly one needs to make this all a bit more precise: What is the correct generalization to time-dependent signals? Do saturation characteristics adapt to different scene statistics, or is there some universal form for the contrast distribution in natural scenes? Does one need to take account of signal-dependent noise in constructing the matched saturation characteristics? These are interesting open questions which can be addressed with existing experimental techniques.

Efficient coding

Almost all sensory systems encode incoming data in trains of identical action potentials. The sensory data is usually a continuous function of time such as sound pressure or light intensity, chemical concentration, ... The sensory neuron performs a discretization of these signals, in which the only available output information is the occurance time of each spike. While a great deal is known about the phenomenology of this encoding in different sensory systems, relatively little is understood about the structure of the code from an information theoretic point of view.

Various authors have suggested that the limited dynamic range of spiking neurons forces the system to throw away information, signaling only features which may be of special interest to the animal. The best-known statement of this approach is contained in the work of Lettvin *et al.* [32], although the basic idea of feature detection appears much earlier in Barlow's work [5]. By the time of the Lettvin paper, and of Hubel and Wiesel's discovery of feature selectivity in mammalian cortex, Barlow had gone on to suggest a very different point of view [7]. Concerned about the fact that photoreceptors vastly outnumber retinal ganglion cells in the cat, he suggested that the array of ganglion cells must provide a more economical encoding of the visual world than would be given by transmitting the receptor voltage signals. This is possible only because signals in the real world have a high degree of redundancy or correlation. The transformations which the nervous system performs before the generation of spikes must serve to reduce redundancy, and qualitatively it is clear that lateral inhibition and light adaptation both serve such a function.

Redundancy reduction cannot be the whole story, since at low signal-to-noise ratios redundancy has the positive effect of reinforcing the real signal against the incoherent noise background. Atick and Redlich formulate the problem of minimizing redundancy while holding fixed the magnitude of the information transfer, which allows for interpolation between the low- and high-SNR regimes [2]. If one identifies this variation in SNR with the variation in background light intensity, it is possible to account for the form of the ganglion cell receptive field as a function of adaptation level. Atick, Li and Redlich extend this approach to discuss the encoding of color, showing that general information theoretic considerations lead to interactions between chromatic and spatial processing which depend on the extent of overlap between the spectral sensitivities of the different receptors [3]. These predictions are in agreement with observations on ganglion cells in primates and goldfish, which represent very different limits of the theory. Atick *et al.* also predict unusual interactions between spatio-chromatic opponency and the temporal aspects of the stimulus, and these predictions may help to unravel some seemingly conflicting experiments in the goldfish.

The work of Atick *et al.* serves to illustrate the general structure of any information theoretic approach to the prediction of neural coding strategies. One starts with the hypothesis that some quantity is being optimized, subject to certain constraints. To relate these notions back to real neurons one needs a model of the coding process and knowledge of the signal and noise characteristics of the receptor cells which feed the encoder neurons. Finally one needs some general statistical data about the source of signals. Unfortunately, relatively little is known about the ensemble of stimuli which animals encounter in the natural setting [25].

As a different example of these ideas, Linsker has explored the consequences of assuming that signal transfer from one layer of neurons to the next serves to maximize information transmission subject to some constraints of architecture [33]. Once again one can predict receptive fields qualitatively like those found at different layers of the visual system, and it possible to show that relatively simple local "learning rules" can lead to the optimization of information transfer, without any global oracle to measure the actual transfer rates. Ruderman, Zee and I have studied a version of the maximum information transmission problem in the low SNR limit, and we find a choice of the constraints which results in a mapping from this optimization problem onto the variational method in quantum mechanics [18]. This formulation makes it especially clear that even the qualitative coding strategies may depend on the ensemble of natural signals. In addition there are multiple solutions to the optimal coding problem, which one might hope eventually to connect with the existence of multiple cell types.

In all of this work, however, the "neurons" are cartoon devices which take analog inputs and deliver analog outputs. One can mutter kind words about firing rates, but nonetheless one has not come to grips with the fundamental fact about sensory neurons, namely that they spike. There are also no experiments in the literature which give convincing measures of the rate at which the spike train of a primary sensory neuron provides information about a time-varying, natural stimulus; for reviews see Refs. [15,38,47].

In a separate contribution to these proceedings, Rieke, Warland and I have taken a direct experimental approach to the questions of information capacity and coding efficiency in sensory neurons [43]. We adopt the point of view of the organism, which has access only to the spikes arriving at times $\{t_i\}$ and must estimate from these data some aspects of the unknown, continuously varying stimulus waveform $s(t)$. In several different systems one can estimate the waveform itself by simple linear filtering,

$$s_{\mathrm{est}}(t) = \sum_i F(t - t_i), \tag{3}$$

where the reconstruction filter F is "learned" (by us) by demanding that the mean-square deviation between $s_{\mathrm{est}}(t)$ and the true $s(t)$ be minimized. Linearity of our reconstruction algorithm does not imply a simple linear response of the neuron, and reconstructing the stimulus waveform *in real time* involves much more than just inverting a conventional stimulus/response relationship — one must actually interpolate between the spikes to recover a continuous function from discrete data at the spike arrival times. These and other issues are discussed at greater length in Refs. [15,17,19,47].

Once we have reconstructed the stimulus waveform we can use conventional methods from signal processing to characterize the quality of the reconstruction, determining the magnitude of the information [52] which the reconstruction provides about the true signal. In experiments on cricket cercal mechanoreceptors, we found that spike trains could provide more than 300 bits per second of information about random displacements of the sensory hair, corresponding to roughly 3 bits per spike. One can (for now) only fantasize about the information which could be conveyed by an entire array of such cells. Similarly impressive bit/spike ratios were obtained in experiments on the bullfrog sacculus, a vibratory sensor in the inner ear.

In 1952 MacKay and McCulloch pointed out that a system which keeps track of spike arrival times or inter-spike intervals could convey several bits per spike, much more than crude rate codes [34]. There is an upper bound to the transmitted information set by the entropy of the spike sequences themselves: The entropy measures the number of distinguishable spike sequences given some timing precision, and the information rate measures the number of distinguishable stimulus waveforms. In both the cricket and the bullfrog, information rates are within a factor of two of the spike train entropy, corresponding to a coding efficiency in excess of 50%. Although the spike sequences are highly variable, most of this variability is thus being used to convey information about the stimulus, a possibility emphasized by Bullock [21,22]. From these results it is natural to return to our starting point and ask if some optimization principles for the code itself can be formulated.

As a first try at the theory of optimal coding in spiking neurons, consider the following model problem. A spiking neuron receives its signal $s(t)$ in a background of Gaussian noise $\eta(t)$. The combination of signal and noise is filtered through a filter $F(t)$, and the output of this filter is encoded in spikes. Thus the signal driving spike generation is

$$y(t) = \int dt' F(t-t')[s(t') + \eta(t')]; \tag{4}$$

a spike is generated each time t_i the signal y crosses a threshold θ with positive slope. This misses the refractoriness of real neurons, as well as the possibility that the cell is driven into a regime of self-sustained oscillation. These effects are negligible if threshold crossings are infrequent, in which case the spikes approximate a Poisson process; experimentally many sensory neurons can be described at least approximately in terms of modulated Poisson spike sequences [19]. The optimal coding problem is to choose the filter $F(t)$ and the threshold θ to maximize the information which the spike arrival times $\{t_i\}$ provide about the stimulus $s(t)$; we should hold the entropy of the spike train fixed to enforce a high coding efficiency. At low signal-to-noise ratio optimizing the information rate is not a well-posed problem, because the information can always be increased by making filters which have longer lasting transient responses. This would mean, however, that the information takes a long time to find its way into the spike train. Presumably there is a penalty for such long delays — it is not much help to be very certain that a tiger was lunging at us one minute ago. The exact structure of the optimal filter depends on the structure of this penalty, but there is a simple choice which leads to a suggestive result: The optimal filters are solutions of a Schrödinger equation, very much analogous to the situation found by Ruderman, Zee and I in the problem of optimizing spatial receptive fields. Again there will be multiple solutions, corresponding to multiple cell types.

Optimization of the threshold leads to the conclusion that, for example, retinal ganglion cells should adapt not only to the background light intensity but also to the variance of the contrast as seen through its spatio-temporal receptive field. There is preliminary evidence for this in recordings from the tiger salamander retina [36], and we have seen similar effects in the cricket cercal system; for low frequency auditory neurons the well-known adaptation to sound intensity (which is just the pressure variance) is the same phenomenon. If the threshold adapts to maintain optimal information transmission, then within this class of models it also turns out that linear decoding, as in Eq. (3), should be a good approximation. This suggests that our two fundamental observations on the structure of the neural code — the success of simple decoding algorithms and the very high information rates — may be tied together by a principle of optimal coding.

Specialized computations

Different organisms are interested in different aspects of their sensory environment and thus perform a variety of specialized computations. Consider the target ranging computations of echolocating bats, specifically the species *E. fuscus* described by Simmons in these proceedings. This bat can discriminate echo delay differences of 10 nanoseconds [53,54]; in noisy backgrounds the threshold for reliable delay discrimination is essentially equal to the theoretical limit imposed by the noise itself: The animal makes optimal use of the available signal to achieve its behavioral performance. We know the mathematical structure of the computations required for optimal processing, but we do not know how these computations are realized in the bat's brain.

Humans are apparently very good at recognizing symmetry and regularity in images. Indeed, as scientists we must take care not to see too much regularity in our data. Barlow and co-workers [8] generated random dot patterns which had a statistical tendency toward bilateral symmetry — some of the randomly placed dots had near-symmetric partners across a reflection axis, while others did not — and asked observers to discriminate between these patterns and a set of random dot patterns with no correlations across the symmetry axis. The observer gets just one look at one randomly chosen pattern, so that there is a well-defined probability of confusion even for a noiseless observer.

If we convert the percent correct discrimination performance into effective signal-to-noise ratios for the recognition of symmetry, human observers achieve signal-to-noise ratios within a factor of two of the limits imposed by randomness in the stimulus itself. This conclusion is robust to many changes in the details of the task, and the results have been extended by Tapiovaara

[58]. Similar results are obtained for the recognition of selected patterns in noisy backgrounds, the detection of variations in density for random dot patterns, and so on.

For every case where the human approaches optimal performance one can ask about the mathematical structure of the optimal computation. For symmetry recognition it seems obvious that the appropriate strategy is to measure the correlation between the two halves of the image. Under the conditions of Tapiovaara's experiments this is in fact not the optimal strategy, and human observers perform better than a simple correlator. Evidently our subjective impressions of symmetry are quantified by subtler computations.

In a natural setting much of biological vision is devoted to the processing of movement signals — animals use their vision to measure their own movement relative to fixed background objects, as well as estimating the trajectories of predators, prey, prospective mates, ... In many insects the visual system produces a movement estimate which is used to control the flight muscles and thereby stabilize desirable flight paths.

In flies one can study visually-guided flight both by measuring flight paths during natural chasing behaviors [30] and by examining the torques produced by tethered flies in response to movements of controlled patterns across the visual field [39,41]. One can also achieve very stable recordings from the photoreceptor cells of the compound eye, allowing us to measure the signal and noise characteristics at the input to the computation of motion. Finally, the output of the visual movement computation is carried by just a handful of identified cells in the lobula complex [27]. Recently de Ruyter van Steveninck, Rieke, Warland and I have reported a series of experiments which characterize the precision of the movement computation in the cell H1, which senses rigid horizontal movements [15,17,44,46,47].

Following the approach described above, we have decoded the spike train of H1 to recover estimates of the trajectory of random patterns moving across the visual field; Fig. 2 shows an example of one segment in these reconstructions. Clearly the spike train of H1 contains enough information to infer — in real time, without averaging — details of the stimulus waveform on times comparable to the typical inter-spike interval, as suggested by the behavioral reaction times. The errors in our reconstruction correspond to an angular displacement noise of $\sim 10^{-4}$ (degrees)^2/Hz. With a behaviorally relevant integration time of 30 msec one can thus use the output of H1 to estimate trajectories to a precision of $\sim$ 0.1 degree, one order of magnitude smaller than the spacing between photoreceptors in the lattice of the compound eye.

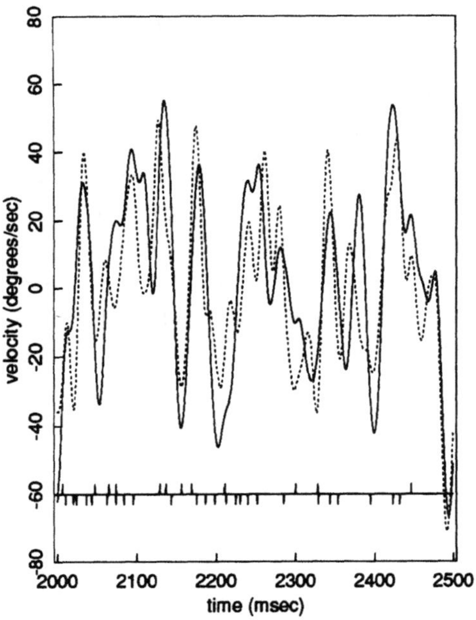

Figure 2: Time-dependent angular velocity signals are reconstructed (solid line) by filtering the spike train of H1, shown at bottom; true stimulus waveform shown as dashed line. Positive spikes represent the response of H1 to the stimulus as shown, negative spikes represent the response to the negative of the stimulus as would be seen by the H1 cell on the opposite side of the head during rigid rotation of the fly. The precision of the reconstructions can be quantified by computing the spectral density of the errors, as described in Refs. [15,17].

This is the regime corresponding to hyperacuity in human vision, and we have confirmed these results in experiments where we ask directly about the discriminability of step displacements using the H1 spike train.

Given the signal and noise characteristics of the photoreceptors [46] we can calculate the limits to hyperacuity on the assumption that the fly is the optimal processor of the voltage signals in the array of photoreceptors. The key point is that in response to a sudden movement the fly can generate a torque within 30 ms. But for small amplitude signals, the photoreceptors need nearly 10 ms to reach their peak response. Even with very large signals there is a delay of roughly 15 ms before H1 will respond. For small displacements, then, we expect that there is a delay of almost 25 ms before H1 can report the full signal corresponding to a displaced pattern. But this leaves only 5 ms during which the fly can be "looking" at the full signal as it tries to estimate the appropriate compensating torque. This is essentially the correlation time in the photoreceptor voltage noise (~ 8 ms), so rather than looking at time-averaged signals the fly must base its behavioral decisions on a single snapshot of the voltage array.

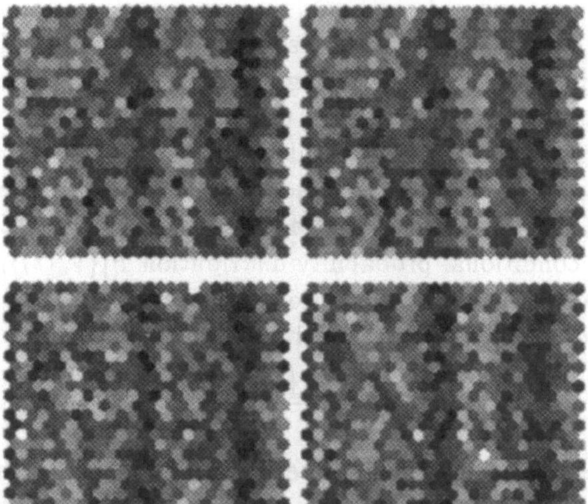

Figure 3: Snapshots of the receptor voltage array in response to small displacements. Left panel, displacement of a random bar pattern by 0.27 receptor spacings (from top to bottom panel). Right panel, displacement of a random bar pattern by 0.36 receptor spacings. In the experiments of Ref. [46], these displacements are discriminated with > 75% correct using just one example of the spike train from H1.

Based on Ref. [46], we can simulate what the fly "sees" when patterns are displaced by small fractions of the photoreceptor spacing. Figure 3 shows an example of these simulations, carried out by N. Socci. This gives us graphic evidence that *on the time scales of relevance to fly behavior the signals represented in the photoreceptor array are very noisy;* as a result the hyperacuity task is hard. Making these observations quantitative we find that the effective noise level of the reconstructions is essentially equal to the fundamental limit imposed by noise in the receptor array. Thus the fly visual system carries out an optimal and nearly noiseless extraction of movement estimates from the receptor signals.

What is the structure of the optimal estimator for angular velocity [15,44]? If the fly is looking at some rigid contrast pattern $C(\phi)$ which moves along a trajectory $\theta(t)$, each receptor produces a voltage

$$V_n(t) = \int d\tau T(\tau) \int d\phi M(\phi - \phi_n) C[\phi - \theta(t-\tau)] + \delta V_n(t), \qquad (5)$$

where $T(\tau)$ is the temporal impulse response to a spatially uniform contrast pulse, and $M(\phi)$ is the normalized angular sensitivity profile of the cell determined by its optics. The ϕ_n are the orientations of the cells' optical axes, which form a regular lattice, and the noise voltages $\delta V_n(t)$ are known experimentally to be Gaussian and additive, characterized by some spectral density $N(\Omega)$. This description of the photoreceptor voltage responses determines the conditional probability distribution $P[\{V_n(t)\}|\theta(\tau)]$, and everything that we know about the trajectory $\theta(\tau)$ by virtue of observing the array of receptor voltages is contained in the distribution

$$P[\theta(\tau)|\{V_n(t)\}] = \frac{P[\{V_n(t)\}|\theta(\tau)]P[\theta(\tau)]}{P[\{V_n(t)\}]}, \qquad (6)$$

which combines the response characteristics of the cells with our *a priori* knowledge of the statistics of $\theta(t)$. In the experiments of Ref. [17,47], for example, the statistics of the trajectories $\theta(t)$ correspond to a simple random walk, or angular diffusion with diffusion constant $D \sim 4\,\text{deg}^2/\text{sec}$. Our task is to understand how one can use all of this information to estimate the angular velocity $\dot{\theta}(t)$ based on the data contained in the $\{V_n(t)\}$.

Let us take as the optimal estimator of $\dot{\theta}(t)$ the conditional mean,

$$\dot{\theta}_{\text{est}}(t) = \langle \dot{\theta}(t) \rangle_{\{V_n(t)\}} = \int [d\theta] \dot{\theta}(t) P[\theta(\tau)|\{V_n(t)\}], \qquad (7)$$

which is the average trajectory that gives rise to our observations. Alternatively one could look for the most likely trajectory, but the theory for the average trajectory is more straightforward. In the low SNR limit, which is relevant to our experiment, the optimal estimator can be written as

$$\begin{aligned} \dot{\theta}_{\text{est}}(t) \quad &\propto \quad \int \frac{d\Omega}{2\pi} \int \frac{d\Omega'}{2\pi} e^{-i(\Omega+\Omega')t} \frac{1}{-i(\Omega+\Omega')} \\ &\times \quad \sum_{nm} \left[\frac{\tilde{T}^*(\Omega)}{N(\Omega)} g_{nm}(\Omega) \tilde{V}_n(\Omega) \right] \left[\frac{\tilde{T}^*(\Omega')}{N(\Omega')} \tilde{V}_n(\Omega') \right], \end{aligned} \qquad (8)$$

where the filters g_{nm} are given by

$$g_{nm}(\Omega) = \int d\tau e^{i\Omega\tau} \frac{\phi_n - \phi_m}{[(\Delta\phi)^2 + D|\tau|]^{3/2}} \exp\left[-\frac{1}{4} \cdot \frac{(\phi_n - \phi_m)^2}{(\Delta\phi)^2 + D|\tau|}\right]; \qquad (9)$$

as before D is the angular diffusion constant of the motion we are trying to estimate and $\Delta\phi$ is the width of the photoreceptor aperture, $M(\phi) \propto \exp[-\phi^2/2(\Delta\phi)^2]$.

The optimal estimator at low signal-to-noise ratios has the form of a correlator, multiplying spatially displaced pairs of photoreceptor voltages which have been passed through different temporal filters. This is a generalization of the scheme proposed by Hassenstein and Reichardt from their experiments on insect optomotor behavior [39,41]. The phenomenological filters of the original discussion now take specific forms related to the characteristics of the photoreceptors and the ensemble of stimuli. There is ample evidence that the cell H1 is approximately described by a Reichardt-like model, qualitatively supporting the existence of neural dynamics as predicted from the principle of optimal processing. Several points should be emphasized:

1. The optimal estimator includes contributions from all pairs of receptors. The peak amplitude of g_{nm} is proportional to $1/(n-m)^2$, so that nearest neighbors dominate, although contributions from more distant neighbors should be observable. Higher-neighbor terms have been seen in experiments where one stimulates individual facets of the compound eye [51].

2. Before entering the filters g_{nm} all photoreceptor signals pass through a filter $\tilde{T}^*(\Omega)/N(\Omega)$. This is in accord with the general arguments described above, where we found that such a filter — matched to the signal and noise characteristics of the photoreceptor but independent of the stimulus ensemble or the computational task — is the universal first step in any optimal estimation task at low signal to noise ratio. By analogy with the bipolar cells in the vertebrate retina, we might imagine that this filtering is achieved in the lamina, the first stage of signal processing in the fly.

3. The filters g_{nm} have dynamics which depend on the statistics of the movement signal. If small amplitude motions are superposed on a constant velocity — as in natural flight — then the dynamics of the g_{nm} depend also on this constant velocity. Thus we predict that the cell's filter characteristics adapt to background motions. This is observed in H1 [48], and the fact that velocity-specific adaptation occurs locally suggests that the filters g_{nm} reside in a portion of the fly visual system which retains the orderly geometrical relations of the retina itself, perhaps the medulla.

4. The outputs of the filters g_{nm} are correlated with signals from neighboring cells, then summed in space and integrated in time. Presumably the summation and integration steps reside in H1 itself. Optimal movement estimation thus involves at least three mathematically distinct stages of computation, which we can tentatively identify with three layers of processing circuitry in the lamina, the medulla, and the lobula complex.
5. In [17,46,47] we have evidence for near-optimal performance under conditions where the total signal to be computed [e.g., in Eq. (8)] involves a sum over roughly 2500 cell pairs, and it is this total signal which exhibits a signal-to-noise ratio near unity. But then the movement signal from each isolated pair of photoreceptors is at a signal-to-noise (amplitude) ratio of just $1/\sqrt{2500} \sim 1/50$. Each multiplication step must therefore be done with about 5 to 6 bits of accuracy ($\log_2 50 = 5.64$) in order that the subsequent averaging among parallel channels be effective in revealing the signal.
Finally, the correlation scheme is *not* optimal at high signal-to-noise ratio. Franceschini *et al.* [26] have probed the dynamics of H1 under conditions far removed from those discussed here — high SNR inputs to just two receptor cells, as opposed to low SNR inputs to thousands of cells. Under these conditions the non-linearities in the pathway to H1 are very different from a simple multiplication. A crucial challenge for the theory is to see if this sort of adaptive variation in computational strategy can be predicted.

Conclusions

As early as 1911, Lorentz realized that the available data on the minimum energy for vision were suggestive of single-photon counting, though it was many years before a solid case could be made [14]. There is a tendency to see such performance as an isolated curiosity, perhaps significant only for its amusement value in the physics community. I hope to have convinced you that this theme of optimality extends through several levels of neural organization and across a wide range of species. It is this universality which makes optimal processing so attractive from a theoretical point of view.

What do we need to establish optimality as a general principle for the early stages of sensory signal processing? One approach is to continue the quantitative comparison of neural performance with optimal performance in the largest possible variety of systems. We would like to move away from the consideration of one cell at a time and think about computation and coding in arrays of cells. Methods are now available to record simultaneously from large numbers of retinal ganglion cells [37], and our decoding algorithms have a natural generalization to the multi-cell case, so I believe that the tools are available to attack these problems.

Ideally we would like to find some dramatic qualitative signatures of optimality. There is a strong hint from the work reviewed here that sensory adaptation must be much richer than usually suspected if optimal performance is to be maintained over a wide range of conditions — filters must adapt their dynamics, spiking neurons must adapt their thresholds in relation to the stimulus variance, and even the qualitative character of non-linear operations must shift as a function of signal-to-noise ratio. This is much more than just shifting the operating point of the system to compensate for static backgrounds, which is the textbook picture of adaptive responses.

As things stand, one is forced to illustrate the different issues in the design of sensory systems by looking at different pieces of different organisms. Obviously it would be nice to make a complete story straight through one system. This naturally leads to the question of whether there is a "simplest case." One candidate is the auditory system of noctuid moths, which mediates complex bat-evading flight control using inputs from just two sensory neurons [45]. Another possibility, which Kruglyak and I have been considering [29], is to step back from the nervous system and look at what may be the simplest systems of all, the chemosensors of single-celled organisms.

From Berg and Purcell [13,40] we know a great deal about the physical constraints under which these systems operate. Essentially the problem is to find the position in three-dimensional space where the concentration of some attractive substance is maximized. But a bacterium is too small to make reliable measurements of a spatial gradient; measurements of the concentration itself must be averaged for a long ($\sim$ 1 sec) time to beat down the diffusive noise. The bacterium has no ability to steer; it can choose only between swimming straight (running) and "tumbling" to pick a new direction at random. Finally, rotational Brownian motion of the entire bacterium prevents straight swimming for periods of longer than a few seconds. Within this environment, the problem of optimizing performance is to optimize the transformations from measured concentration to the switching between runs and tumbles. This is much like the optimal filtering problems which we have solved in the case of neural signal processing, and the resulting filters have a strong resemblance to the available data [50], but problems remain in the interpretation of these experiments. It would clearly be very attractive if the same general physical principles are relevant to the design of sensory systems at such different levels of the evolutionary tree.

Acknowledgements

I have already thanked my many collaborators who helped make this work so much fun. Conversations with many others have (hopefully) helped to make the ideas more precise: J. Atick, D. Baylor, S. Block, N. Franceschini, D. Glaser, A. Libchaber, M. Meister, J. Simmons, and G. Zweig.

References

1. A.-C. Aho, K. Donner, C. Hydén, L. O. Larsen, and T. Reuter, "Low retinal noise in animals with low body temperature allows high visual sensitivity," *Nature* **324,** 348 (1988).
2. J. J. Atick and A. N. Redlich, "Towards a theory of early visual processing," *Neural Comp.* **2,** 308 (1990).
3. J. J. Atick, Z. Li, and A. N. Redlich, "Color coding and its interaction with spatiotemporal processing in the retina, preprint.
4. H. B. Barlow, "The size of ommatidia in apposition eyes," *J. Exp. Biol.* **29,** 667-674 (1952).
5. H. B. Barlow, "Summation and inhibition in the frog's retina," *J. Physiol.* **119,** 69 (1953).
6. H. B. Barlow, "Retinal noise and absolute threshold," *J. Opt. Soc. Am.* **46,** 634 (1956).
7. H. B. Barlow, "Possible principles underlying the transformation of sensory messages," in *Sensory Communication,* W. Rosenblith, ed. p. 217-234 (MIT Press, Boston MA, 1961).
8. H. B. Barlow, "The absolute efficiency of perceptual decisions," *Philos. Trans. R. Soc. London Ser. B* **290,** 71 (1980).
9. H. B. Barlow, "Thermal limit to seeing," *Nature* **334,** 296 (1988).
10. D. A. Baylor, T. D. Lamb, and K.-W. Yau, "Responses of retinal rods to single photons," *J. Physiol.* **288,** 613-634 (1979).
11. D. A. Baylor, G. Matthews, and K.-W. Yau, "Two components of electrical dark noise in toad retinal rod outer segments," *J. Physiol.* **309,** 591 (1980).

12. D. A. Baylor, B. J. Nunn, and J. F. Schnapf, "The photocurrent, noise and spectral sensitivity of rods of the monkey *Macaca fascicularis, J. Physiol.* **357,** 575 (1984).

13. H. C. Berg and E. M. Purcell, "Physics of chemoreception," *Biophys. J.* **20,** 193 (1975).

14. W. Bialek, "Physical limits to sensation and perception," *Ann. Rev. Biophys. Biophys. Chem.* **16,** 455-478 (1987).

15. W. Bialek, "Theoretical physics meets experimental neurobiology," in *1989 Lectures in Complex Systems, SFI Studies in the Sciences of Complexity, Lect. Vol. II,* E. Jen, ed., pp. 513-595 (Addison-Wesley, Menlo Park CA, 1990).

16. W. Bialek and W. G. Owen, "Temporal filtering in retinal bipolar cells: Elements of an optimal computation?," *Biophys. J.* **58,** 1227-1233 (1990).

17. W. Bialek, F. Rieke, R. R. de Ruyter van Steveninck, and D. Warland, "Reading a neural code," *Science* **252,** 1854-1857 (1991).

18. W. Bialek, D. L. Ruderman, and A. Zee, "Optimal sampling of natural images: A design principle for the visual system?," in *Advances in Neural Information Processing 3,* R. Lippman, J. Moody, and D. Touretzky, eds., pp. 363-369 (Morgan Kaufmann, San Mateo CA, 1991).

19. W. Bialek and A. Zee, "Coding and computation with neural spike trains," *J. Stat. Phys.* **59,** 103-115 (1990).

20. E. de Boer and R. MacKay, "Reflections on reflections," *J. Acoust. Soc. Am.* **67,** 882 (1980).

21. T. H. Bullock, "The reliability of neurons," *J. Gen. Physiol.* **55,** 584-656 (1970).

22. T. H. Bullock, "Redundancy and noise in the nervous system: Does the model based on unreliable neurons sell nature sort?," in *Electrobiology of Nerve, Synapse, and Muscle,* ed. J. Reuben, D. Purpura, M. V. L. Bennett, and E. Kandel, p. 179 (Raven Press, 1976).

23. K. Donner, "The absolute sensitivity of vision: Can a frog become a perfect detector of light induced and dark rod events?," *Phys. Scr.* **39,** 133 (1989).

24. R. P. Feynman, R. Leighton, and M. Sands, *The Feynman Lectures on Physics, Volume I.* (Addison-Wesley, Reading MA, 1963).

25. D. Field, "Relations between the statistics of natural images and the response properties of cortical cells," *J. Opt. Soc. Am. A* **4,** 2379 (1987).

26. N. Franceschini, A. Riehle, and A. le Nestour, "Directionally selective motion detection by insect neurons," in *Facets of Vision,* R. C. Hardie and D. G. Stavenga, eds. pp. 360-390 (Springer-Verlag, Berlin, 1989).

27. K. Hausen, 'The lobular complex of the fly: structure, function and significance in behavior," in *Photoreception and vision in invertebrates,"* M. A. Ali, ed., pp. 523-559 (Plenum, New York, 1984).

28. S. Hecht, S. Shlaer, and M. Pirenne, "Energy, quanta, and vision," *J. Gen. Physiol.* **25,** 819-840 (1942).

29. L. Kruglyak and W. Bialek, "Optimal filtering in bacterial chemotaxis," in preparation.

30. M. F. Land and T. S. Collett, "Chasing behavior of houseflies (*Fannia canicularis*): A description and analysis," *J. Comp. Physiol.* **89,** 331 (1974).

31. S. B. Laughlin, "A simple coding procedure enhances a neuron's information capacity," *Z. Naturforsch.* **36c,** 910-912 (1981).

32. J. Y. Lettvin, H. R. Maturana, W. S. McCulloch, and W. H. Pitts, "What the frog's eye tells the frog's brain," *Proc. I. R. E.* **47,** 1940-1951 (1959).

33. R. Linsker, "An application of the principle of maximum information preservation to linear systems," in *Advances in Neural Information Processing 1,* ed. D. Touretzky, p. 186 (Morgan Kaufmann, San Mateo, CA, 1989).

34. D. MacKay and W. S. McCulloch, "The limiting information capacity of a neuronal link," *Bull. Math. Biophys.* **14,** 127 (1952).

35. A. Mallock, "Insect sight and the defining power of compound eyes," *Proc. R. Soc. Lond. Ser. B* **55,** 85-90 (1894).

36. M. Meister, personal communication.

37. M. Meister, R. O. L. Wong, D. A. Baylor, and C. J. Shatz, "Synchronous bursts of action potentials in the ganglion cells of the developing mammalian retina," *Science* **252,** 939-943 (1991).

38. D. H. Perkel and T. H. Bullock, "Neural coding," *Neurosci. Res. Prog. Sum.* **6,** 221 (1968).

39. T. Poggio and W. Reichardt, "Visual control of orientation behavior in the fly. Part II. Towards the underlying neural interactions," *Q. Rev. Biophys.* **9,** 377 (1976).

40. E. M. Purcell,"Life at low Reynolds number," *Am. J. Phys.* **45,** 3-11 (1977).

41. W. Reichardt and T. Poggio,, "Visual control of orientation behavior in the fly. Part I. A quantitative analysis," *Q. Rev. Biophys.* **9,** 311 (1976).

42. F. Rieke, W. G. Owen, and W. Bialek, "Optimal filtering in the salamander retina," in *Advances in Neural Information Processing 3,* R. Lippman, J. Moody, and D. Touretzky, eds., pp. 377-383 (Morgan Kaufmann, San Mateo CA, 1991).

43. F. Rieke, D. Warland, and W. Bialek,"Coding efficiency and information capacity in sensory neurons," preprint.

44. F. Rieke, D. Warland, R. R. de Ruyter van Steveninck, and W. Bialek, "Optimal real-time processing of visual movement signals: Theory and experiments in the blowfly," in preparation.

45. K. D. Roeder and R. S. Payne, "Acoustic orientation of a moth in flight by means of two sense cells." *Symp. Soc. Exp. Biol.* **20,** 251-272 (1966).

46. R. R. de Ruyter van Steveninck, *Real-time Performance of a Movement-Sensitive Neuron in the Blowfly Visual System,* Academish Proefschrift, Rijksuniversiteit Groningen (1986).

47. R. de Ruyter van Steveninck and W. Bialek, "Real-time performance of a movement sensitive neuron in the blowfly visual system: Coding and information transfer in short spike sequences," *Proc. R. Soc. London Ser. B* **234,** 379-414 (1988).

48. R. R. de Ruyter van Steveninck, W. H. Zaagman, and H. A. K. Mastebroek, "Adaptation of transient responses of a movement-sensitive neuron in the visual system of the blowfly *Calliphora erythrocephala*," *Biol. Cybern.* **54,** 223-236 (1986).

49. B. Sakitt, "Counting every quantum," *J. Physiol. (Lond.)* **223,** 131-150 (1972).

50. J. Segall, S. M. Block, and H. C. Berg, "Temporal comparisons in bacterial chemotaxis," *Proc. Nat. Acad. Sci. USA* **83,** 8987 (1986).

51. F. Schuling, H. Mastebroek, R. Bult, and B. Lenting, "Properties of elementary movement detectors in the fly *Calliphora erythrocephala*," *J. Comp. Physiol. A* **165,** 179 (1989).

52. C. E. Shannon, "Communication in the presence of noise," *Proc. I. R. E.* **37,** 10-21 (1949).

53. J. A. Simmons, "A view of the world through the bat's ear: The formation of acoustic images in echolocation," *Cognition* **33,** 155-199 (1989).

54. J. A. Simmons, M. Ferragamo, C. F. Moss, S. B. Stevenson, and R. A. Altes, "Discrimination of jittered sonar echoes by the echolocating bat, *Eptesicus fuscus*: The shape of target images in echolocation," *J. Comp. Physiol. A* **167,** 589-616 (1990).

55. A. W. Snyder, "Acuity of compound eyes: Physical limitations and design," *J. Comp. Physiol. A* **116,** 161-182 (1977).

56. A. W. Snyder, S. B. Laughlin, and D. G. Stavenga, "Information capacity of eyes," *Vision Res.* **17,** 1163-1175 (1977).

57. A. W. Snyder, D. S. Stavenga, and S. B. Laughlin, "Information capacity of compound eyes," *J. Comp. Physiol.* **116,** 183-207 (1977).

58. M. Tapiovaara, "Ideal observer and absolute efficiency of detecting mirror symmetry in random images," *J. Opt. Soc. Am. A* **7,** 2245-2253 (1990).

59. N. Wiener, *Extrapolation, Interpolation and Smoothing of Time Series.* (Wiley, New York, 1949).

60. G. Zweig, "Basilar membrane motion," *Cold Spring Harbor Symp. Quant. Biol.* **40,** 619 (1975).

61. G. Zweig, R. Lipes, and J. R. Pierce, "The cochlear compromise," *J. Acoust. Soc. Am.* **59,** 975 (1976).

Measuring the coding efficiency of sensory neurons

Fred Rieke,* David Warland, and William Bialek*

* NEC Research Institute
4 Independence Way
Princeton, New Jersey 08540

Departments of Physics and Molecular and Cell Biology
University of California
Berkeley, California 94720

Abstract

All of an organism's knowledge about the sensory world comes from observation of the spike trains in its own sensory cells. How much information is carried in these spike trains, and how efficient is the coding? We have recently developed an approach to neural coding which allows us to measure the information a spike train carries about a sensory stimulus without assumptions about how the information is coded [1]. Comparing this measured information with an upper bound to the information rate determined by the spiking statistics results in a direct measure of the efficiency of coding. For two different mechanoreceptors, from the cricket cercal system and from the bullfrog sacculus, we find information rates close to 3 bits per spike (corresponding to 300 bits per sec in the cricket, 160 bits per sec in the frog), and coding efficiencies greater than 50%.

Introduction

Living creatures receive information from the environment at tremendous rates. Much of this information, especially that which is crucial to the animal's survival, must be processed and transmitted quickly. How do sensory systems cope with these tremendous data rates? One idea is that these systems extract certain 'relevant features' of the input while discarding large amounts of 'irrelevant' information [2].* Another idea is that the initial stages of sensory processing serve to convert the incoming signals into a form where they can be efficiently coded [5, 6, 7]. A number of unanswered experimental questions lie at the center of the theoretical debate between these different views of sensory processing: How much information is carried in a neural spike train? How efficient is the coding of sensory information? Is the coding process itself an 'informational bottleneck,' limiting the ability of the system to transmit information?

Here we measure information rates and coding efficiencies for two very different mechanoreceptors — filiform hair afferents from the cercal system of the cricket *Acheta domestica* and auditory nerve afferents from the sacculus of the bullfrog *Rana catesbeiana*. As far as we know these are the first measurements of information rates in primary sensory neurons which do not depend on assumptions about how the information is coded in the spike trains. The surprisingly high information rates and coding efficiencies we measure in these systems serve to sharpen the theoretical issues surrounding neural coding.

Measuring the information rate

To measure the information which the spikes carry about the stimulus we need to understand something about how sensory stimuli are represented in neural spike trains. What are some of the salient properties of the coding process? First of all the coding is in real time. An organism makes decisions about events in the world based on single examples of the spike trains in its sensory cells; it does not have the luxury of averaging neural responses over many repetitions of the stimulus. Second, the coding is nonlinear. The most obvious nonlinearity is the all-or-nothing generation of an action potential.

---

*Feature detection was also proposed by Barlow [3], however, his definition of what constituted a feature was much broader than the current definition. For a historical account of development of feature detection see Ref. [4].

There are also a number of 'softer' nonlinearities, such as saturation and adaptation.

What do these properties mean in terms of how we approach the study of neural coding? The fact that the coding is real-time means that we should study the code from the point of view of the organism. The nonlinear property means that we should be careful in our choice of stimuli — in particular we should study the coding of complex, natural stimuli. These two considerations are the focus of our approach to neural coding [1]: Can we decode neural spike trains in real-time to estimate continuous naturalistic stimuli?

Motivated by theoretical studies of a model neuron [8], we write our estimate of the stimulus as a filtered version of the spike train:

$$\begin{aligned} s_{\text{est}}(t) &= \int d\tau F_1(\tau) \sum_i \delta(t - t_i - \tau) \\ &+ \int d\tau \int d\tau' F_2(\tau, \tau') \sum_{i,j} \delta(t - t_i - \tau)\delta(t - t_j - \tau') + \cdots, \end{aligned} \tag{1}$$

where the spikes occur at times $\{t_i\}$. We choose the filters F_{n} to minimize the mean square error between the true stimulus and our estimate, $\chi^2 = \int dt |s(t) - s_{\text{est}}(t)|^2$. In a number of systems — a movement sensitive cell in the visual system of the blowfly *Calliphora erythrocephela* [1, 9], primary afferents from the cercal system of the cricket (*Acheta domestica*) [10, 11], and sacculus afferents in the bullfrog *Rana catesbeiana* [12] — we find that the series (1) is dominated by the first term. Thus we can estimate the sensory stimuli in these systems by linearly filtering the spike trains; only in the case of the bullfrog sacculus does the second order term make significant contributions to the estimate. Fig. 1 shows examples of the reconstructions, spike trains and original stimuli for the bullfrog and cricket mechanoreceptors.

It is apparent from Fig. 1 that our reconstructions are not perfect. The errors determine the amount of information the reconstruction provides about the original stimulus. The reconstructions consist of a piece which is deterministically related to the stimulus and some inevitable random noise. To separate these we introduce a frequency dependent gain $g(\omega)$ and write

$$\tilde{s}_{\text{est}}(\omega) = g(\omega) \left[\tilde{s}(\omega) + \tilde{n}_{\text{eff}}(\omega)\right], \tag{2}$$

where g corrects for systematic errors and n_{eff} is a measure of the random errors written as an effective input noise. Experimentally we determine $g(\omega)$

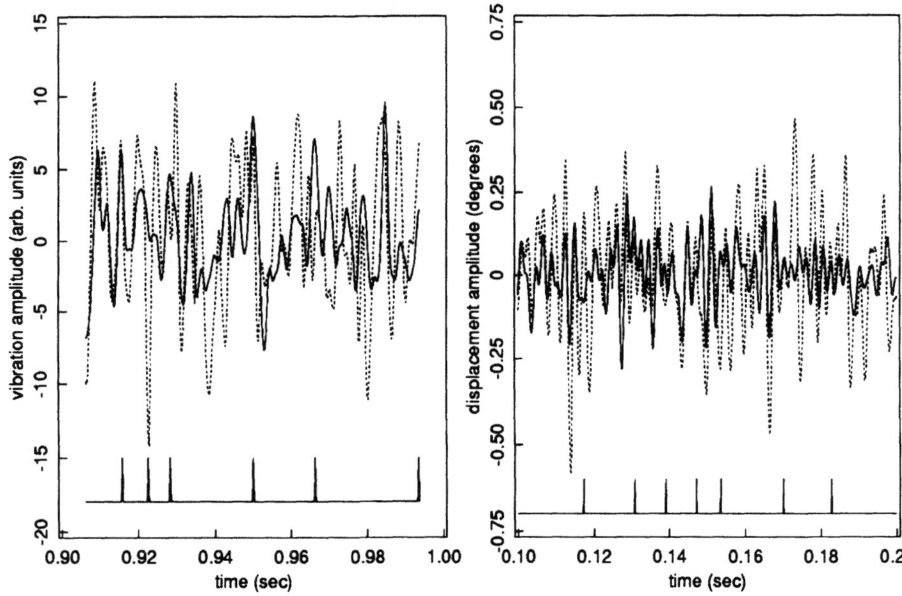

Figure 1: Estimated stimulus, spike train and original stimulus for bullfrog (left) and cricket (right) mechanoreceptors. The reconstructions are able to interpolate between the discrete spikes to estimate the continuous stimulus waveform. Note the time scales; the reconstructions capture information on time scales as short as milliseconds.

as the slope of the best fit line in a plot of $\tilde{s}(\omega)$ versus $\tilde{s}_{\mathrm{est}}(\omega)$, and $\tilde{n}_{\mathrm{eff}}(\omega)$ as the scatter about this line.

The distribution of errors in these decoding experiments turns out to be nearly Gaussian, as shown in Fig. 2. Thus we have succeeded in mapping the (discrete!) spikes into a continuous waveform which is deterministically related to the original stimulus plus additive Gaussian noise. If the original stimulus is itself Gaussian the rate at which the reconstructions provide information about the stimulus (in bits per second) is [13]:

$$I = \frac{1}{2}\int_{-\infty}^{\infty}\frac{d\omega}{2\pi}\log_2\left[1+\frac{S(\omega)}{N_{\mathrm{eff}}(\omega)}\right], \tag{3}$$

where the signal and noise power spectra are, respectively, $S(\omega) = \langle|\tilde{s}(\omega)|^2\rangle$ and $N_{\mathrm{eff}}(\omega) = \langle|\tilde{n}_{\mathrm{eff}}(\omega)|^2\rangle$ (Fig. 2). An important point about this measure

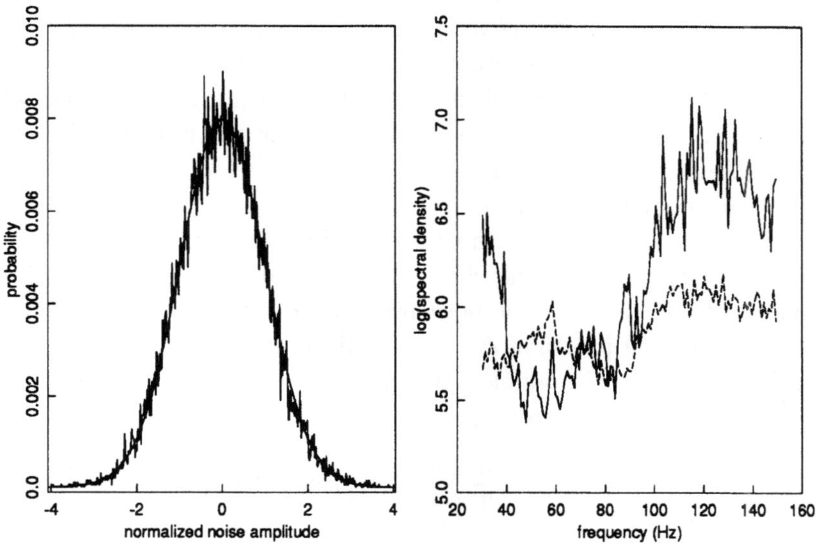

Figure 2: (left) Histogram of the values of n_{eff} from the different segments of the reconstruction in the bullfrog. To plot all frequencies together we scale each $\tilde{n}_{\text{eff}}(\omega)$ by the standard deviation $[\langle|\tilde{n}_{\text{eff}}(\omega)|^2\rangle/2]^{1/2}$. With this scaling, Gaussian noise produces a histogram corresponding to a Gaussian of zero mean and unit variance, which is shown as a solid line. The agreement is excellent, indicating that the effective noise n_{eff} in our reconstructions is indeed Gaussian. (right) Effective noise spectrum $N_{\text{eff}} = \langle|\tilde{n}_{\text{eff}}(\omega)|^2\rangle$ and stimulus spectrum $S(\omega) = \langle|\tilde{s}(\omega)|^2\rangle$ (dashed line) for the bullfrog reconstructions.

of information is that I is not some hypothetical measure based on assumptions about the relevant features of the spike trains; instead I is the rate at which we are able to extract information, in real time, about the stimulus from the spikes. For the reconstructions shown in Fig. 1 we find information rates of 155 ± 3 bits per sec (bullfrog) and 294 ± 6 bits per sec (cricket), corresponding to 2.6 ± 0.05 and 3.2 ± 0.07 bits per spike, respectively.

The coding efficiency

Can we put these measured information rates on an absolute scale? Under a given set of experimental conditions, with stimuli chosen from a particular ensemble, the spike train has some entropy per unit time S, which is a measure of the number of degrees of freedom in the spike train. This entropy is determined by the statistics of the spikes alone, without reference to the stimulus. For a given set of spiking statistics the entropy sets an absolute upper bound to the rate at which the spikes can carry information about the stimulus. The coding efficiency $\epsilon \equiv I/S$ thus measures the fraction of the available degrees of freedom in the spike train which are used to convey information about the stimulus.

Up to now we have used the full experimental precision in specifying the spike times; we have treated the spike arrival times $\{t_i\}$ as continuous variables with no errors. Suppose instead we view the spike train with some limited timing precision $\Delta\tau$. With this limited timing resolution there are a finite number of possible spike patterns in a time T. This number corresponds to a finite entropy per unit time:

$$S(\Delta\tau) = -\frac{1}{T}\sum_{\{t_i\}} P(\{t_i\})\log_2 P(\{t_i\}), \tag{4}$$

where $P(\{t_i\})$ is the probability of a particular spike sequence $\{t_i\}$, the spike times are measured with a precision $\Delta\tau$, and we sum over all possible spike sequences in the time T.

How do we estimate this entropy in a finite experiment? Since correlations in the timing of the spikes can only lower the entropy, a natural series of upper bounds to $S(\Delta\tau)$ is provided by successively including higher order correlations. It is important to realize that these upper bounds are determined by spike statistics measured during presentation of the same stimuli as the measured information rates.

- First order — rate fixed: The firing rate $\bar{r}$ determines a probability per unit time for spiking. Thus in a time $\Delta\tau$ there is a probability $p = \bar{r}\Delta\tau$ of finding a spike, and a probability of $(1-p)$ of no spike. With no other restrictions this gives us an entropy per unit time (in bits per sec) of

$$S^{(1)}(\Delta\tau) = -\frac{1}{\Delta\tau}\left[p\log_2 p + (1-p)\log_2(1-p)\right]. \tag{5}$$

  This is essentially same calculation as MacKay and McCulloch [14].

- Second order — interval distribution fixed:

$$S^{(2)}(\Delta\tau) = -\frac{\sum_n P(\tau_n)\log_2 P(\tau_n)}{\sum_n \tau_n P(\tau_n)}, \tag{6}$$

where the interval distribution $P(\tau_n)$ measures the probability of an interval τ_n in the spike train viewed with a timing precision $\Delta\tau$.

- Third order — joint interval distribution fixed:

$$S^{(3)}(\Delta\tau) = -\frac{\sum_{n,m} P(\tau_n, \tau_m) \log_2 [P(\tau_n, \tau_m)/P(\tau_n)]}{\sum_n \tau_n P(\tau_n)}, \qquad (7)$$

where τ_n and τ_m are successive intervals, and $P(\tau_n, \tau_m)$ is measured for the spike train with timing precision $\Delta\tau$.

Measuring the entropy with a finite timing precision $\Delta\tau$ is equivalent to 'coarsening' the spike train so that the spike times t_i are determined to an accuracy $\Delta\tau$. As above, we reconstruct the stimulus using this coarsened spike train to measure the information $I(\Delta\tau)$ as a function of timing precision.

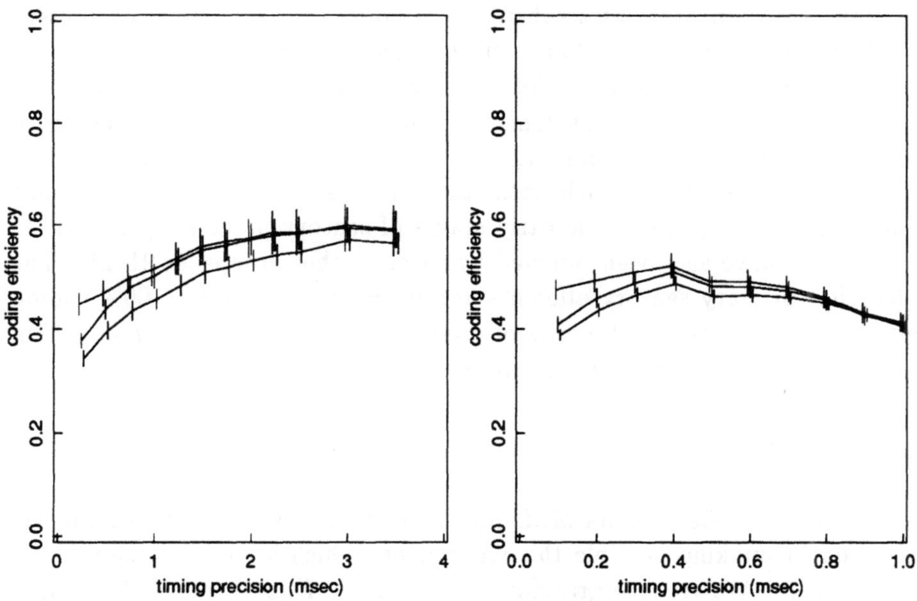

Figure 3: Coding efficiency in the bullfrog (left) and cricket (right). Successively higher coding efficiencies correspond to successively lower bounds to the entropy, $S^{(0)}$ to $S^{(2)}$.

We now have all the ingredients to measure the coding efficiency $\epsilon(\Delta\tau) = I(\Delta\tau)/S(\Delta\tau)$ The results are shown in Fig. 3. It is important to realize that these results are lower bounds to the 'true' coding efficiency. There are certainly other considerations that may reveal higher efficiencies than we measure. For example, neural systems may encode natural signals more efficiently than the band-limited white noise used here. Another consideration is that the information $I(\Delta\tau)$ is the information we are able to extract about a particular stimulus using our simple decoding approach; this measured information may fall short of the total information the spike train carries.

Implications

What does all this mean? The information rates we measure are remarkably high — if single sensory neurons are are carrying information at rates of 100–300 bits per second many of the traditional neural coding problems (e.g. the coding of vowels in the auditory nerve) do not seem so hard. These information rates indicate that a substantial amount of the information present at the receptor level is coded and sent to higher processing centers, and thus that less pre-processing in the form of 'feature selection' is required.

There are now several systems in which there is quantitative evidence for optimal signal processing [1, 15–17]; for a brief review see Ref. [18]. We have suggested that such optimality may used to predict the relevant computational aspects of the early stages of sensory processing [19, 20, 7]; these ideas have also been pursued by several other groups [6, 21–25]. The idea that primary sensory afferents approach perfect coding (100% coding efficiency) suggests that theories of pre-processing based on optimal coding are of direct relevance to real neurons.

Acknowledgements

We thank our collaborators M. Landolfa, E. R. Lewis, J. P. Miller, and W. Yamada for making possible the experiments which led to the analysis described here. We are also grateful to J. Atick, G. Kuperberg, K. Moortgat, E. Posner, and D. Tank for helpful discussions. W. B. thanks the Aspen Center for Physics for providing a congenial environment in which to talk about these ideas. Work at Berkeley was supported in part by the National Science Foundation through a Presidential Young Investigator Award (to

W. B.), supplemented by funds from Sun Microsystems, Cray Research and the NEC Research Institute, and through a Graduate Fellowship (to F. R.). D. W. was supported in part by the Systems and Integrative Biology Training Program of the National Institutes of Health. Additional support was provided by an INCOR grant through the Los Alamos National Laboratory.

References

1. W. Bialek, F. Rieke, R. R. de Ruyter van Steveninck, and D. Warland. *Science*, 252:1854–1857, 1991.

2. J. Y. Lettvin, H. R. Maturana, W. S. McCulloch, and W. H. Pitts. *Proc. Inst. Radio Engr.*, 47:1940–1951, 1959.

3. H. B. Barlow. *J. Physiol.*, 119:69, 1953.

4. H. B. Barlow. In D. G. Albrecht, editor, *Recognition of Pattern and Form*, page 4. Springer-Verlag, Berlin, 1979.

5. H. B. Barlow. In W. A. Rosenblith, editor, *Sensory Communication*, page 217. MIT Press, Cambridge, Mass, 1961.

6. J. J. Atick and N. Redlich. *Neural Computation*, 2:308, 1990.

7. W. Bialek, D. L. Ruderman, and A. Zee. In D. Touretzky and J. Moody, editors, *Advances in Neural Information Processing Systems 3*, pages 363–369. Morgan Kaufmann, San Mateo, Ca., 1991.

8. W. Bialek and A. Zee. *J. Stat. Phys.*, 59:103–115, 1990.

9. W. Bialek, F. Rieke, R. R. de Ruyter van Steveninck, and D. Warland. In D. Touretzky, editor, *Advances in Neural Information Processing Systems 2*, pages 36–43. Morgan Kaufmann, San Mateo, Ca., 1990.

10. D. Warland, M. A. Landolfa, J. P. Miller, and W. Bialek. In F. Eeckman, editor, *Analysis and Modeling of Neural Systems 1*. Kluwer, Norwell, Mass., 1991.

11. D. Warland. *Reading Between the Spikes: Real-time Processing in Neural Systems.* University of California Doctoral Thesis, 1991.

12. F. Rieke, W. Yamada, K. Moortgat, E. R. Lewis, and W. Bialek. In Y. Cazals, L. Demany, and K. Horner, editors, *Auditory Physiology and Perception: Proceedings of the 1991 European Hearing Workshop.* in press, 1991.

13. C. E. Shannon. *Proc. Inst. Radio Engr.*, 37:10–21, 1949.

14. D. MacKay and W. McCulloch. *Bull. Math. Biophys.*, 14:127, 1952.

15. J. A. Simmons. *Cognition*, 33:155–199, 1989.

16. A.-C. Aho, K. Donner, C. Hydèn, L. O. Larsen, and T. Reuter. *Nature*, 324:348, 1988.

17. W. Bialek. In E. Jen, editor, *1989 Lectures in Complex Systems, SFI Studies in the Sciences of Complexity*, volume 2, pages 513–595. Addison-Wesley, Reading, Mass., 1990.

18. W. Bialek. *in preparation*, 1991.

19. W. Bialek and W. G. Owen. *Biophys. J.*, 58:1227–1233, 1990.

20. F. Rieke, W. G. Owen, and W. Bialek. In D. Touretzky and J. Moody, editors, *Advances in Neural Information Processing Systems 3*, pages 377–383. Morgan Kaufmann, San Mateo, Ca., 1991.

21. M. V. Srinivasan, S. B. Laughlin, and A. Dubs. *Proc. R. Soc. Lond. B*, 216:427–459, 1982.

22. A. W. Snyder, D. G. Stavenga, and S. B. Laughlin. *J. Comp. Physiol.*, 116:183–207, 1977.

23. B. Sakitt and H. B. Barlow. *Biol. Cybernetics*, 43:97–108, 1982.

24. R. Linsker. In D. Touretzky, editor, *Advances in Neural Information Processing 1*, page 186. Morgan Kaufmann, San Mateo, CA, 1989.

25. R. Linsker. *Annu. Rev. Neurosci.*, 13:257–281, 1990.

Non-linear Analysis of Models for Biological Pattern Formation: Application to Ocular Dominance Stripes

M. J. Lyons and L. G. Harrison
Physics Dept. (MJL) and Chemistry Dept. (LGH)
University of British Columbia
Vancouver, BC, Canada V6T 2A6
e-mail: lyons@physics.ubc.ca

Abstract

We present a technique for the analysis of pattern formation by a class of models for the formation of ocular dominance stripes in the striate cortex of some mammals. The method, which employs the adiabatic approximation to derive a set of ordinary differential equations for patterning modes, has been successfully applied to reaction-diffusion models for striped patterns [1]. Models of ocular dominance stripes have been studied [2,3] by computation, or by linearization of the model equations. These techniques do not provide a rationale for the origin of the stripes. We show here that stripe formation is a non-linear property of the models. Our analysis indicates that stripe selection is closely linked to a property in the dynamics of the models which arises from a symmetry between ipsilateral and contralateral synapses to the visual cortex of a given hemisphere.

Introduction

Much of biological pattern formation occurs in two-dimensional domains, such as single sheets of cells, surfaces of large single cells, or single layers of many nuclei within a single cell, such as the widely-studied ***Drosophila*** blastoderm. In many instances, the patterns formed are in essence a set of parallel stripes. For such cases as mammalian coat markings (*e.g.* zebra), while the general nature of the striped pattern may serve a biological purpose, a fair amount of variation from individual to individual may be tolerable in both the total number of stripes, their precise orientation, and the occurrence of defects such as branching. In other cases, such as the onset of insect segmentation, a pattern of a precisely-determined number of stripes, properly oriented and with no tendencies to break into spots or to form branches, is essential to development of the body plan. Most mutantions which disturb this pattern are lethal at an early developmental stage. Our work on striped patterns had insect segmentation as its initial motivation.

The mechanism of stripe formation remains in all cases unknown. Theoreticians have found reaction-diffusion theory, as initiated by Turing [12], a promising possibility. There are, however, some instances in which the mechanism is likely to be quite different, though the morphological phenomenon is the same. A well-known example of this kind is the formation of ocular dominance stripes. These are spatial patterns of synaptic connections observed in the striate cortex of some mammals, especially primates. In these highest vertebrates (and in contrast to some lower ones) each side of the brain receives roughly equal numbers of connections from both eyes. A stripe consists of cortical cells (in layer IVc) all of which receive connections from the same eye. The stripes bear a strong similarity to zebra stripes.

Ocular dominance patterns also closely resemble the striped patterns which we obtained by computation using a particular class of reaction-diffusion models [1]. This class has a particular dynamical symmetry. In addition to the computations, we showed analytically, using the adiabatic approximation, that this particular symmetry should generate stripes. Here, we apply the same technique to the ocular dominance models.

Ocular Dominance Models

Two well known models for OD patterns [2,3] take the form of integro-differential equations. We consider a phenomenological model originally proposed by Swindale [2], based on a mechanism suggested by Hubel and Wiesel [4]. The densities of left and right synapses $n_L(\vec{r},t)$ and $n_R(\vec{r},t)$, which are functions of a two dimensional spatial coordinate $\vec{r}$ on the cortex and time, develop according to:

$$\frac{\partial n_R}{\partial t} = f(n_R)\int(n_R(\vec{r}-\vec{r}\,')w_{RR}(\vec{r}\,')+n_L(\vec{r}-\vec{r}\,')w_{LR}(\vec{r}\,'))d\vec{r}\,' , \quad (1)$$

$$\frac{\partial n_L}{\partial t} = f(n_L)\int(n_R(\vec{r}-\vec{r}\,')w_{RL}(\vec{r}\,')+n_L(\vec{r}-\vec{r}\,')w_{LL}(\vec{r}\,'))d\vec{r}\,' . \quad (2)$$

The integrals sum competitive and cooperative interactions between like and opposite terminals over the entire patterned region, weighted by the intracortical interactions $w(\vec{r}\,')$ (shown in Figure 1). The function $f(n)$ takes into account the non-linear effects limiting the growth of the synaptic densities. We take a simple logistic form $f(n) = n(N-n)$ where N is the maximum density of synapses. Equations (1,2) may be reduced to a single equation for the difference left and right densities with the assumption that the total density of synapses

$N = n_L + n_R$ is constant [2]:

$$\frac{\partial a}{\partial t} = (a * w + K)(N - a)(N + a)\,. \tag{3}$$

$a = n_R - n_L$, $w = 1/4(w_{RR} + w_{LL})$, $K = (N/4) * (w_{RR} - w_{LL})$ and $*$ indicates convolution. (3) represents a minimal model for OD stripes which has been shown by computation to possess the essential patterning capabilities exhibited by more complex models [2,3,5].

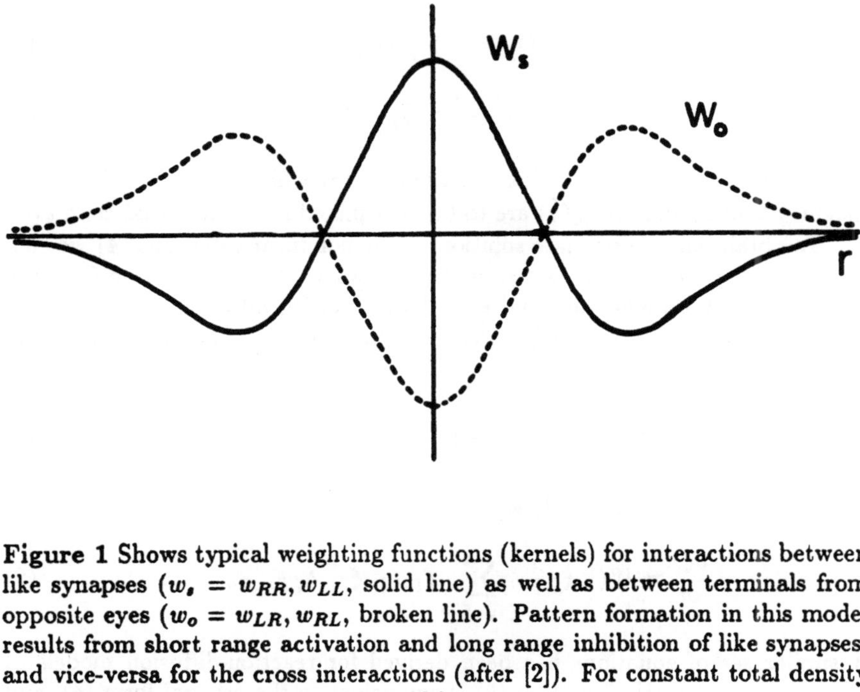

Figure 1 Shows typical weighting functions (kernels) for interactions between like synapses ($w_s = w_{RR}, w_{LL}$, solid line) as well as between terminals from opposite eyes ($w_o = w_{LR}, w_{RL}$, broken line). Pattern formation in this model results from short range activation and long range inhibition of like synapses, and vice-versa for the cross interactions (after [2]). For constant total density of synapses $N = n_L + n_R$, $w_s = -w_o$ [2].

Non-linear Analysis

We are primarily interested in models with symmetry between left and right terminals, *i.e.* $w_{RR} = w_{LL}$ so $K = 0$. This implies that equation (3) is unchanged under the operation $a \to -a$. We analyze the simplest possible pattern-forming system having this symmetry:

$$\frac{\partial a}{\partial t} = a * w - a^3\,. \tag{4}$$

The coefficient of the cubic term can be set to unity without loss of generality by changing the dimensions of the ocular dominance a. Other constants may be absorbed into w. (4) has homogeneous steady state solution given by $a = 0$ which corresponds to an unpatterned state consisting of an equal mixture of left and right terminals. Close to this state the linear part of the equation predominates:

$$\frac{\partial a}{\partial t} = a * w \, . \tag{5}$$

Equation (5) is exactly soluble by Fourier analysis. For the moment, we treat a finite domain, but otherwise leave the boundary conditions unspecified. The general solution is a superposition of plane waves with amplitudes $\xi_k(t)$

$$a(\vec{r},t) = \sum_{\vec{k}} \xi_k(t) e^{i\vec{k}\cdot\vec{r}} \, , \tag{6}$$

$$\xi_{\vec{k},t} = \xi_0(\vec{k}) e^{W(k)t} \, . \tag{7}$$

$W(k)$ is the Fourier tranform of the kernel $w(r)$ which depends only on the magnitude of $\vec{k}$, and the $\xi_0(\vec{k})$ are to be determined from the initial condition.

To obtain an approximate solution to the nonlinear equations (4) we continue to write the solution as a sum of plane waves (6), and derive differential equations for the mode amplitudes. Substituting (6) into (4), multiplying on the left by $e^{i\vec{k}\cdot\vec{r}}$ and integrating over the domain using the orthogonality of plane waves:

$$\frac{1}{A}\int e^{i(\vec{k}-\vec{k}')\cdot\vec{r}} dA = \delta_{\vec{k},\vec{k}'} \, , \; A = \text{ domain area} \, , \tag{8}$$

we obtain the mode amplitude equations:

$$\frac{\partial \xi_{\vec{k}}}{\partial t} = W(k)\xi_{\vec{k}} - \sum_{\vec{k}'\vec{k}''\vec{k}'''} \xi_{\vec{k}'}\xi_{\vec{k}''}\xi_{\vec{k}'''}\delta_{\vec{k},\vec{k}'+\vec{k}''+\vec{k}'''} \, . \tag{9}$$

A similar set of equations has been derived for reaction-diffusion models of pattern formation [1]. Equation (9) is equivalent to the original form (4), and is intractable. It is possible to understand some of the behaviour of model (4) by studying it with a special form for $w(r)$ which allows an approximate solution of equations (9) by the adiabatic method. Briefly, the conditions for the approximation to hold are:

$$W(k_c) > 0 \text{ for one value of } k_c \, , \tag{10}$$

$$W(k) < 0 \text{ for } k \neq k_c \, , \tag{11}$$

$$|W(k)| \gg |W(k_c)| \, . \tag{12}$$

If these conditions are satisfied then the relaxation times of the stable modes are much shorter than that for the unstable modes. The stable mode amplitudes then adjust rapidly to the current values of the ξ_{k_c}. In the adiabatic approximation [6] all of the stable modes ($k \neq k_c$) are then eliminated from the dynamics leaving a set of ODEs for the unstable mode amplitudes, $k = k_c$ at all possible relative orientations ϕ:

$$\dot{\xi}_\phi = W(k_c)\xi_\phi - 3\xi_\phi\xi^*_\phi\xi_\phi - 6\sum_{\phi'}\xi_\phi\xi^*_{\phi'}\xi_{\phi'},\ 0 < \phi, \phi' < \pi,\ \phi' \neq \phi\,. \tag{13}$$

For details on the derivation of equation (13) from (9) we suggest the reader consults [1], where a similar equation was found for a class of reaction-diffusion models which make striped patterns (see Figure 2).

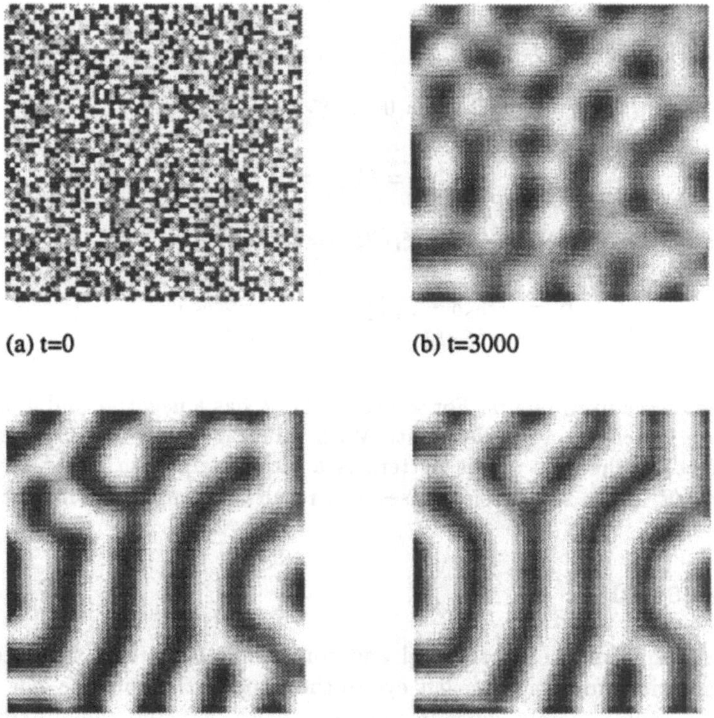

(a) t=0 (b) t=3000

(c) t=5000 (d) t=15000

Figure 2 Time evolution of the concentration profile for one of the morphogens in a reaction-diffusion system with the same dynamical symmetry as equation (4)[1,7,8]. Times are iterations of a finite-difference algorithm. The domain is square with no-flux boundary conditions to mimic the behaviour observed at the boundary of area 17 of the cortex where OD stripes run perpendicular to the edge. The similarity of the RD patterns to OD stripes led us to investigate the properties of OD models.

It may be shown for any number of unstable modes (same k_c, variable ϕ) that the only stable solutions of (13) are stripes [1]. We illustrate this for the simple case of a square domain where there are only four unstable modes corresponding to striped patterns in mutually perpendicular directions. The unstable modes have $\phi = 0, \pi/2, \pi, 3\pi/2$ and $k_c = 2\pi l_c, l_c \in Z^+$ This constrains l_c^2 to integers which cannot be written as the sum of two squares. Then eqns. (13) are:

$$\begin{aligned} \dot{\xi}_0 &= W\xi_0 - 3\xi_0\xi_0^*\xi_0 - 6\xi_0\xi_{\frac{\pi}{2}}^*\xi_{\frac{\pi}{2}} \,, \\ \dot{\xi}_{\frac{\pi}{2}} &= W\xi_{\frac{\pi}{2}} - 3\xi_{\frac{\pi}{2}}\xi_{\frac{\pi}{2}}^*\xi_{\frac{\pi}{2}} - 6\xi_{\frac{\pi}{2}}\xi_0^*\xi_0 \,. \end{aligned} \tag{14}$$

Eqn (18) has fixed points:

$$(I)\ \xi_0^*\xi_0 = W/3, \xi_{\frac{\pi}{2}} = 0 \ \ or \ \ \xi_{\frac{\pi}{2}}^*\xi_{\frac{\pi}{2}} = W/3, \xi_0 = 0 \,, \tag{15}$$

$$(II)\ \xi_0^*\xi_0 = \xi_{\frac{\pi}{2}}^*\xi_{\frac{\pi}{2}} = W/9 \,. \tag{16}$$

The linear stability matrix $A_{ij} = \partial\dot{\xi}_i/\partial\xi_j$ is:

$$\mathbf{A} = \begin{pmatrix} W - 6(\xi_0^*\xi_0 + \xi_{\frac{\pi}{2}}^*\xi_{\frac{\pi}{2}}) & -6\xi_0\xi_{\frac{\pi}{2}}^* \\ -6\xi_{\frac{\pi}{2}}\xi_0^* & W - 6(\xi_0^*\xi_0 + \xi_{\frac{\pi}{2}}^*\xi_{\frac{\pi}{2}}) \end{pmatrix} \tag{17}$$

evaluated at the fixed points. For solution (II), **A** has a negative and a positive eigenvalue indicating a saddle point. With solution (I) we find two negative eigenvalues, so the only stable pattern is a set of stripes, in either direction $\phi = 0$ or $\pi/2$. Mode suppression arises as a result of 'competitive'interactions.

Discussion

In the macaque monkey, ipsilateral and contralateral afferents to the cortex (*i.e.* connections from right or left eye to the same or the opposite side of the brain) are approximately equal in numbers. Also, left and right dominance stripes in layer IVc are similar in form and width (Figure 3(a)). The only difference between left and right dominated bands appears to be their connection to different eyes. This symmetry between ipsilateral and contralateral inputs is related to the selection of stripes rather than any other spatial pattern. In monocular deprivation experiments, this symmetry is disrupted, leading to a spotted pattern of ocular dominance in which terminals from the eye receiving visual information are more numerous than ones from the blindfolded eye. Binocular deprivation restores the left/right symmetry and stripes are observed when neither eye receives input during development.

In the cat, there is normally a quantitative asymmetry between the ipsilateral (about 30%) and contralateral (about 70%) inputs to the cortex [9]. A splotchy pattern of ocular dominance is found (Figure 3(b)). Binocular deprivation gives a pattern quite unlike stripes. To examine the effect on ocular dominance patterns of varying degrees of asymmetry we suggest a comparative survey of various species of mammals with different ratios of ipsilateral/contralateral afferents.

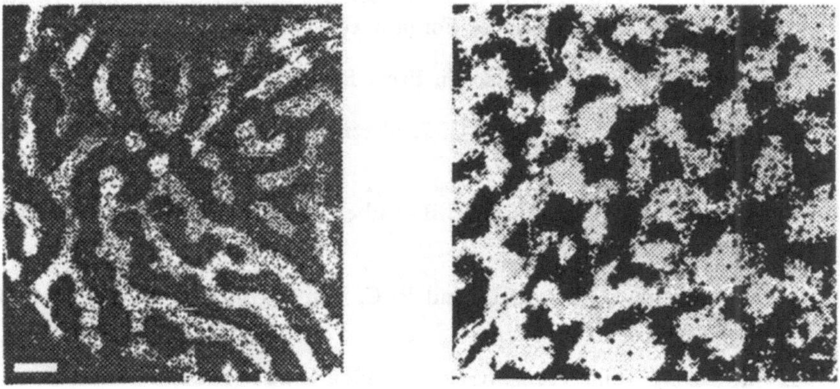

(a) (b)

Figure 3 (**a**) shows the pattern of ocular dominance in layer IVc of a macaque monkey's primary visual cortex. (**b**) shows the pattern for the visual cortex of a cat. (a) was adapted from a digitally scanned image of figure 5 of [10], (b) was adapted from figure 7b of [11].

Acknowledgements

We thank Ken Miller, Nick Swindale, Martin Zuckermann, and Thurston Lacalli for helpful discussions. Thanks also to Florence Tam for drawing Figure 1, and Leah Keshet for useful comments on the manuscript. This work was made possible by financial support from NSERC Canada.

References

[1] M. J. Lyons and L. G. Harrison, Chemical Physics Letters, (accepted for publication).

[2] N. V. Swindale, Proc. R. Soc. Lond. B **208** (1980) 243.

[3] K.D. Miller, J.B. Keller, and Michael P. Stryker, Science **245** (1989) 605.

[4] D. H. Hubel and T. N. Wiesel, Proc. R. Soc. Lond. B **198** (1977) 1.

[5] J. R. Thomson, Z. Zhang, Wm. Cowan, M. Grant, J. A. Hertz, and M. J. Zuckermann, Physica Scripta, **T33** (1990) 102.

[6] H. Haken, Synergetics, 3rd ed. (Springer-Verlag, Berlin, 1983).

[7] L. G. Harrison and M. J. Lyons, Proceedings of the Les Houches Winter Workshop, D. Beysens and G. Forgacs, eds. 1991.

[8] L. G. Harrison and T. C. Lacalli, Proc. R. Soc. Lond. B **202** (1978) 361.

[9] S. Levay, M. P. Stryker, and C. J. Shatz, J. Comp. Neurol. **179** (1978) 223.

[10] S. LeVay, T. N. Wiesel, and D. H. Hubel, J. Comp. Neurol. **191** (1980) 1.

[11] P. A. Anderson, J. Olavarria, and R. C. Van Sluyters, J. Neuroscience **8** (1988) 2183.

[12] A. M. Turing, Phil. Trans. R. Soc. London B **237** (1952) 37.

A Hierarchical Sensory-Motor Architecture of Oscillating Cortical Area Subnetworks

Bill Baird
Dept Mathematics and Dept Molecular and Cell Biology,
U.C.Berkeley, Berkeley, Ca. 94720, baird@math.berkeley.edu

Frank Eeckman
O - Division, Lawrence Livermore National Laboratory,
P.O. Box 808 (L-270), Livermore, Ca. 94550

Abstract

We show how hierarchical networks may be constructed of interconnected oscillatory network modules developed previously as models of olfactory cortex, or caricatures of "patches" of neocortex. The architecture is such that the larger system is itself a special case of the type of network of the submodules, and can be analysed with the same tools used to design the subnetwork modules. A particular subnetwork is formed by a set of neural populations whose interconnections also contain higher order synapses. These synapses determine attractors for that subnetwork independent of other subnetworks. Each subnetwork module assumes only minimal coupling justified by known anatomy. An N node module can be shown to function as an associative memory for up to $N/2$ oscillatory and $N/3$ chaotic memory attractors.

The modules can learn connection weights between themselves which will cause the system to evolve under a clocked "machine cycle" by a sequence of transitions of attractors within the modules, much as a digital computer evolves by transitions of its binary flip-flop states. Thus the architecture employs the principle of "computing with attractors" used by macroscopic systems for reliable computation in the presence of noise. Clocking is done by rhythmic variation of certain bifurcation parameters which hold sensory modules clamped at their attractors while motor states change, and then clamp motor states while sensory states are released to take new states based on input from external motor output and internal feedback.

Simulations show robust storage of oscillatory attractor transition sequences in a system with a sinusoidal clock and continuous oscillatory intermodule driving. The phase-locking or "binding" which occurs rapidly between coupled attractors of similar resonant frequency in different modules is important for reliable transitions. An oscillating systm with phase-locking modules is more robust to perturbation by additive Gaussian noise than an otherwise identical system with static attractors (frequencies set to zero). We show analytically how modular networks with more fault tolerant and biologically plausible distributed patterns can be built from "spreading activation" style networks which use single node representations.

1 Introduction

Patterns of 40 to 80 Hz oscillation have been observed in the large scale activity (local field potentials) of olfactory cortex [11] and visual neocortex [12, 13, 6, 7], and shown to predict the olfactory and visual pattern recognition responses of a trained animal. Similar observations of 40 Hz oscillation in auditory and motor cortex, and in the retina and EMG have been reported. It thus appears that cortical computation in general may occur by dynamical interaction of resonant modes, as has been thought to be the case in the olfactory system [9]. Given the sensitivity of neurons to the location and arrival times of dendritic input [17], the sucessive volleys of pulses (the "wave packet" of Freeman) that are generated by the collective oscillation of a neural net may be ideal for reliable longe range transmission of the collective activity of one cortical area to another.

The oscillation can serve a macroscopic clocking function and entrain or "bind" the relevant microscopic activity of disparate cortical regions into a well defined phase coherent collective state or "gestalt". This can overide irrelevant microscopic activity, produce coordinated motor output, and determine what goes into long term memory. Coherent ensemble activity may thus be an important medium of cortical computation. If this view is correct, then oscillatory network modules form the actual cortical substrate of the diverse sensory, motor, and cognitive operations now studied in static networks. It must ultimately be shown how those functions can be accomplished with oscillatory dynamics.

With this intent, we have constructed a parallel distributed processing architecture that is designed to model, however simplistically, the architecture of cerebral cortex. The goal is to accomplish real tasks that brains can do, using ordinary differential equations, in networks that are as faithful as possible to the known dynamics and anatomical structure of cortex. In this effort we hope to uncover novel design principles that may advance the art of computation.

The system may be viewed as a discrete-time symbol processing architecture, with analog input and oscillatory subsymbolic representations, constructed from a system of continuous nonlinear ordinary differential equations. Single network modules of the architecture have been sucessfully used for recognition of handwritten characters, and the capabilities of networks of coupled modules are being explored by application to the problem of word recognition from character sequences. Other approaches to oscillatory memory may be found in [8, 24, 16, 19].

1.1 Computing with attractors

In the design of this system, we follow an approach inspired by a particular concept of the physical structure required of macroscopic computational systems in general for reliable computation in the presence of noise. We view a computational medium as a set of structurally stable subsystems which can be coupled to form a larger system. By "structurally stable" we mean that the dynamical behavior of each subsystem is to a large extent immune to small perturbations due to noise or parameter changes. We assume that the dynamics of each subsystem is organized into attractor basins. As the overall system evolves in time (which could in principle proceed indefinitely), each subsystem passes through a sequence of attractors. *These sequences of attractors constitute the "computation" of the system.*

Present day digital computers are such systems. They are built of "flip-flops", which at the level of their transistors are continuous, structurally stable dissipative dynamical systems with two stable equilibrium states underlying the symbols we call "0" and "1". These states are attractors, i. e. returned to after small perturbations, because the flip-flops are "dissipative open systems" - each with its own source of power, and each continually dissipating energy to maintain its own equilibrium state.

We argue that the neuron itself is not the required structurally stable subsystem - it is not functionally a bistable element (flip-flop) whose activity is buffered against the effects of noise, even though it is commonly modeled as such an element in many neural networks. The output of most active neurons is a Poisson distribution of pulse intervals which suggests a firing threshold buried in noise. There are no such things as two-state neurons which go into hard saturation on either side of a threshold and stay there. An action potential is a brief pulse, not a step change of state. If repetitive firing is considered the "on" state, then the observed variation of mean firing rate with different levels of suprathreshold stimulation is unaccounted for.

We consider an intermediate "network" level of organization between the neuron and the whole brain - the "cortical area" to be the necessary structurally stable subsystem of the computational system that is the brain. We take a local population of neurons, possibly as large as a minicolumn, to be a node of the network. Its activation level is its average cell voltage or pulse output density. A network module then spans a "patch" of cortex which might be as small as a hypercolumn in primary visual cortex, or as large as a cortical area, such as inferotemporal or prepyriform cortex. The patch then functions as a multiflop with many oscillatory attractors.

The computations of the brain that occur on the timescale of seconds are hypothesized to consist of transitions of attractors in these cortical area subnetworks that occur by interactions of the areas through learned (programmed) connections, as in cellular automata or Von Neuman machines. The 200 millisecond time period required between successive voluntary eyeball movements or alterations of a motor action implies an internal clock interval at which new sensory data is sampled and motor output given. This interval allows for sucessive relaxations of internal subnetwork modules to their attractors, as is required for digital circuits.

In a digital computer, the binary flip-flop elements are made to bifurcate, say from the 1 equilibrium state being stable to the 0 state being stable, by changes in the signal or clock inputs. We hypothesize that the associative networks of the neocortex are clocked to change oscillatory attractors by the alpha rhythm at a 10 Hz "framing rate" under thalamic attention control — much as the olfactory system appears to be clocked to change attractors between 40 Hz "bursts" by the 3 - 8 Hz respiratory rhythm[2].

There is a demonstrable role for analog signal processing and motor control in the brain, but "computation with attractors" is also a necessity. Somewhere in the loop from sensation to action, an animal filters out many details of a perception to make a response choice that is invariant to those details. The hungry cat, seeing a mouse, engages in stalking behavior that is unaffected by most details of the mouse's appearance. Without the categories that can be established by basins of attraction, the world never repeats. The ability of humans and primates to manipulate internal visual and auditory symbols, as a subset of their capabilities, may be considered evidence that neural attractors are employed in some domains.

2 Biological Subnetwork Module

We have determined a biologically "minimal" model that is intended to assume the least anatomically justified coupling sufficient to allow function as an oscillatory associative memory. The network is meant only as a cartoon of the real biology, which is designed to reveal the general mathematical principles and mechanisms by which the actual system might function.

Long range excitatory to excitatory connections are well known as "associational" connections in olfactory cortex[15] and cortico-cortico connections in neocortex. Since our units are neural populations, we can expect that some density of full cross-coupling exists in the system[15], and our weights are taken to be the average synaptic strengths of these connections. Local inhibitory "interneurons" are a ubiquitous feature of the anatomy of cortex [13, 7]. It is unlikely that they make long range connections (> 1 mm) by themselves. These connections, and even the debated interconnections between them, are therefore left out of a minimal coupling model. The resulting network is a fair cartoon of the well studied circuitry of olfactory (pyriform) cortex. Since almost all of

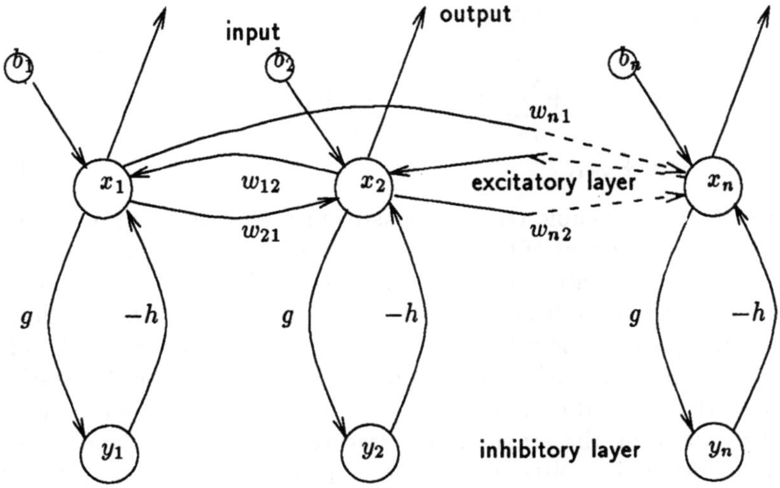

Figure 1: Biological subnetwork of excitatory cell populations x_i, inhibitory cell populations y_i, inputs b_i, adaptive excitatory to excitatory connections W_{ij}, and constant local inhibitory feedback connections g and $-h$.

cortex has this type of structure in the brains of amphibia and reptiles, our super-network of these submodules has the potential to become a reasonable caricature of the full cortical architecture in these animals. Although the neocortex of mammals is more complicated, we expect the model to provide useful suggestions about the principles of oscillatory computation there as well.

For an N dimensional system, this minimal coupling structure is described mathematically by the matrix T.

$$T = \begin{bmatrix} W & -hI \\ gI & 0 \end{bmatrix} \tag{1}$$

W is the $N/2 \times N/2$ matrix of excitatory interconnections, and gI and hI are $N/2 \times N/2$ identity matrices multiplied by the positive scalars g, and h. These give the strength of coupling around local inhibitory feedback loops. A state vector is composed of local average cell voltages for $N/2$ excitatory neuron populations $\vec{x}$ and $N/2$ inhibitory neuron populations $\vec{y}$. Intuitively, since the inhibitory units receive no direct input and give no direct output, they act as hidden units that create oscillation for the amplitude patterns stored in the excitatory cross-connections W. This may perhaps be viewed as a specific structural addition to a recurrent analog higher order network architecture to convert its static attractors to periodic attractors. Here the symmetric sigmoid functions of such a network are Taylor expanded up to cubic terms with third order weights (quadratic terms are killed by the symmetry). Network equations with the first order coupling (1 shown above, plus these higher order excitatory synapses, are shown below, in component form.

$$\dot{x}_i = -\tau x_i - hy_i + \sum_{j=1}^{N/2} W_{ij}x_j - \sum_{jkl=1}^{N/2} W_{ijkl}x_jx_kx_l + b_i, \quad (2)$$

$$\dot{y}_i = -\tau y_i + gx_i, \quad (3)$$

The competitive (negative) cubic terms of constitute a directly programmable nonlinearity that is independent of the linear terms. Normal form theory shows that these cubics are the essential nonlinear terms required to store oscillations, because of the (odd) phase shift symmetry required in the vector field. They serve to create multiple periodic attractors by causing the oscillatory modes of the linear term to compete, much as the sigmoidal nonlinearity does for static modes in a network with static attractors [1, 3]. Intuitively, these terms may be thought of as sculpting the maxima of a "saturation" (energy) landscape, into which the modes with positive eigenvalues expand, and positioning them to lie in the directions specified by the eigenvectors to make them stable. A Liapunov function for this landscape may be explicitly constructed in a special polar coordinate system [1, 3]. We use this network directly as our biological model. From a physiological point of view, (2) may be considered a model of a biological network which is operating in the linear region of the known *axonal* sigmoid nonlinearities, and contains instead these higher order *synaptic* nonlinearities.

Adding the higher order weights corresponds, in connectionist language, to increasing the complexity of the neural population nodes to become "higher order" or "sigma-pi" units. Clusters of synapses within a population unit can *locally* compute products of the activities on incomming primary connections W_{ij}, during higher order Hebbian learning, to establish a weight W_{ijkl} (see figure 3.1). These secondary higher order synapses are then used in addition to the synapses W_{ij}, during operation of the overall network, to weight the effect of triple products of inputs in the output summation of the population.

Using only the long range excitatory *connections* W_{ij} available, some number of the higher order synaptic weights W_{ijkl} could be realized locally within a neural population in the axo-dendritic interconnection plexus known as "neuropil" [2]. Only $(N/2)^2$ of these $(N/2)^4$ possible higher order weights are required in principle to approximate the performance of the projection algorithm [4]. The size of our cortical patches is limited by this number, and is itself motivation for modularity.

3 Learning by the Projection Algorithm

A key feature of a net constructed by this algorithm is that the underlying dynamics is explicitly isomorphic to any of a class of standard, well understood nonlinear dynamical systems - a "normal form" [14]. This system can be chosen in advance, independent of both the patterns to be stored and the learning algorithm to be used. This control over the dynamics permits the design of important aspects of the network dynamics independent of the particular patterns to be stored. Stability, basin geometry, and rates of convergence to attractors can be programmed in the standard dynamical system.

Here we use the normal form for the Hopf bifurcation [14] as a simple recurrent competitive k-winner-take-all network with a cubic nonlinearity shown here in Cartesian coordinate form.

$$\dot{v}_i = \sum_{j=1}^{N} J_{ij} v_j - v_i \sum_{j=1}^{N} A_{ij} v_j^2 \tag{4}$$

This network lies in what might considered diagonalized, or "mode," or "overlap," or "memory coordinates" (one memory per k nodes). For temporal patterns, these nodes come in complex conjugate pairs (with eigenvalues specified in J) which supply Fourier components for trajectories to be learned. Chaotic dynamics may be created by specific programming of the interaction of two pairs of these nodes.

The rule for learning desired distributed spatial or spatio-temporal patterns can be shown to be equivalent to the operation of "projecting" sets of these nodes into "network" coordinates (the standard basis) using the desired vectors as corresponding columns of a transformation matrix P. The differential equations of the recurrent network itself may be viewed as linearly transformed or "projected", leading to new recurrent network equations with general coupling T_{ij} for the linear terms, and general higher order weights T_{ijkl} for the cubic terms [1]. The projection learning rule is,

$$T = PJP^{-1}\ , \quad and \quad T_{ijkl} = \sum_{mn=1}^{n} P_{im} \tilde{A}_{mn} P_{mj}^{-1} P_{nk}^{-1} P_{nl}^{-1}. \tag{5}$$

Thus, in the general case, the normal form (4) is projected to become,

$$\dot{x}_i = -\tau x_i + \sum_{j=1}^{N} T_{ij} x_j - \sum_{jkl}^{N} T_{ijkl} x_j x_k x_l. \tag{6}$$

The minimal network coupling (1) for T results from the projection learning rule (5) when a specific biological form is chosen, in the columns s of P, for the patterns to be stored. Only the higher order weights W_{ijkl} between excitatory populations shown in the biological module (2) are required for approximate pattern storage [4]. This special complex form for P^s and the corresponding asymptotic solutions $X^s(t)$ established are,

$$P^s = \begin{bmatrix} |\vec{x}^s| e^{i\vec{\theta}_x^s} \\ \sqrt{\frac{g}{h}} |\vec{x}^s| e^{i\vec{\theta}_y^s} \end{bmatrix} \quad \Rightarrow \quad X^s(t) = \begin{bmatrix} |\vec{x}^s| e^{i\vec{\theta}_x^s + i\omega^s t} \\ \sqrt{\frac{g}{h}} |\vec{x}^s| e^{i\vec{\theta}_y^s + i\omega^s t} \end{bmatrix}. \tag{7}$$

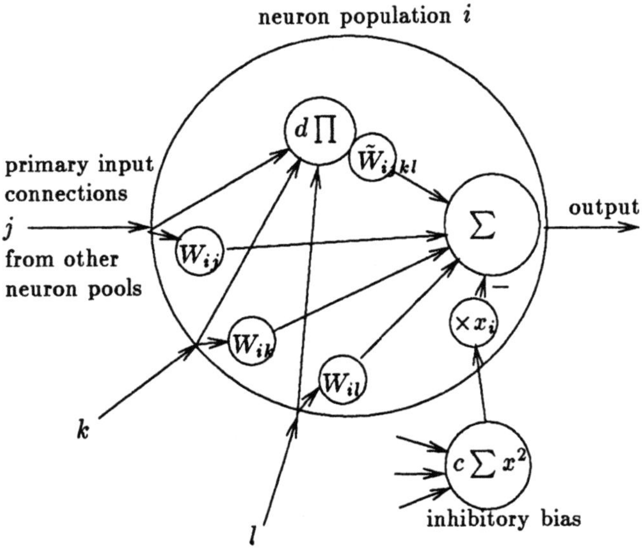

Figure 2: Neural population subnetwork acting as a sigma-pi unit. It uses secondary higher order synaptic weights $\tilde{W}_{ijkl}$ on products of the activities of incoming primary connections W_{ij}, and receives a global inhibitory bias.

The phase θ_x, θ_y is constant over the components of each kind of neural population x and y, and differs only between them. This is basically what is observed in the olfactory bulb (primary olfactory cortex) and prepyriform cortex[2]. The phase of inhibitory components θ_y in the bulb lags the phase of the excitatory components θ_x by approximately 90 degrees.

3.1 Hebbian Learning

A Hebbian learning rule may be derived from the projection learning rule which allows a network to learn its attractor categories by local self organization of synapses and synaptic clusters according to pre and post synaptic activities experienced during external input forcing. For orthonormal static patterns $\vec{x}^s$, $P^{-1} = P^T$, and the projection rule for the W matrix reduces to an outer product, or "Hebb" rule, and the projection for the higher order weights becomes a *multiple* outer product rule:

$$W_{ij} = \sum_s^{N/2} \alpha^s x_i^s x_j^s \;, \qquad W_{ijkl} = c\delta_{ij}\delta_{kl} - d\sum_{s=1}^{N/2} x_i^s x_j^s x_k^s x_l^s \;. \tag{8}$$

When the Hebbian learning rule (8) is used, the higher order weights W_{ijkl}

of the network model (2) can be decomposed so that (2) becomes,

$$\dot{x}_i = -\tau x_i - hy_i + \sum_{j=1}^{N/2} W_{ij}x_j + d \sum_{jkl=1}^{N/2} \tilde{W}_{ijkl}x_jx_kx_l - cx_i \sum_{j=1}^{N/2} x_j^2 + b_i \tag{9}$$

where $\tilde{W}_{ijkl}$ comes from the multiple outer product in (8), and $-cx_i \sum_{j=1}^{N/2} x_j^2$ comes from the $c\delta_{ij}\delta_{jk}$ term in (8). This single weight negative term corresponds to a shunting inhibitory bias which depends on the global excitatory activity of the network. "Shunting" here means multiplied by the current average cell voltage x_i of population i. This is an input which is identical for all excitatory neural populations and could be calculated by a single node of the network which receives input from all excitatory populations, as shown in figure 3.1. Such a node might correspond to one of the nuclei which lie below the prepyriform cortex. These send and receive the diffuse projections required to and from prepyriform cortex. The constants c and d of eq. (9) give the magnitude of the inhibitory bias c and the average higher order weight d. These constants are derived from entries in the normal form matrix A, and, as we will discuss below, $c > d$ guarantees stability of all stored patterns. The greater the bias c relative to d, the greater the "competition" between stored patterns, the more robust is the stability of the present attractor, and vice versa. This is the mechanism employed later in the biological sensory-motor architecture for central control of attractor transitions within modules.

3.2 Programming the Normal Form Network

The key to the power of the projection algorithm to program these systems lies in the freedom to chose a well understood normal form for the dynamics, independent of the patterns to be learned. The Hopf normal form used here is especially easy to work with for programming periodic attractors, but handles fixed points as well. J is a matrix with real eigenvalues for determining static attractors, or complex conjugate eignevalue pairs in blocks along the diagonal for periodic attractors. The real parts are positive, and cause initial states to move away from the origin, until the competitive (negative) cubic terms dominate at some distance, and cause the flow to be inward from all points beyond. The off-diagonal cubic terms cause competition between directions of flow within a spherical middle region and thus create multiple attractors and basins. The larger the eigenvalues in J and off-diagonal weights in A, the faster the convergence to attractors in this region.

It is easy to choose blocks of coupling along the diagonal of the A matrix to produce different kinds of attractors, static, periodic, or chaotic, in different coordinate subspaces of the network. The sizes of the subspaces can be programmed by the sizes of the blocks. The basin of attraction of an attractor determined within a subspace is guaranteed to contain the subspace [3]. Thus basins can be programmed, and "spurious" attractors can be ruled out when all subspaces have been included in a programmed block [1, 3].

This can be accomplished simply by choosing the A matrix entries outside the blocks on the diagonal (which determine coupling of variables within a subspace) to be greater (more negative) than those within the blocks. The principle is that this makes the subspaces defined by the blocks compete exhaustively, since intersubspace competition is greater than subspace self-damping. Within the middle region, the flow is forced to converge laterally to enter the subspaces programmed by the blocks.

An simple example is a matrix of the form,

$$A = \begin{bmatrix} d & & & & & & & \\ & d & & & & (g) & & \\ & & \begin{bmatrix} d & c \\ c & d \end{bmatrix} & & & & & \\ & & & \begin{bmatrix} d & d & c & c \\ d & d & c & c \\ c & c & d & d \\ c & c & d & d \end{bmatrix} & & & & \\ & & (g) & & & & \ddots \end{bmatrix},$$

where $0 < c < d < g$. There is a static attractor on each axis (in each one dimensional subspace) corresponding to the first two entries on the diagonal, by the argument above. In the first two dimensional subspace block there is a single fixed point in the interior of the subspace on the main diagonal, because the off-diagonal entries within the block are symmetric and less negative than those on the diagonal. The components do not compete, but rather combine. Nevertheless, the flow from outside is into the subspace, because the entries outside the subspace are more negative than those within it.

The last subspace contains entries appropriate to guarantee the stability of a periodic attractor with two frequencies (Fourier components) chosen in the J matrix. The doubling of the entries is because these components come in complex conjugate pairs (in the J matrix blocks) which get identical A matrix coupling. Again, these pairs are combined by the lesser off-diagonal coupling within the block to form a single limit cycle attractor. A large subspace can store a complicated continuous periodic spatio-temporal sequence with many component frequencies [3].

The discrete Fourier transform (DFT) of a set of samples of such a sequence in space and time can be input directly to the P matrix as a set of complex columns corresponding to the frequencies in J and the subspace programmed in A. A total of $N/2$ such DFT samples of N dimensional time varying spatial vectors may be placed in the P matrix, and parsed by the A matrix into $M < N/2$ separate sequences as desired. Separate basins of attraction are then guaranteed for these sequences by the projection theorem[3]. For a symmetric A matrix, there is a Liapunov function, in the amplitude equations of a polar coordinate version of the normal form, which governs the approach of initial states to stored trajectories.

3.3 Chaotic Attractors

Chaotic attractors may be created in this normal form, with sigmoid nonlinearities added to the right hand side, $v_i \rightarrow tanh(v_i)$. The sigmoids yield a spectrum of higher order terms that break the phase shift symmetry of the system. Two oscillatory pairs of nodes like those programmed in the block above can then be programmed to interact chaotically. In our simulations, for example, if we set the upper block of d entries to -1, and the lower to 1, and replace the upper c entries with 4.0, and the lower with -0.4, we get a chaotic attractor of dimension less than four, but greater than three [5].

This is "weak" or "phase coherent" chaos that is still nearly periodic. It is created by the broken symmetry, when a homoclinic tangle occurs to break up an invariant 3-torus in the flow [14]. This is the Ruelle-Takens route to chaos and has been observed in Taylor-Couette flow when both cylinders are rotated. Experiments of Freeman have suggested that chaotic attractors of the above dimension occur in the olfactory system [10]. These might most naturally occur by the interaction of oscillatory modes. We have demonstrated in simulations as well that a network with multiple Lorenz attractors can be created simply by adding off-diagonal normal form competitive terms to couple sets of the three Lorenz equations [5].

In normal form, or in a biological network, these chaotic attractors have a basin of attraction in the N dimensional state space that constitues a category, just like any other attractor in this system. They are, however, "fuzzy" attractors, and there may be computational advantages to the basins of attraction (categories) produced by chaotic attractors, or to the effects their outputs have as fuzzy inputs to other network modules. The particular N dimensional spatio-temporal patterns learned for the four components of these chaotically paired modes may be considered a coordinate specific "encoding" of the strange attractor, which may constitute a recognizable input to another network, if it falls within some learned basin of attraction. While the details of the trajectory of a strange attractor in any real physical continuous dynamical system are lost in the noise, there is still a particular statistical structure to the attractor which is a recognizable "signature".

4 Architecture of Subnetwork Modules

Because we have developed this class of mathematically well-understood associative memory networks, we can take a constructive approach to building a cortical architecture, using these networks as modules, in the same way that digital computers may be designed from well behaved flip-flop circuits. The operation of the architecture can be precisely specified, even though the outcome of a particular computation set up by the programmed connections may be unknown.

Each module is described in normal form or "mode" coordinates as the k-winner-take-all network discussed above, where the winning set of units may have static, periodic or chaotic dynamics. By choosing modules to have only two attractors, networks can be built which are similar to networks using binary units. There can be fully recurrent connections between modules. The entire super-network of connected modules, however, is itself a polynomial network that can be projected into standard network coordinates like those for the biological model. The attractors within the modules may then be distributed patterns like those described for the biological model (7), and observed experimentally in the olfactory system [11]. The system is still equivalent to the architecture of modules in normal form, however, and may easily be designed, simulated, and theoretically evaluated in these coordinates.

We show here how normal form subnetworks maybe coupled with connections learned by a sequence learning algorithm, and this larger network itself may be projected without destroying the identity of the subnetworks. The A matrix for a network module like that described above may itself become a block on the diagonal in the matrix A^* of a larger "super-network". The overall network matrix A^* is then like a large normal form A matrix with zero elements in the off-diagonal region outside the diagonal A blocks that define multiple attractors for the modules. The modules neither compete nor combine states, in the absence of A^* matrix coupling between them, but take states independently based on their inputs to each other through the weights in a matrix J^*. What is new here is that J^* is allowed to have nonzero off-diagonal elements that provide coupling between the blocks J along the diagonal that define the modules. There can be fully recurrent connections J^*_{ij} *between* module J blocks, but not *within*, so that the modules themselves remain in normal form - with J matrices having real or complex conjugate eigenvalues on the diagonal.

For simplicity, we first consider the case of modules with static attractors. For a three subnetwork system, of three N dimensional modules, for example, the matrix P^* for the supernet contains diagonal N dimensional block matrices B, C, and D with distributed attractor patterns in their columns to be assigned to the corresponding three subnetworks. Input might be wired into the units of the first subnetwork B, and output taken from the the third module D, leaving C as a "hidden" subnetwork which takes, after learning, the attractor states required to produce desired output sequences for given input sequences. Here, for simplicity, we choose identical postive real eigenvalues $\alpha = 1$ in the J matrix blocks which determine static attractors for the modules. Thus the matrix J^* of the supernet appears with the identity I (actually αI) in the diagonal blocks, but has further weights in the matrices F, G, H, K, L, and M in the off-diagonal grid of blocks that define interconnections between the subnetwork blocks,

$$P^* = \begin{bmatrix} B & & \\ & C & \\ & & D \end{bmatrix}, \; J^* = \begin{bmatrix} I & F & G \\ H & I & K \\ L & M & I \end{bmatrix}, \; P^{*-1} = \begin{bmatrix} B^{-1} & & \\ & C^{-1} & \\ & & D^{-1} \end{bmatrix}.$$

Now J^* can be decomposed, $J^* = I + J^{*'}$,

$$where\ J^{*'} = \begin{bmatrix} 0 & F & G \\ H & 0 & K \\ L & M & 0 \end{bmatrix},\ so\ P^*J^*P^{*-1} = P^*IP^{*-1} + P^*J^{*'}P^{*-1},$$

$$hence\quad T^* = I + P^*J^{*'}P^{*-1} = \begin{bmatrix} I & BFC^{-1} & BGD^{-1} \\ CHB^{-1} & I & CKD^{-1} \\ DLB^{-1} & DMC^{-1} & I \end{bmatrix}.$$

Thus the diagonal and the off-diagonal blocks can be transformed independently - they do not affect each other's transform. The *higher order terms* for each subnet transform independently as well, and no terms cross-connecting subnets are generated. For all terms T^*_{ijkl} which have any index i, j, k, or l outside of a subnet block, the corresponding term of P^* or P^{*-1} is zero, and zeros all terms of the summation over m and n to guarantee that T^*_{ijkl} must be zero.

The identity of the individual subnetwork patches is preserved by the transformation (projection) from normal form to network coordinates. The connections *within* patches in network coordinates determine the particular distributed patterns that correspond to nodes in the normal form coordinates. The between-patch connections in network coordinates are transformed from those learned to effect transitions between winning nodes in normal form coordinates; they now effect transitions between the corresponding distributed patterns in the patches.

For an architecture with biological network modules, the J^* and P^* matrices contain blocks with complex conjugate eigenvalues and eigenvectors respectively. When each block of P^* has the form (7) shown previously for oscillatory attractors, then the connections within each diagonal block of T^* have the form (1) shown for the biological module. There are then nonzero entries in the upper left quadrants of the grid of blocks in the off-diagonal region which contain weights for the connections between excitatory elements of the subnetworks.

The higher order terms for each subnet are independent, as before, and take the same form as they would without being embedded in a super-network. Pairs of the complex conjugate coordinate axes for oscillators in normal form coordinates can be set up by the coupling for a module within an A^* matrix block to produce chaotic attractors as described above. For these systems with oscillatory or chaotic attractors, the weights between subnetworks implicitly encode frequency and vector valued amplitude and phase pattern information in order to effect transitions to the proper dynamic attractors.

A network can thus be designed or analysed in normal form coordinates as a "spreading activation" style network (only one node within a subnet is "activated" over others) where "concepts" correspond to single nodes, but can be transformed to operate with any of these more fault tolerant and biologically plausible attractors, where concepts (symbols) are distributed (possibly dynamic) patterns. For example, a super-network with can be constructed with modules in normal form coordinates that correspond to the "cortical patches" with competitive winner take all dynamics used in an architecture due to Schmidhuber [22]. The intermodule connections can then be trained in these coordinates by his "neural bucket brigade algorithm", for which simulation results demonstrating sequence learning are available. Basically, we can substitue our modules for his, use his connections, and his results apply to our system. This is true even when we use oscillatory attractors and project to get distributed activity patterns, because our network is still functionally isomorphic to his. Many other network models, such as the McClelland-Rummelhart letter recognition model[20], employ this basic structure of cooperatively connected subsets of competitive elements, and can be translated into our system to explore the effects of more complex dynamics on its operation.

4.1 Sensory-Motor Clocking

The neural bucket brigade algorithm and discrete time temporal learning algorithms like the temporal Hebbian rule, $w_{ij}(t) = w_{ij}(t-1) + \alpha x_i(t) x_j(t-1)$, require that at least the previous time step activations at each node be saved in order to calculate the weight update after the transition to the next state. Since our transition intervals are several hundred milliseconds long, it is hard to point to a specific biological process within a local neural population that might retain memory of such an activation, and feed it into the weight changing process, without interfering with the on going dynamical relaxation to the new state.

Without claiming positive biological evidence, but to get started and explore one network level solution to this problem, we introduce a "machine cycle" for our system. This is implemented by reciprocally varying the d "competition" (bifurcation) parameters of the sensory and motor patches, to alternately clamp one set of attractors, while the other is allowed to make attractor transitions prescribed by the present sensory states acting through the intermodule connections with weights specified by the J^* matrix [2]. Motor states feed back to sensory states to cause transitions, so that internal activity can be self-sustaining, as in any recurrent network; but this internal activity can also be guided by a sequence of sensory inputs to produce a learned sequence of motor outputs. When the inputs from one module to the next are given as impulses that establish initial conditions, the relaxation dynamics of a module is exactly predicted by the projection theorem[1]. Discrete time clocking is then done essentially by pulsing the states of the sensory modules into the motor modules and vice versa.

This alternation of sensory and motor states is inspired to some degree by the structure of cerebral cortex, with its segregation and pairing of functionally related sensory and motor areas accross the central sulcus, and because it accomplishes the task of distinguishing temporal order by "action - reaction". This allows the implementation of weight changes between time steps without requiring that neurons "store" their last activation states. Temporal transitions are always between sensory and motor areas, and the activations preceeding any motor state are always available in the present sensory state, and vice-versa. A hierarchy of temporal "chunks" can be created in the system, by allowing transitions of higher level modules only after many transitions of lower level modules. Such high level modules may thus encode "plans" that activate long sequences of low level behaviors - a notion which is also being explored by other researchers [21]. Even the attractors of the motor modules which produce final output in our system might be thought of as supplying "plan vectors" in premotor cortex to lower motor control networks, like those of Jordan [18], which generate smoothed and coordinated analog muscle activations.

5 Simulations

Preliminary simulations of this architecture show robust storage of oscillatory attractor transition sequences in a system with a sinusoidal clock and continuous oscillatory intermodule driving. Coupled attractors of similar resonant frequency in different modules rapidly become phase locked. This is similar to the apparent "binding" of sensory features by phase-locking that has been observed in visual cortex [13, 7], and used in other oscillatory memory models [23, 24, 16]. In this system, it appears that oscillating modules within the sensory set must phase-lock to the modules of the motor set in order to reliably drive them through the transitions of attractors that have been programmed by the intermodule connections. This perhaps models a further function of binding at a temporal level of processing.

A set of four modules, each with three oscillatory attractors, was divided into two sensory and two motor modules. Each module consisted of six differential equations of the form (4), with identical complex conjugate eigenvalues in the corresponding diagonal blocks of the J^* matrix, and a six by six dimensional A^* matrix coupling block, of the form discussed earlier that guarantees three single frequency oscillatory attractors. The super-network of coupled modules thus consisted of 24 differential equations.

Two further equations of the same form were added with a two dimensional A^* matrix block on the diagonal that formed a single stable oscillator. This was used as an autonomous clock to vary the bifurcation parameter d that allowed sensory-motor transitions. The value of one of its variables was subtracted from the off-diagonal A^* matrix coupling d of the sensory modules and added to the value of that for the motor modules. The amplitude of the oscillation was chosen to be such that, at maximum excursion, the off-diagonal A matrix coupling d equalled the diagonal coupling c in one set of modules, and was much greater in the other. At zero difference, as discussed above, there is no internal "competition" between learned attractors within a module, and a new input vector can relocate the state of a module to a new attractor basin. This causes the desired alternation of attractor transitions within the sensory and motor modules as the the oscillator variable reaches positive and negative peaks.

The period of the clock in our simulations was chosen to be five times that of the modules, to simulate the 8Hz rate of respiration in olfaction (or the cortical alpha rythum) relative to the 40Hz cortical oscillation. The convergence rate of the modules was adjusted, by setting the maximum difference of the off diagonal A^* matrix coupling (competition), to a value sufficient for modules to relax to their attractors within a clock cycle (sniff).

Particular sequences were programmed into the system by adding to the J^* matrix specific off-diagonal coupling between modules. A cyclic sequence through all attractor combinations (states of the supernet) was programmed by hand. In another experiment, the J^* matrix weights determining several non-overlapping cyclic sequences of lengths from three to seven were "learned" by a heteroassociative Hebb rule between pre-chosen states. Input (starting states) were given to the network as handwritten digits, using sensory modules previously trained to recognize digits [5]. A distributed amplitude pattern was assigned to each attractor, and placed in a P^* matrix block so that the system could be transformed into network coordinates. A menu switch allowed the activity of the system to be viewed either in mode coordinates or in terms of the oscillatory activity patterns, distributed over three excitatory and three inhibitory neural populations (90 degrees out of phase), that would be observed in the biological model.

Another menu switch set frequencies ω to zero in the J^* matrix to convert the modules of the system to static attractors (in two dimensional subspaces). A set of slide pots on the screen allowed on-line variation of the operating parameters of the simulation to allow real time interactive control and observation of the system dynamics. The clock cycle of variation in intramodule competition levels could be effected by hand by moving a slider up and down. The simulation's free-running internal oscillator would take over this control when this slide was exactly centered. A slide pot controling the overall magnitude of intermodule coupling was configured so that when it was set to zero, the sensory-motor cycle operated in "pulse" mode. Here the states of one set of modules were sent through the connection matrix at the peaks of the clock variation to establish initial conditions in the other set of modules for subsequent relaxation. There was a slide control as well to scale the overall range of competition levels, and one to vary the real part of the eigenvalues of the modules. At larger positive values, this eigenvalue control increases attractor oscillation amplitudes and convergence rates. At negative values, it "resets" all modules to have only a single static attractor at the origin. There were also sliders controling the magnitude of the variance of zero mean Gaussian perturbations added to the states of the modules on each numerical integration time step, and added as well to the frequencies of all attractors at any time the frequency dispersion slider was moved. Finally, the speed of the simulation could be varied to observe particular dynamical transitions in detail.

The major results of our observations to date are the following. The supra-system relaxed from unprogrammed starting states to the preprogrammed cycles of transitions. These were unpreturbed by additive Gaussian noise of a variance of up to 25% of the peak amplitude of the oscillations of the attractors. Spurious transitions were made, however, when the resonant frequencies of the attractors were randomly distributed from each other by greater than about 7% of the mean frequency, and phase locking was lost. This indicated the importance of phase locking for reliable transitions in sinusoidally driven modules. When the system was switched to run with static attractors which provided constant ("clamped") inputs between modules, spurious transitions were observed with noise perturbations of as little as 10% of the maximum of activity at the attractor. This activity was the same as the peak oscillation amplitude, since these systems were identical in all respects save the absence of frequency terms ω in this static case.

When either the static or oscillating system was clocked in the pulse mode by sending the states of the sensory modules into the motor modules and vice versa to establish initial conditions, there was, of course, no longer any sensitivity to frequency dispersion. However, convergence times were then slower, and transitions were more sensitive to the additive Gaussian noise. Spurious transitions were observed at a variances of only 8%. These results suggest that noise tolerance is conferred on this system by phase locked oscillatory intermodule forcing, and point to a specific advantage that oscillatory dynamics may provide for biological systems.

5.1 Resonant Attractor Transitions

The observation here that phase locking between the driving and the driven modules is necessary to effect reliable transitions in the presence of noise, suggests an important computational role in *this* system, at least, for the kind of selective phase locking observed in visual cortex. We have some understanding of the mechanism of these transitions, in the case of phase-locked forcing, because it is similar to the process of attractor transition which we can observe in detail in two dimensional phase portraits for two static attractors with clamped inputs.

We must first view the process of the machine cycle in more detail. The larger the difference of off-diagonal A^* matrix coupling within a module created by the clock, the greater the competition between attractors and the faster the convergence to a chosen attractor. At high values of competition the module is "clamped" at its current attractor, and the effect of the changing input it receives by feedback from the modules undergoing transition is negligable. The feedback between sensory and motor modules is therefore effectively cut. The super-system can thus be viewed as operating in discrete time by transitions between a finite set of states. This kind of clocking and "buffering" of some states while other states relax is essential to the reliable operation of digital architectures. In our simulations, if we clock all modules to transition at once, the programmed sequences lose stability, and we get transitions to unprogrammed fixed points and simple limit cycles for the whole system. This is a primary justification for the use of the "machine cycle".

By this reasoning, when one set of modules is clamped at their periodic attractors, the other modules are externally "forced" by these inputs, and theoretical analysis of the transition process becomes possible. When a module is driven at the same frequency as the intrinsic resonant frequency of all its attractors, and has phase locked to the input, then we can look at its dynamics in terms of a strobed series of snapshots at a particular point of the cycle. This lets us freeze the periodic motion and construct a vector field for the competition between amplitudes of the periodic attractors independent of their phase. This is the phase portrait on a Poincare section of the system and looks like a vector field for static attractors, since equilibria in this system now describe the asymptotic states of the underlying oscillations. In the absence of input, this is exactly the vector field of the amplitudes of the polar coordinate version of the normal form equations [1, 3]. The amplitudes of the driving inputs appear in this picture as clamped static inputs, and the series of phase portraits we have calculated for static attractor transitions applies to describe the transitions between oscillatory attractors [4].

At zero difference between diagonal and off-diagonal A matrix coupling, in these studies of static attractor transitions, there is no internal "competition" between attractors within a module. The learned internal dynamics is effectively turned off, and the state vector is determined entirely by a clamped input vector. There is only a single attractor within a module then that is a scalar multiple of whatever input vector is applied. The previous attracting state of a module is completely erased at this point. As the competition increases again, this attractor is forced to a new position in state space determined by the nearest learned attractor. The other learned attractors reappear by saddle-node bifurcations at a distance from this attractor and create other basins of attraction. Even though this story can be applied to oscillating systems under the special circumstances described above, this picture is still a projection of the true higher dimensional periodic flow. Something about that flow gives the oscillating networks the greater robustness to noise perturbation during attractor transitions that we see in simulations. The focus of future investigation is to understand this dynamical difference.

A simulation similar to that described above was carried out using modules with Lorenz attractors. There is always some probability of a spurious transition in this system, since trajectories of this attractor pass arbitraily close to zero, and other attractors may take over dominance when they are released from supression by this spontaneous self-destruction. Many variations of the basic system configuration, and operation, and various training algorithms, such as reinforcement learning with adaptive critics, are being explored. The ability to operate as an finite automaton with oscillatory states is an important benchmark for this architecture, but only a subset of its capabilities. At low competition, as we have pointed out, the supra-system reverts to a continuous dynamical system. We expect that this kind of variation of the operational regime, including chaotic dynamics, though unreliable for habitual behaviors, may nontheless be important in other areas such as the the search process of reinforcement learning.

5.2 Acknowledgments

Supported by AFOSR-87-0317, and a grant from LLNL. It is a pleasure to acknowledge the invaluable assistance of Morris Hirsch, Walter Freeman, and Todd Troyer.

References

[1] B. Baird. A bifurcation theory approach to vector field programming for periodic attractors. In *Proc. Int. Joint Conf. on Neural Networks, Wash. D.C.*, pages 1:381–388, June 1989.

[2] B. Baird. Bifurcation and learning in network models of oscillating cortex. In S. Forest, editor, *Emergent Computation*, pages 365–384. North Holland, 1990. also in Physica D, 42.

[3] B. Baird. A learning rule for cam storage of continuous periodic sequences. In *Proc. Int. Joint Conf. on Neural Networks, San Diego*, pages 3: 493–498, June 1990.

[4] B. Baird. *Bifurcation Theory Approach to the Analysis and Synthesis of Neural Networks for Engineering and Biological Modeling.* Research Notes in Neural Computing. Springer, 1992. to appear.

[5] B. Baird, W. Freeman, F. Eeckman, and Y. Yao. Applications of chaotic neurodynamics in pattern recognition. In *SPIE Proceedings Vol. 1469*, 1991. in press.

[6] R. Eckhorn, R. Bauer, W. Jordan, M. Brosch, W. Kruse, M. Munk, and H. Reitboeck. Coherent oscillations: A mechanism of feature linking in the visual cortex? *Biological Cybernetics*, 60:121, 1988.

[7] A. K. Engel, P. Konig, C. Gray, and W. Singer. Synchronization of oscillatory responses: A mechanism for stimulus-dependent assembly formation in cat visual cortex. In R. Eckmiller, editor, *Parallel Processing in Neural Systems and Computers*, pages 105–108. Elsevier, 1990.

[8] J. Freeman, W., Y. Yao, and B. Burke. Central pattern generating and recognizing in olfactory bulb: A correllation learning rule. *Neural Networks*, 1(4):277, 1988.

[9] W. Freeman. *Mass Action in the Nervous System.* Academic Press, New York, 1975.

[10] W. Freeman. Simulation of chaotic eeg patterns with a dynamic model of the olfactory system. *Biological Cybernetics*, 56:139, 1987.

[11] W. Freeman and B. Baird. Relation of olfactory eeg to behavior: Spatial analysis. *Behavioral Neuroscience*, 101:393–408, 1987.

[12] W. J. Freeman and B. W. van Dijk. Spatial patterns of visual cortical eeg during conditioned reflex in a rhesus monkey. *Brain Research*, 422:267, 1987.

[13] C. Gray, P. Konig, A. Engel, and W. Singer. Oscillatory responses in cat visual cortex exhibit intercolumnar synchronization which reflects global stimulus properties. *Nature(London)*, 338:334–337, 1989.

[14] J. Guckenheimer and D. Holmes. *Nonlinear Oscillations, Dynamical Systems, and Bifurcations of Vector Fields.* Springer, New York, 1983.

[15] L. B. Haberly and J. M. Bower. Olfactory cortex: model circuit for study of associative memory? *Trends in Neuroscience*, 12(7):258, 1989.

[16] D. Horn and M. Usher. Parallel activation of memories in an oscillatory neural network. *Neural Computation*, 3(1):31–43, 1991.

[17] J. Jack, D. Noble, and R. Tsien. *Electric Current Flow in Excitable Cells.* Clarendon Press, Oxford, 1983.

[18] M. Jordan. Motor learning and the degrees of freedom problem. In M. Jeanerod, editor, *Attention and Performance, XIII*, Hillsdale, NJ, 1990. Laurence Erlbaum.

[19] Z. Li and J. Hopfield. Modeling the olfactory bulb and its neural oscillatory processings. *Biological Cybernetics*, 61:379, 1989.

[20] J. McClelland and D. Rummelhart. An interactive activation model of context effects in letter perception. part 1: an account of basic findings. *Psychological Review*, 5:375–407, 1981.

[21] M. Mozer. Connectionist music composition based on melodic, stylistic, and psychophysical constraints. *Tech Report CU-CS-495-90, Univ. of Colorado at Boulder*, 1991.

[22] J. H. Schmidhuber. A local learning algorithm for dynamic feedforward and reccurent networks. *Connection Science*, 1:719–722, 1990.

[23] C. von der Malsburg. How are nervous structures organized? In E. Basar, H. Flor, H. Haken, and A. Mandell, editors, *Synergetics of the Brain*, pages 238–249. Springer, 1983.

[24] D. Wang, J. Buhmann, and C. von der Malsburg. Pattern segmentation in associative memory. *Neural Computation*, 2(1):94–106, 1990.

A Computationally Efficient Spike Initiator Model that Produces a Wide Variety of Neural Responses

B.R. Parnas
E.R. Lewis

Department of EECS
University of California
Berkeley, CA 94720

We have developed an integrate-and-fire spike-initiation model based on that of Hill (1936), which comprises a pair of coupled, linear differential equations describing membrane potential and a threshold variable. In our version, we add an intrinsic noise source and explicit representation of the spike and its effect on threshold. The equations of the Hill model describe the time evolution of membrane potential and threshold in the subthreshold region. The model has three time constants. We demonstrate that, with suitable choice of these time constants and the mean value of the intrinsic noise, the model can mimic the spike-response characteristics of many classes of neurons. Specifically, with respect to auditory brainstem neurons, we produce the responses characteristic of the primary-type neuron, several varieties of chopper cell and onset cells, all of which are found in the mammalian cochlear nucleus. We demonstrate that the model computes simulated spike responses quickly.

1. Introduction

Models for the spike initiation function of neurons have ranged from simple logic elements (e.g., McCulloch & Pitts, 1943) to complicated biophysical models (e.g., Hodgkin and Huxley, 1952). Artificial neural network research has yielded a model in which inputs are linearly summed and this is passed through a nonlinear "squashing" function resulting in an analog output which loosely represents mean spike rate. Biophysical studies have employed complex compartment neuron models with nonlinear, multi-state spike generation functions. While the logic element representation lacks certain features of real neurons, such as individual unit accommodation, and ignores interspike timing, a signal parameter shown to be important in signal coding (Optican & Richmond, 1987), detailed biophysical models are too unwieldy to be used in models of neural ensembles. The model that we present here represents a compromise between these two ideas. It includes the features that we believe to be relevant to the formation of spikes and the statistics of the spike trains, but does not attempt to reflect the underlying biophysical mechanisms.

The test of any model is to explore its capabilities and verify that it does accurately portray the desired phenomena. We demonstrate that this simple model can generate the neural response types necessary for modeling the mammalian auditory system, the ultimate goal of our work. The simplicity of the system allows efficient computation but there is yet enough complexity to represent the necessary behavior. We show that this model is well-suited for the simulation of large-scale neural arrays.

2. Model Details

Our spike initiator is described by the two time constant model of Hill. The equations of this model govern the time evolution of the membrane potential and a threshold variable. The equations have the following form:

$$\frac{dV}{dt} = -\frac{V - V_0}{k} + I \tag{1}$$

$$\frac{dU}{dt} = \frac{V - V_0}{\beta} - \frac{U - U_0}{\lambda} \tag{2}$$

where V is the membrane potential at the spike initiator, U is the threshold variable, I is the stimulus input, k is the membrane time constant, β represents the time constant of coupling between U and V and λ is the time constant for the threshold variable. The parameters k, β, the ratio β/λ and the resting threshold value may be altered at runtime. In Hill's formulation λ is set equal to β so that, with sustained subthreshold depolarization, $U_\infty - V_\infty = U_0 - V_0$. We have allowed these values to be varied independently as this plays an important role in our ability to generate the desired neural response characteristics. This simple set of linear equations models the behavior of the spike initiator quite well in the subthreshold region. It does not reflect spike activity as this is a highly nonlinear process. We have chosen to simply "paste on" spikes when the

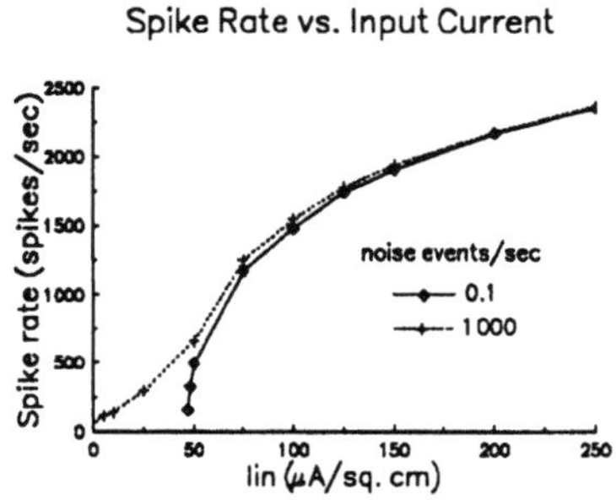

Figure 1. Spike rate versus input amplitude for modified Hill model spike initiator. The solid curve indicates the behavior without noise and exhibits the typical jump from zero to finite spike rate at a critical current value. The dashed curve shows the effects of noise in removing this jump phenomenon. The model parameters are k = 0.4 ms, β = 8 ms, β/λ = 5.

membrane potential crosses the (variable) threshold, adding an appropriate spike value to *V*. This allows us to indicate spike activity and timing and to adjust the membrane potential properly so that the threshold variable receives the proper input (see equation 2). In this way a spike brings about a relative refractory period by causing a large increase in the threshold variable.

Noise is included in the model in the form of variably-sized noise events which occur at random intervals governed by a Poisson process with mean value λ (not to be confused with the time constant λ of equation 2). The details of the noise generation process have been published previously (Parnas, et al, 1991). Figure 1 shows the spike rate versus DC input current curve for our model in the presence and absence of noise. It can be seen that the noise serves to linearize the nonlinear spike generation process in a probabilistic sense (Yu and Lewis, 1989). The noise also provides a background firing rate for the spike initiator.

3. Response Characteristics

This simple model is capable of generating a wide variety of neuron response types. This is accomplished through variation in the parameters of the model: the noise parameter λ, the coupling time constant β, and the ratio of β to the threshold time constant λ. We will show the effects of varying each of these parameters independently. The membrane time constant, *k*, can also be varied to reflect the actual geometry and membrane properties of specific cell types, but this will not be explored here. Response types are described by their peri-stimulus time histograms (PSTH) and fall into several categories. Primary-type units show a single strong peak at the onset of the stimulus and remain relatively quiet thereafter. Primary-type with notch units have this peak and, after a quiet period, some low level activity returns. Choppers fire at a frequency related to the amplitude, and not the frequency, of the stimulus. These are often subclassified based on their coefficient of variation (cv) - the standard deviation σ of the inter-spike interval histogram (ISIH) divided by the mean μ of the ISIH. Larger values of the cv correspond to transient choppers while smaller values classify sustained choppers. Onset_c (onset chopper) units display characteristics between those of primary-type and chopper units.

3.1 Parameter: Noise. The effects of varying the noise parameter λ can be seen in figure 2. Here the PSTH responses are shown for four different values of λ, from 0.1 to 5000 events per second. While the qualitative appearance of the response becomes worse, the regularity of discharge, as determined by the coefficient of variation does not change much.

3.2 Parameter: β. Next we vary the coupling time constant between the membrane and threshold potentials, β. This determines the strength of the effect of the membrane potential on the threshold potential. For small values of β the threshold rises rapidly with increasing membrane potential providing more of a relative refractory period. The effects of varying β are shown in figure 3 where we have plotted PSTH responses for four values of β, from 8 ms to 1 ms.

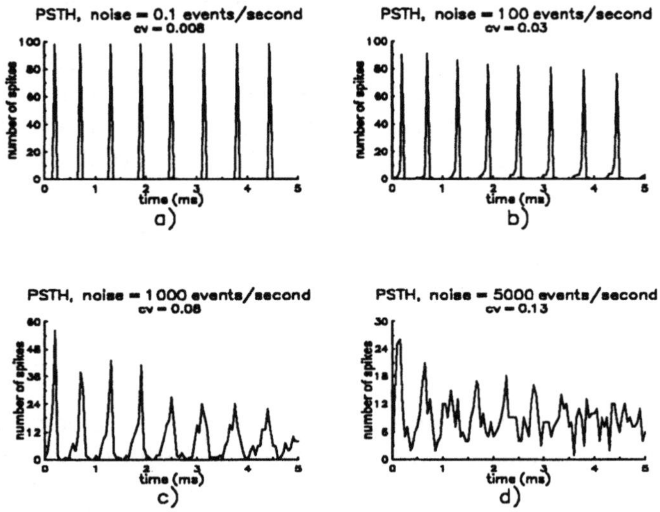

Figure 2. Effects of noise on PSTH plot. a) 0.1 noise events per second. b) 100 noise events per second. c) 1000 noise events per second. d) 5000 noise events per second. In all plots $\beta = 8$ ms, $\beta/\lambda = 5$ and the input is a 100 Hz square wave of current at 100 $\mu A/cm^2$. The cv for each plot is indicated above.

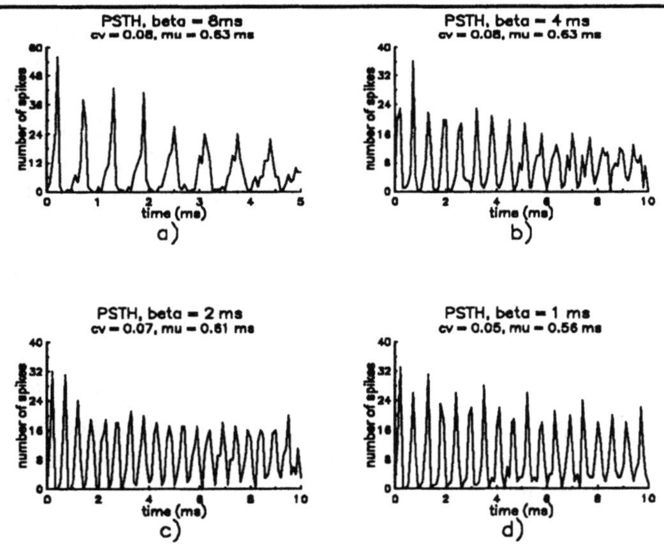

Figure 3. Effects of varying β on PSTH plot. a) $\beta = 8$ ms. This plot is the same as figure 2c; the input frequency is 100 Hz. b) $\beta = 4$ ms. c) $\beta = 2$ ms. d) $\beta = 1$ ms. In all plots noise parameter = 1000 events per second, $\beta/\lambda = 5$ and the input is a 50 Hz square wave of current (except for a) at 100 $\mu A/cm^2$. The cv and mean interspike interval, μ, are indicated above each plot.

3.3 Parameter: β/λ. The parameter β/λ affects the relative contributions of V and U on the threshold. For large values of β/λ the threshold recovers quickly and the effect of spikes is reduced. For small values, the membrane potential dominates and threshold recovers more slowly. The increased weight of the spikes on the threshold serves to make the response more irregular. The effects of varying β/λ are shown in figure 4, for values ranging from 5 to 1.5.

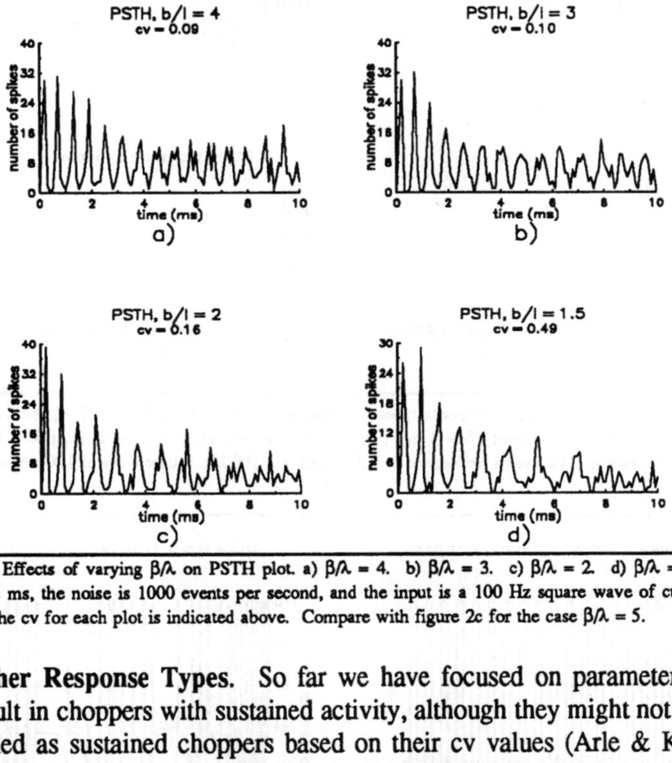

Figure 4. Effects of varying β/λ on PSTH plot. a) β/λ = 4. b) β/λ = 3. c) β/λ = 2. d) β/λ = 1.5. In all plots β = 8 ms, the noise is 1000 events per second, and the input is a 100 Hz square wave of current at 100 $\mu A/cm^2$. The cv for each plot is indicated above. Compare with figure 2c for the case β/λ = 5.

3.4 Other Response Types. So far we have focused on parameter variations which result in choppers with sustained activity, although they might not technically be classified as sustained choppers based on their cv values (Arle & Kim, 1991). This type of response is easiest to generate with our model due to the linearity of the system. A steady state between threshold and membrane potential variations is reached which determines an inherently stable firing pattern for a large range of the system parameters. With suitable manipulation of the parameters, however, other response types may be obtained. A sample of these response types are displayed in figure 5.

3.5 Spiketrain waveforms. Figure 6 displays sample spiketrain waveforms. These plots show regular and irregular chopping patterns. The "pasted on" nature of the spikes is evident in their rectangular shape and consistent form. The effects they have on timing and on the threshold waveform are not dependent upon the exact form of the spike. For this reason, the short rectangular pulses are a quite adequate representation of the spike phenomenon.

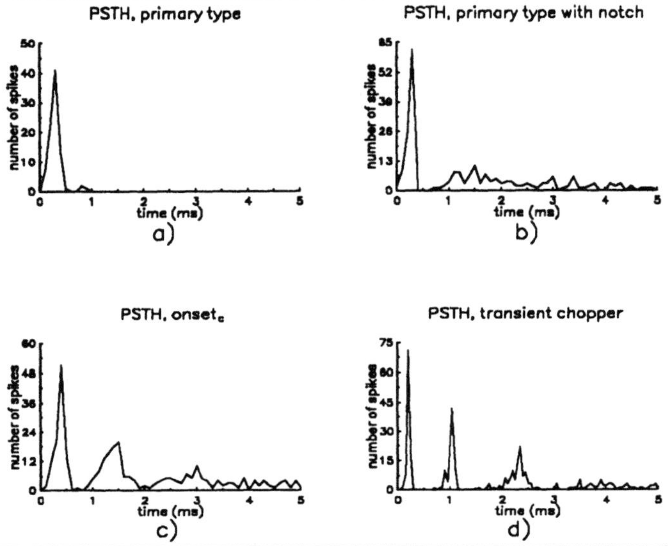

Figure 5. PSTH for different neural response types. a) Primary type, $\beta = 0.8$ ms, $\beta/\lambda = 0.7$. b) Primary type with notch, $\beta = 2$ ms, $\beta/\lambda = 1$. c) Onset$_c$, $\beta = 4$ ms, $\beta/\lambda = 1$. d) Transient chopper, $\beta = 2$ ms, $\beta/\lambda = 1.5$. In all plots noise parameter is 1000 events per second and the input is a 100 Hz. square wave at 100 $\mu A/cm^2$.

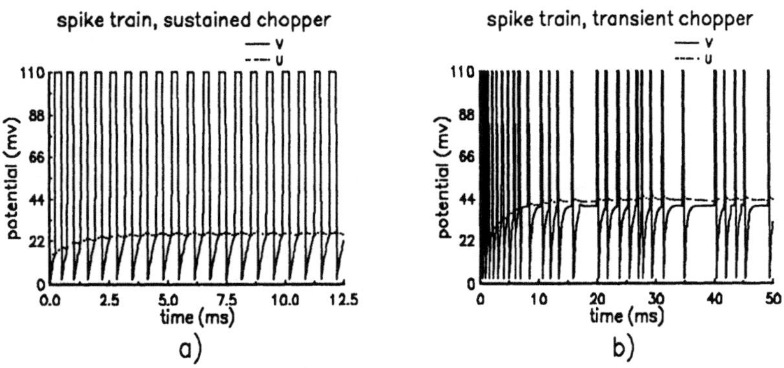

Figure 6. Sample spike trains for different values of system parameters. a) $\beta = 8$ ms, $\beta/\lambda = 5$, no noise. This plot corresponds to the PSTH of figure 2a. b) $\beta = 8$ ms, $\beta/\lambda = 1.5$, noise equals 1000 events per second. This plot corresponds to the PSTH of figure 4d. The input is a current step at 100 $\mu A/cm^2$.

4. Discussion

Figures 2-4 give an indication of the effects of the system parameters on the neural response characteristics. The value of the noise parameter λ affects the appearance of the waveform, but does not substantially affect the regularity of discharge. In varying λ from 0.1 to 5000 events/second, a factor of 50000, the cv changes only from 0.008 to 0.13, a factor of about 16. The chopping signal does not appear as clean, due to the noise-generated background spikes, but the pattern is still quite regular. Figure 7 shows the ISIH curves for the no noise and high noise situations. It can be seen that, while the interval distribution spreads out, it is still quite regular. The mean inter-spike interval (ISI) changes by only 13% with this large increase in the noise parameter.

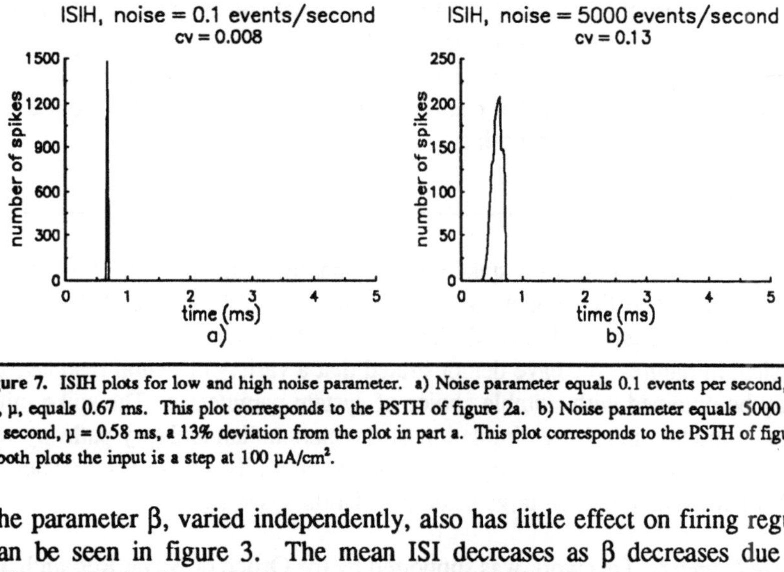

Figure 7. ISIH plots for low and high noise parameter. a) Noise parameter equals 0.1 events per second, mean ISI, μ, equals 0.67 ms. This plot corresponds to the PSTH of figure 2a. b) Noise parameter equals 5000 events per second, $\mu = 0.58$ ms, a 13% deviation from the plot in part a. This plot corresponds to the PSTH of figure 2d. In both plots the input is a step at 100 $\mu A/cm^2$.

The parameter β, varied independently, also has little effect on firing regularity, as can be seen in figure 3. The mean ISI decreases as β decreases due to the speeding up of the threshold timecourse; the threshold system recovers more quickly and is ready to spike again. This parameter is useful for fine-tuning the response of choppers in order to generate specific frequency vs. input curves, but is not effective in modifying the response type.

As mentioned earlier, the ratio β/λ has a more profound effect on the spiking regularity. As this parameter gets smaller, the chopping pattern becomes more irregular, as is evident from figure 4. With large values of β/λ the patterns are all choppers, more or less sustained. A wider variety of response types is obtained as the ratio decreases. The response types shown in figure 5 all take advantage of this and derive their characteristics from the small ratio. Since the threshold variable has no measurable analog in the real neural system it is certainly reasonable to choose values for β and β/λ which elicit the desired responses. Also, there is no attempt to be biophysically precise in this model. The goal is to find the combination of parameters that produces the desired response type. The qualitative behavior quite closely resembles PSTH plots made from real neurons. In the interests of space, however, data from physiological studies have not been included here for comparison with our results. These data may be found in the neurophysiological literature (e.g., Rhode, et al., 1983; Smith and Rhode, 1989; Blackburn and Sachs; 1989).

As we stated at the beginning, the model is computationally efficient. The lack of detailed nonlinear dynamics and the use of only two state variables per spike initiator endow this model with its efficiency. In comparisons with a Hodgkin-Huxley simulation, this model was found to require only 57% of the computation time (Parnas & Lewis, 1991). Because the model is linear, this computation time disparity can be made even greater by using a larger timestep in the integration routine. Since numerical stability is not an issue the choice of timestep does not present a hazard from the standpoint of computational convergence. The tradeoff in this situation is the resolution available from the system in the timing of spike arrivals and the sampling frequency we choose for our sound waveforms. We would like to achieve good temporal and spatial resolution from this system and this requires a fine timestep for spike occurrence timing.

5. Conclusions

We have developed a model for spike initiation based on that of Hill. It is essentially a leaky integrate-and-fire model with a variable threshold, noise and the effects of pikes explicitly included. The model is simple, computationally efficient. It is not biophysical in nature. We have demonstrated that a variety of neural response types that may be obtained with suitable choice of system parameters. The spike initiator serves as a module in a large simulation of the mammalian auditory pathway which we are developing. The simplicity and computational efficiency of the model make it ideal for inclusion in such large-scale simulation systems.

Acknowledgements. This work was supported by the Office of Naval Research, ONR grant # N00014-91-J-1333.

6. References

Arle, J.E. and D.O. Kim (1991), Neural modeling of intrinsic and spike-discharge properties of cochlear nucleus neurons. *Biol. Cybern.*, **64**: 273-283.

Blackburn C and M Sachs (1989), Classification of unit types in the anteroventral cochlear nucleus: PST histograms and regularity analysis. *J. Neurophysiol.*, **62**: 1303-1329.

Hill, A.V. (1936), Excitation and accommodation in nerve. *Proceed. Royal Soc., Series B*, **119**:305-355.

Hodgkin, A.L. and A.F. Huxley (1952), A quantitative description of membrane current and its application to conduction and excitation in nerve. *J. Physiol.*, **117**: 500-544.

McCulloch, W.S. and W. Pitts (1943), A logical calculus of the ideas immanent in nervous activity. *Bull. Math. Biophys.*, **5**: 115-133.

Optican, L.M. and B.J. Richmond (1987), Temporal encoding of two-dimensional patterns by single units in primate inferior temporal cortex. III. information theoretic analysis. *J. Neurophys.*, **57**: 162-178.

Parnas, B.R., K.S. Gangnes, D.A. Feld, W.L. Lee and E.R. Lewis (1991), A parallel neural model for auditory front end processing, in F.H. Eeckman (ed) **Analysis and Modeling of Neural Systems**, pp. 283-288.

Parnas, B.R. and E.R. Lewis (1991), On biological neural networks: criteria for selection of spike initiator models, Proceedings of the IEEE EMBS Conference, *Oct. 31 - Nov. 3, 1991*.(In Press)

Rhode, W.S., P.H. Smith and D. Oertel (1983), Physiological response properties of cells labeled intracellularly with horseradish peroxidase in cat dorsal cochlear nucleus. *J. Comp. Neurol.*, **213**: 426-447.

Smith, P.H. and W.S. Rhode (1989), Structural and functional properties distinguish two types of multipolar cells in the ventral cochlear nucleus. *J. Comp. Neurol.*, **282**: 595-616.

Yu, X. and E.R. Lewis (1989), Studies with spike initiators: linearization by noise allows continuous signal modulation in neural networks. *IEEE Trans. Biomed. Eng.*, **BME-36**: 36-43.

Linearization by Noise and/or Additional Shunting Current of a Modified FitzHugh Nagumo Spiking Model.

Frédéric E. Theunissen[1], Frank H. Eeckman[2] and, John P. Miller[1].
[1]Biophysics Group and Dept. of Mol. and Cell Biology. University of California, Berkeley, CA 94720
[2] O-division, Lawrence Livermore Lab, Livermore CA 94550

Abstract

We have developed a simplified description of voltage dependent channels to describe the currents involved in spike generation. We included this spiking model in a compartmental neuron model in order to obtain the correct scaling factor between the current actually injected in an experiment and the current arriving at the spike initiating zone. After fixing the parameters for our spiking model to match the spike shape of the real neuron, the model had a strong bifurcation in its intensity versus rate characteristic. The rate of firing would abruptly jump from 0 to a value close to the saturating rate as the intensity of the stimulus was increased. To eliminate this artificial singularity we looked at the effect of both an additional shunting current and of current noise as suggested by previous researchers. We found that either the shunt or the noise alone could get rid of the bifurcation but that only a specific combination of shunt and noise would also yield the correct spiking variability as seen in interspike histograms.

Realistic modeling of spiking neurons requires not only an accurate description of the passive electroanatomy of the cell (as it relates to the effective integration of presynaptic inputs) but also an accurate description of the spiking mechanism, as the spike train is the ultimate output of the unit. An accurate spiking model is of particular importance when the modeled cell is capable of repetitive firing and possibly "codes" by modulating its spike rate or statistics. Many such neurons (in particular sensory and motor neurons) are capable of firing rates which vary continuously from 0 to their saturating value. For example, we can record from interneurons in the cricket cercal sensory system which respond to increasing steps of current by firing repetitively with steady rates which vary continuously from 0 to 200 spikes per second. Similar behaviors have been observed in crustacean motor neurons (Connor *et al*, 1977), cat motor neurons (Calvin & Stevens, 1968) and in auditory sensory neurons (Yu & Lewis, 1989) to mention a few. One goal of an effective spiking model must be to match not only this relationship between spike rate and current (FI plot) but also the variability as described by the inter spike interval (isi) histogram obtained at each specific current.

Spiking models at various levels of abstraction have been more or less successful at this task. At one extreme, models which are single point stochastic processes (ie. stochastic versions of the integrate and fire models) are capable of matching both the mean rate and variability found in the spike trains (Stein, 1965; Calvin & Stevens, 1968). These models are useful in determining the nature and therefore the probable origin of the variability, but they give little or no insight into the mechanism of spike generation. In this sense, they might not be sufficient to predict a behavior other than the one that was matched.

At the other extreme, models based on the exact kinetics of each of the different classes of channels, such as the Hodgkin-Huxley (HH) model, can successfully match the spike shape (and the currents involved) and also predict some of the complex spiking behavior (relative refractory period, etc). However, the classic HH equations (and many of their reduced versions) exhibit a strict non-linearity as the neuron is driven into repetitive firing, with the firing rate jumping from zero to a value very close to the saturating level. This phenomenon is understood mathematically and is called a Hopf bifurcation (Rinzel & Ermentrout, 1989).

Connor *et al.* (1977), modified the HH model to match the kinetics of Na and K channels observed in the crustacean motor neuron and added a description for an additional channel which carried a transient potassium current ("A" current). With their model, they were able to not only match the currents of their specific channels, but were also able to match the low firing rates that were observed experimentally. Such models, however, are cumbersome in the number of necessary parameters, and are useful only in situations where the modeler has a knowledge of all the channel kinetics and is not trying to empirically fit a more macroscopic expression of the channel currents (like spike shape or the spiking statistics).

To allow modeling and theoretical analysis at a more abstract level, theoreticians (FitzHugh, 1961; Abbott and Kepler, 1990; Rinzel & Ermentrout, 1989) have worked on reducing the number of variables of HH-like equations by effectively grouping variables which have similar time constants. Models with two variables are often chosen, since they display most of the complex observed phenomena and lend themselves to phase plane analysis. Rinzel and Ermentrout (1989) used such a reduced model to show how both the Hopf bifurcation and an arbitrarily slow firing rate (Such as Connor *et al.*) can be explained mathematically. Nothing, however, in the deterministic HH or reduced models can predict the variability observed in real cells.

In another line of studies, Yu and Lewis (1989) showed that the Hopf bifurcation in the HH model dynamics could be linearized simply by adding noise to the spiking model. Using the full HH description with an additional Gaussian current noise, they were able to match the large dynamic firing range of a frog saccular neuron as well as its statistics as described by the cycle histograms.

With these problems (and solutions) in mind, we built a reduced HH model and tested the effect of both a shunting outward current ("A-like" current) and of noise on spike generation. Our goal was to match both the FI characteristics and the spike statistics of interneurons in the cricket cercal sensory system. In addition, we embedded our spiking model in a compartmental model and required the spike shape to match the experimental spike. By matching the spike shape, we obtained more realistic values for the magnitude and the time course of the currents involved in spike generation and by consequence of the magnitude of the Hopf bifurcation. The use of a compartmental model allows one to take into account the passive part of the cell which might constitute a significant additional impedance which is not taken into account in models normalized to a patch of membrane. In this paper, we did not actually investigate the extent of such an effect but we wanted to set up a model where this could be explicitly considered. The compartmental model did, however, allow us to "inject" exactly the same amount of current (measured in nA) as in a set of electrophysiological experiments, and therefore limit the extent of any additional assumptions on the attenuation from such a current injection. To do so, we imposed two additional constraints: 1. To match the model and experimentally measured input resistances at the injection/recording site 2. To match the modeled and observed attenuated shape of the action potential at the recording/injection site. The second step also implied that we had to match the shape of the action potential as recorded from the axon.

Methods

Experimental Data.

We obtained FI plots by intracellular injections of current steps into giant interneurons in the cercal sensory system. Currents up to 5 nA were injected with 0.5 nA increments, and voltage traces were recorded simultaneously. From the size of the action potential we could deduce that the electrode was not in the axon of the cell but still relatively close to the SIZ. The voltage trace was recorded for about 6 s for each of the different values of the current. A small amount of adaptation was measured but did not significantly change the results for the mean spike rate or for the spike statistics. Also we minimized the synaptic input to the interneuron by isolating the sensory input to the cercal system. The cercal system is an air current detection system, and the recordings were performed in a sound proof chamber. Additional recordings were obtained from the axon of the cell, to obtain a more exact shape (ie.: not attenuated) of the action potential.

Model

a. Passive compartments.
Our modeling was done using a compartmental representation of the current injection experiment. We built a 25 compartment cable of about 200 μ of length. The current was injected at one end of the cable, the active (ie. spiking) compartment being at the other end. The diameter of the cable was set at 5 μ, which is the average diameter of the axon in the interneuron. The length of the cable was adjusted so that the amplitude of the attenuated shape of the spike at the injection site matched the observed amplitude. All other electrophysiological parameters for the passive properties were set at "standard" values : Rm = 1000 Ω . cm^2; Ri = 100 Ω . cm;Cm = 1μF/cm^2. An additional shunt of 0.15 mS was added at the current injection extremity to match the input resistance of the cell as deduced experimentally in the same experiment (About 5.5 MΩ).

b. Spike-generating compartment.
The spiking compartment had additional excitatory and inhibitory currents which were simplified versions of HH sodium and potassium currents. The HH model was simplified to minimize the number of parameters we needed in order to fit our model spike to the experimental data. Our model was inspired by the similar reduction found in the FitzHugh-Nagumo model but the reduction was done here to a lesser degree.

The excitatory current is described by :

$$I_\varepsilon = G_m (\alpha_\varepsilon)(\beta_\varepsilon)(V - V_{Na}) ,$$

where α_ε is the activation factor and β_ε is the inactivation factor. The overall conductance is expressed as a factor of G_m , the passive membrane conductance (1/Rm), so as to gain a qualitative idea of the magnitude of the currents involved.

The activation coefficient is given by :

$\alpha_{\varepsilon} = (0.1)\, V_m \; (V_m - V_{Th})$,

where V_{Th} is called the threshold voltage. The activation is set to be instantaneous and the typical sigmoid curve for m, which is raised to the third power (m^3) in the classical HH model, was replaced by a quadratic. (This is similar in spirit to the FitzHugh Nagumo reduction). Increasing the threshold V_{Th} effectively raises the threshold of excitability and is equivalent to shifting the sigmoid curve for m to the right on the voltage axis. Note that the α_{ε} coefficient does not go from 0 to 1. The maximum value for α_{ε} is around 1000 which means that in our model, during an action potential the sodium conductance is 1000 times greater than the leakage conductance. The inactivation coefficient is given by

$$\frac{d\,\beta\varepsilon(t)}{dt} = \frac{\beta_{\varepsilon}\,\infty - \beta\varepsilon\,(t)}{\tau_{\beta}}$$, where

$$\beta_{\varepsilon}\infty = \frac{1}{1 + (\frac{V}{Vhalf})^{-ksl}} \quad \text{and} \quad ksl = \frac{\log\,(Slope)}{\log(Vhalf/Vmax)}$$

Written in this manner the sigmoid for inactivating could be shifted along the V axis by defining Vhalf, the voltage point for which $\beta_{\varepsilon}\infty = 0.5$ and could be made more or less steep by changing the value of Slope. To fit our spike shape we used Vhalf = 40 mV, Vmax = 115 mV, Slope = 40 and $\tau_{\beta} = 0.8$ ms.

The inhibitory current was represented as

$I_{\varphi} = G_m\, (\alpha_{\varphi})\, (\, V - V_K)$,

where the activation coefficient is given by

$$\frac{d\,\alpha_{\varphi}(t)}{dt} = \frac{\alpha_{\varphi}\,\infty - \alpha_{\varphi}(t)}{\tau_{\varphi}}$$, where

$\alpha_{\varphi}\,\infty = 4$ V for $V > 0$ and $\alpha_{\varphi}\,\infty = 0$ for $V < 0$.

τ_{φ} was set to 3.0 ms to match the shape of our action potential (mostly of the inhibitory undershoot).

c. Shunting current.

An additional current was added to the model for the purpose of "linearizing" the FI characteristics, ie: to eliminate the bifurcation. For this purpose we chose a potassium current which would induce a transient outward current for each action potential. This current effectively created an additional shunt after each spike and thus extended the interspike interval. Even though our mathematical representation was different that the one for the A current described in Connor's the effects were similar. The additional shunting current was distributed along 10 compartments of the cable and was described by :

$$I_s = G_m (\alpha_\sigma) (V - V_K),$$

where the activation coefficient is given by

$$\frac{d\,\alpha_\sigma(t)}{dt} = \frac{\alpha_\sigma \infty - \alpha_\sigma (t)}{\tau_\sigma} \text{ , where}$$

$$\alpha_\sigma \infty = Ksh \frac{1}{1 + (\frac{V}{Vhalf})^{-ksl}} \quad \text{and} \quad ksl = \frac{\log (Slope)}{\log(Vhalf/Vmax)}$$

τ_σ was set at 8 ms, Vhalf at 20 mV , Vmax at 25 mV and Slope at 40. In this manner the activation sigmoid had a steep voltage dependance around 20 mV but which was delayed by 8 ms. By distributing this current in 10 compartments, and because ts was relatively long, its addition did not affect the spike shape. We were able to change the value of the scaling factor Ksh to vary this current shunt in our simulations.

d. Synaptic Noise.
In some of the simulations and to finally match the spike statistics in the real data, we had to add some variability to our otherwise deterministic model. We chose to add current noise at the spiking compartment which had the same characteristics as the voltage recorded from real neurons when no current was injected. In such voltage traces, one could observe small depolarizations corresponding to random synaptic events. We "injected" scaled values of the current corresponding to such voltage fluctuations and loked at its effect on the spike statistics. Noise of synaptic origin was chosen since was it clearly visible in our recordings and since other studies have also found ISI variability of synaptic origin (Calvin & Stevens, 1967).

e. Simulator.
The simulations were done using the program Nemosys (this volume) which is a graphical compartmental neuron simulator. The simulator allows the user to add any custom current at each step in the integration method by specifying the conductance change and the reversal potential of each current. The program uses an implicit first order integration method for the determination of the voltage at each node. It is stable

even when conductances are updated at each step. (ie. the differential equations for the conductances in the active part of the cell are calculated explicitly and are in this manner uncoupled from the system of equations for the voltages at each compartment).

Results

a. Matching the spike shape at the axon.

The choice of the values for almost all the parameters in the equations describing the active currents in the spiking compartment were set by matching the spike shape at the spiking compartment to the spike shape of the real neurons as recorded from the axons. This fit dictated the relative values of the magnitude of the inward and outward current as well as the delays for the inactivation of the inward current and the activation of the outward current. The fit is shown in Figure 1 where the model spike is compared to one of the spikes recorded in or very close to the axon. In this particular case, the spike as recorded from the real neuron is as lightly attenuated (and low-passed) version of the model spike. Such an attenuation is to be expected from a recording electrode at some distance from the spike initiating zone (SIZ), and different recordings would give slightly different levels of attenuation. Moreover the exact height of the spike did not change the FI characteristics of the cell. The time course of the spike is much more critical and clearly affects the FI characteristic. In our case by making the model spike (with its hyperpolarization) as narrow as in the observed data, our spiking model (with no shunting current) had kinetics which made the saturating rate quite high (greater than 300 spikes/s) and which made, for the same reasons, the jump at the bifurcation quite severe (jumping from 0 to 100 spikes/s). In other words, by matching the spike kinetics, we made our goal of linearizing the bifurcation harder than if we had used any arbitrary spiking model with slower dynamics (such as the unmodified HH).

b. Matching the spike shape at the recording site.

As mentioned previously, the passive parameters in our model neuron were set so as to approximate the input resistance at the injection site and also the transfer resistance between the spiking compartment (modeling the SIZ) and the injection site. Figure 2 shows the level of this approximation. The figure shows the output of a real cell and of our model to the same step of current (5 nA). Both the steady state depolarization reflecting the input resistance at the site and the size of the attenuated spike reflecting the transfer resistance between the SIZ and the recording site are similar for the model and the experimental data. The main objective was to assure that the current injections in our spiking model were of the right magnitude and therefore that it was significant to try to match the FI plots directly between the data and the model. Clearly, some of the complex transfer impedance is not accounted for,

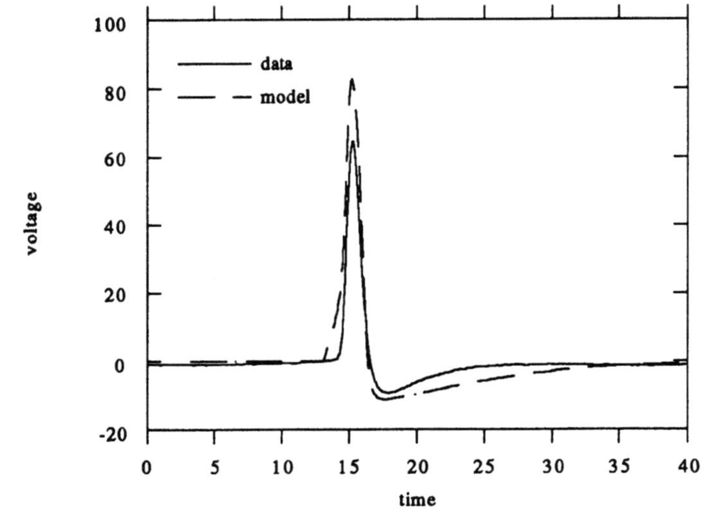

Figure 1: Comparison between the spike shape recorded in the axon and the voltage trace in the spiking compartment of the model.

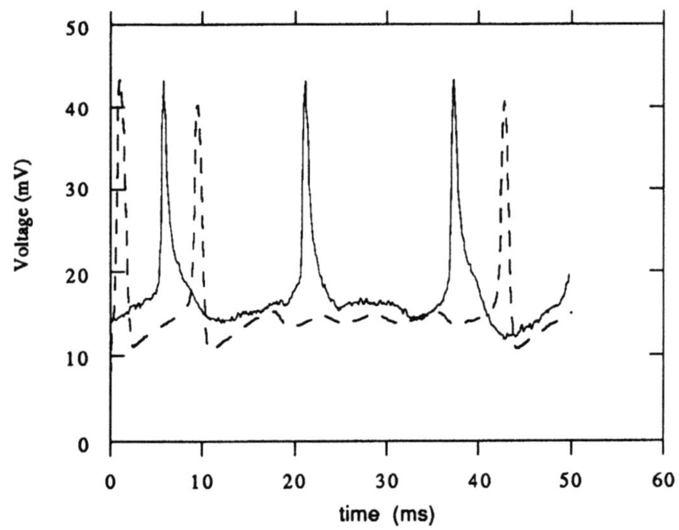

Figure 2: Comparison between the voltage traces recorded at the injection site, some distance away from the spike initiating zone (SIZ) and the first compartment of the model neuron.

since the undershoot seems to be completely filtered out in the experimental data but not in the model. This is understandable since in our model we only fitted the resistance. Note that the complex part of the transfer impedance could be matched by including more of the complex anatomical structure of a real neuron or similarly by adding a complex impedance to the shunt resistance of the recording site.

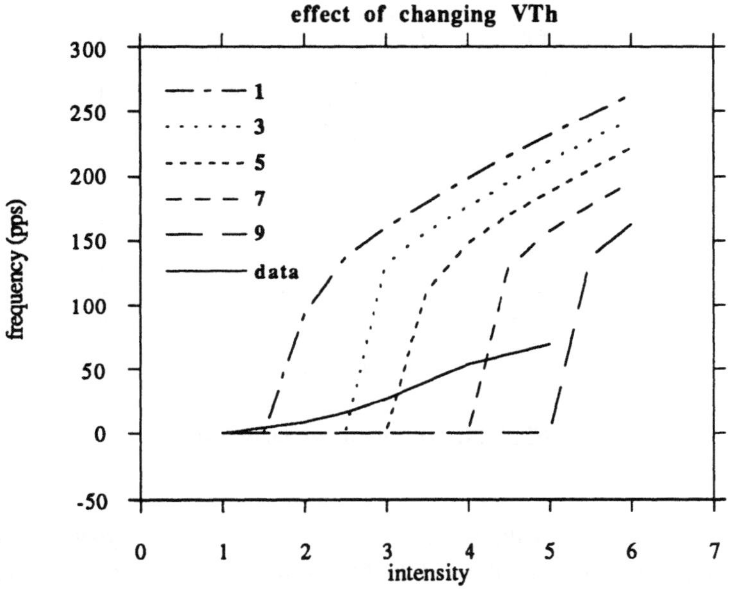

Figure 3: FI plot for different values of V_{Th}

c. Matching the FI plot.

The first step we undertook was to study how changing the remaining free parameters in our model would affect the FI plot. Three parameters in the model could be changed without significantly changing the shape of the action potential or the transfer resistance from the injection/recording site to the SIZ. The parameters are : 1. The threshold of the activation of the inward current (V_{Th}); 2. The amount of shunting current (K_{sh}); 3. The amount of synaptic noise.

Figure 3 shows the effect of changing V_{Th} . As V_{Th} was increased, the slope and the singularity at the bifurcation of the FI curve remained the same but the entire curve is shifted to the right along the input current axis. Note that in the cases shown in Figure 3, with no shunting current and no noise, the FI plot in the simulations is far from the observed FI plot.

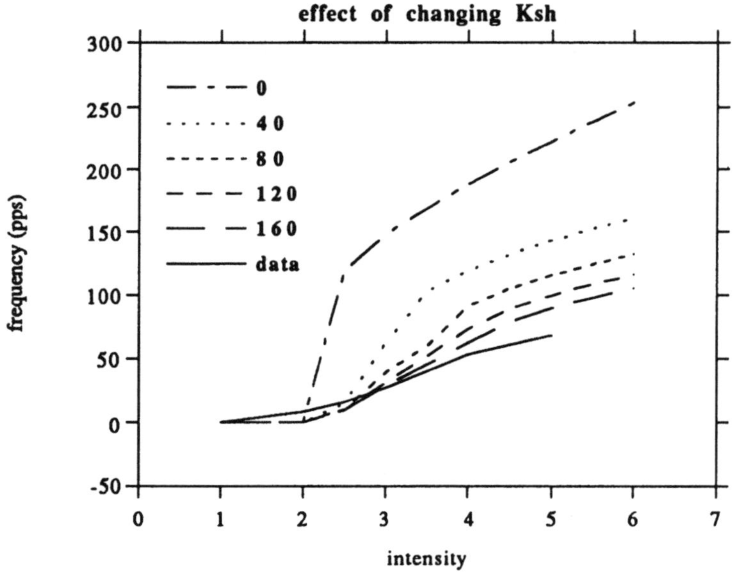

Figure 4: FI plot for different values of K_{sh}

Figure 4 shows the effect of changing K_{sh} for a fixed V_{Th}, with no synaptic noise. Increasing the shunt decreases the slope of the FI plot and also progressively reduces the singularity at the bifurcation.

Figure 5 shows the linearizing effect of adding synaptic noise. The addition of noise doesn't shift the curve or change its final slope but eliminates the singularity at the bifurcation. As the noise level is increased, the slope of the FI plot around the original singularity is decreased.

From Figures 3, 4, and 5 one can see that there will be more that one set of voltage threshold, aditional shunt and noise witch will match the experimental data. In fact for almost any value of Vth one can find the right noise and shunt combination to fit the data. A particular fit is shown on Figure 6. The remaining degree of freedom can be resolved only by matching the spike statistics as described by their interspike histograms.

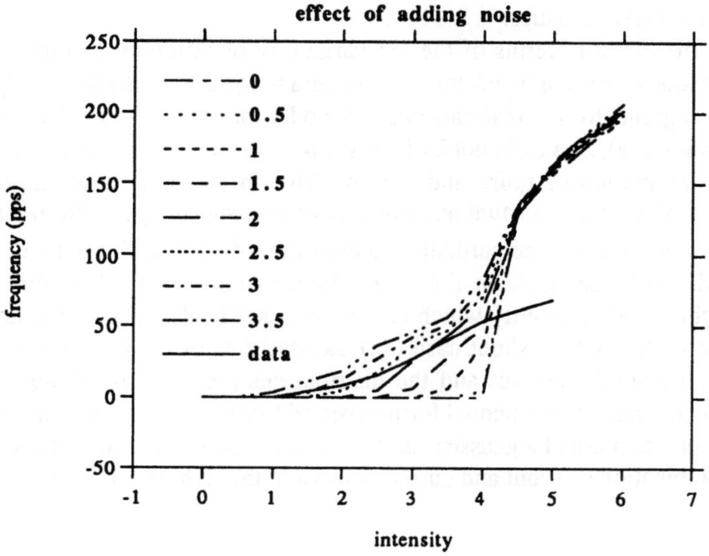

Figure 5: FI plot for different values of added synaptic noise.

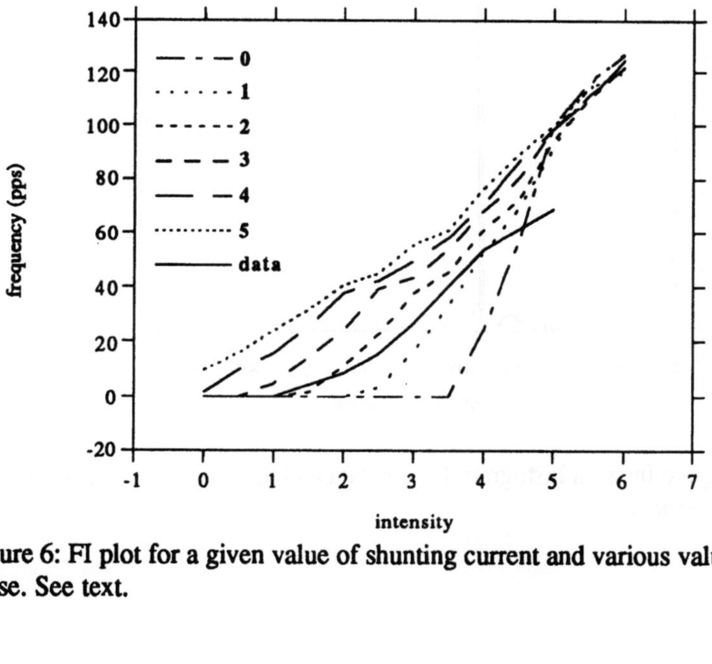

Figure 6: FI plot for a given value of shunting current and various values of synaptic noise. See text.

d. Matching the spike statistics.

To illustrate the effect in terms of the ISI variability of different combinations of noise/shunt values, we displayed the ISI histogram for three situations. Figure 7 shows the histograms for the real data and the model for cases where 1) we did not include any noise (7a), 2) we did not include any additional current shunt (7c), and 3) we used a combination of noise and current (7b). In all three cases the chosen parameters for V_{Th} , current shunt and noise gave a reasonably good fit in terms of the FI plot. However, only one particular combination of noise and shunt will match the observed spike statistics. In the high shunt/no noise case (7a), the model distribution shows no variability which clearly does not fit the observed data (solid line). For the high noise/no shunt case (7c) the statistics of the model are clearly swamped by the noise statistics and the distributions are radically different. The model distribution has an exponential form expected from Poisson like events and the data distribution has more of a gaussian shape. There is a correct middle ground where an unique combination of shunt and current will yield the right distribution as shown in figure 7b.

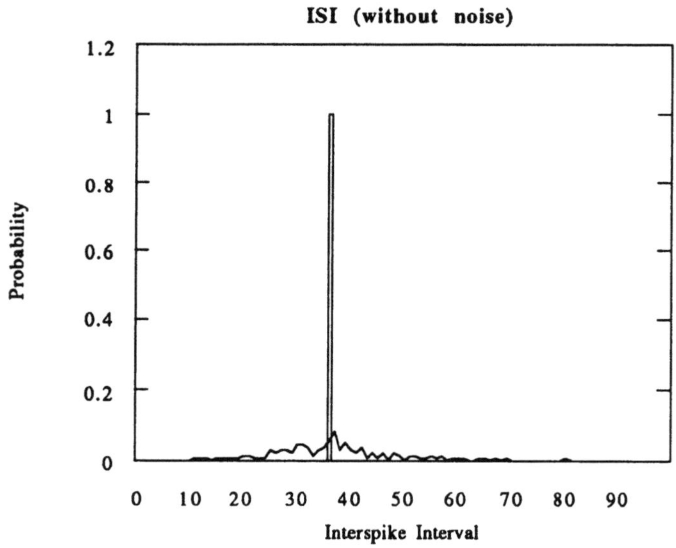

Figure 7a: Interspike Interval histogram for the data and the model in the case where no synaptic noise is added.

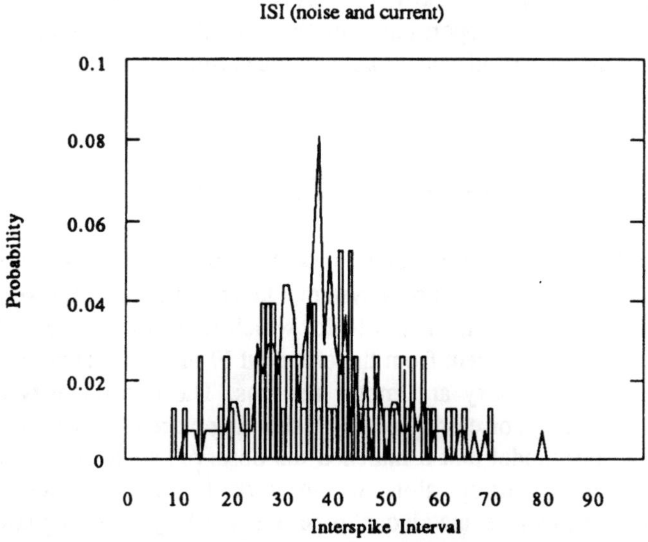

Figure 7b: Interspike Interval histogram for the data and the model in the case where both synaptic noise and shunting current are used. This particular combination gives the best match to the data.

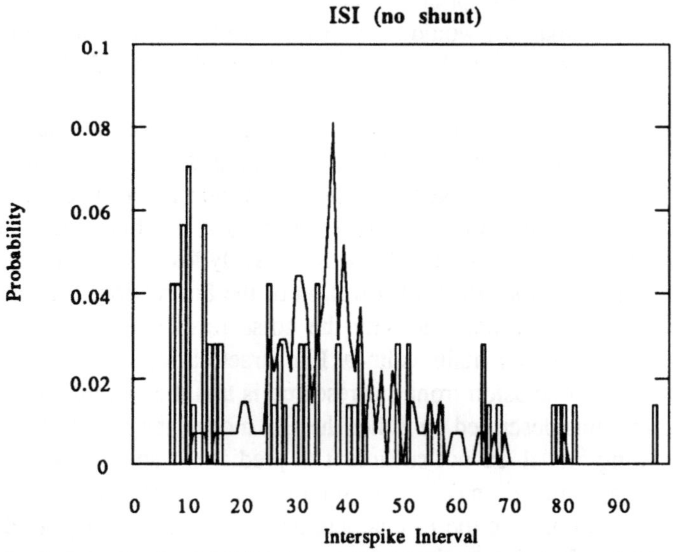

Figure 7c: Interspike Interval histogram for the data and the model in the case where only synaptic noise is used.

Thus to fit both the FI plot and the variability we had to include both noise and shunting current in the right proportions even though either the addition of noise or shunt alone would have been sufficient to obtain the correct FI plot.

Discussion & Conclusion

We used a simplified version of a channel model to match the shapes of spikes in interneurons in the cricket cercal sensory system. Our spiking model showed a strong non-linearity (Hopf Bifurcation) in its FI plot which is characteristic of HH-type models but which is very different from the observed FI plot of cricket interneurons as well as many other sensory and motor neurons. The non-linearity could be linearized either by adding noise or by adding a shunting current. However, when we also required from our model that it matched the observed spike statistics, neither noise alone nor shunting current alone would work. In particular, we could not eliminate the non-linearity *and* match our spike statistics by just adding noise to our model, as other researchers have suggested (Yu and Lewis,1990). To obtain the low frequency firing with the correct spiking statistics we had to add an additional current which produced a shunt. The model description of that current need not be identified by any particular channel in the real neurons. The point is that there must be a mechanism beyond the spike generating inward and outward currents which modulates the interspike interval. The description we chose allowed us to fit our data, but it is possible that other formulations would work just as well. A clear extension of our work would be to investigate the specific nature of this additional shunt experimentally.

After adding the shunting current, the variability in the interspike interval as shown in the interspike histograms was easily fitted by setting the right "magnitude" of noise level in our model. Note that this right magnitude could also be verified experimentally by recording the noise directly at the SIZ. In this work, using the model, we found that the noise level needed was relatively low and it is possible that noise is not an essential factor in the linearization of the bifurcation. It is our guess that by eliminating all presynaptic activity the noise level would decrease even further but the neuron would maintain its linear FI characteristic.

In this sense, the main conclusion from our research is that the inward and outward spike generating currents described by either the HH model, the FitzHugh-Nagumo model, or our spiking model are not sufficient to predict the spiking behavior and that an additional deterministic mechanism such as an A current must be present. Noise is also shown to linearize the FI characteristic but seems to play a relatively small role in our case. It is interesting to see that Yu and Lewis were capable of matching the spike behavior and the statistics of saccular sensory neurons by just adding noise to the HH model. These researchers suggest that noise is actually beneficial in terms of the information capacity of the neurons since it allows for the large observed dynamic range. This is particulary true if the system has many component cells (as it is the case in the frog saccular system) and can effectively nullify the variability in their responses by averaging across all cells. On the other hand, the cricket cercal system consists of only 10 pairs of the type of interneurons examined in this study and the benefit of the linearization by noise would quickly be reversed by the added variability. For these interneurons, it is much more efficient to increase the dynamic range by another mechanism and this seems to be the case.

References

1. L. F. Abbott and T. B. Kepler. (1990) Model Neurons: From Hodgkin to Hopfield. In: *Statistical Mechanics of Neural Networks.* Garrido L. (ed). Springer Verlag, Berlin.
2. W. Calvin and C. F. Stevens. (1968) Synaptic noise and other sources of randomness in motorneuron interspike interval. *J. Neurophys.* **31** : 574-587.
3. J. A. Connor, D. Walter and R. McKown. (1977) Neural Repetitive Firing. *Biophysical Journal* **18** : 81-102.
4. R. FitzHugh. (1961) Impulses and physiological states in theoretical models of nerve membrane. *Biophysical Journal* **1**: 445:466.
5. J. Rinzel and G. B. Ermentrout. (1988) Analysis of neural excitability and oscillations in *Methods in neural modeling.* Koch and Segev (eds). MIT Press Cambridge MA.
6. R. Stein. (1965) A theoretical analysis of neuronal variability. *Biophysical Journal* **5**: 173-194.
7. X. Yu and E. R. Lewis. (1989) Studies with spike initiators: Linearization by noise allows continuous signal modulation in neural networks. *IEEE Trans.on Biomed. Eng* **36** :36-43.

SECTION 2

MODELING

The modeling section is comprised of the papers presented at the Satellite Symposium on Software Development for Single Cell Compartmental Modeling. Major developers of simulation packages were invited to present their software at the symposium. The discussion was focussed around issues of implementation, representation, numerical techniques and user-interface design. The papers are in alphabetical order by first author.

Upinder Bhalla and Jim Bower introduce GENESIS, a package that allows one to construct simulations at many levels of detail. GENESIS has sufficient flexibility and user extendibility to make it possible to work on any type of neural simulation problem. GENESIS was conceived as a research, communication, and teaching tool for the neuroscience community.

Edward Blum and Peyvand Khademi present CAJAL-91. CAJAL-91 is a program for simulating general biological neural networks at a detailed compartmental level. The paper describes the models of dendrites, axons, and synapses used to model neurons and networks of arbitrary architecture. Careful attention is given to numerical accuracy of simulation results.

Erik De Schutter is the author of NODUS, a user friendly program for simulation and development of compartmental models of neurons and small networks on Apple Macintosh computers. The chapter describes the program functions and some issues in developing a user friendly interface.

Frank Eeckman et al. describe NEMOSYS, a software package developed by physiologists for physiologists. NEMOSYS tries to closely mimic the experimental paradigms used in laboratories, studying the behavior of single neurons. A great deal of attention was paid to speed, numerical accuracy and user interface design.

Michael Hines developed NEURON, a nerve simulation program designed around the notion of continuous cable "sections" which are connected together to form any kind of branched cable and which are endowed with properties which vary with position along the section. The design goal is to keep entirely separate the physical properties of the neuron from the numerical issue of size of spatial compartments. Simulations are controlled using an interpreter with c-like syntax. New membrane channels or mechanisms are described using a high level model description language which allows the expression of models in terms of kinetic schemes or sets of simultaneous equations. Membrane voltage and states are computed efficiently using an implicit integration method optimized for branched structures.

Genesis : a neuronal simulation system.

Upinder S. Bhalla and James M. Bower
Division of Biology
California Institute of Technology
Pasadena, CA 91125

Introduction

The use of numerical simulation techniques to study neurobiology has grown over the past few years to encompass systems ranging from molecular level processes occurring in real neurons to systems of many thousands of highly abstract neuron-like elements (Touretzky, 1990, Koch and Segev, 1989; Zornetzer et al 1990). It is increasingly apparent that all these levels of analysis are interrelated and may be essential to the understanding of the functioning of the nervous system (Bower, 1991)

In principle, numerical simulation techniques can provide a means of performing an analysis of neurobiological structure / function relationships over many levels of biological investigation. Such an effort is likely to benefit from the availability of a simulation environment capable of representing neurobiological assemblies at levels of detail ranging from molecules to complete systems. This is one of the principal design objectives of the GENESIS system.

Design Philosophy.

We have developed Genesis (GEneral NEuronal SImulation System) and its graphical interface Xodus (X-based Output and Display Utility for Simulators) primarily as a means of constructing biologically realistic neuronal simulations. There are three basic elements of our design philosophy. One, the simulator must be capable of addressing problems at many levels of detail. Two, the system should facilitate communication between groups working on many kinds of modeling problems. Three, the system should be open-ended : it should not place limits on the kinds of problems that can be addressed and should allow the incorporation of user extensions. The following specific features of the system have evolved based on these criteria.

Modular Design.

As a key principle conferring both flexibility and organization on Genesis, a highly object-oriented programming approach has been adopted. Simulations are constructed of 'building blocks' or modules, each of which perform well-defined functions and have standardized means for communicating with each other. The level of detail of a simulation is then determined by the nature of the building blocks one chooses. For example, one could use more realistic (and computationally expensive) Hodgkin-Huxley type conductances to describe the spiking properties of a neuron, or just simple threshold elements. The user has complete freedom to assemble a model with any set of modules, so the system does not constrain the kind of simulation developed. Object-oriented design also enables the user to easily add new modules to extend Genesis for a particular application.

Flexible interface.

The Genesis user interface consists of two parts. The underlying level is the Script Language Interface, or SLI. This is an interpretive programming language similar to the UNIX shell, with an extensive set of commands related to building simulations. The interpreter can read SLI programs (scripts) either interactively from the keyboard, or from files.

The graphical interface, Xodus, provides a higher level and user-friendly means for developing simulations and monitoring their execution. Xodus consists of a set of graphical modules which are exactly the same as the computational modules from the user's point of view, except that they perform graphical functions. Like the computational modules, Xodus modules can be set up in any manner that the user chooses to display or enter data. Furthermore, the graphical modules can call functions from the script language, so the full power of the SLI is available through the graphical interface.

These two aspects of the Genesis user interface give the user complete freedom to specify the structure of a simulation, including graphical input and output.

Device independence.

In order to make the simulator portable and available to the maximum number of users, the system has been designed to run under UNIX and X-windows. The code has been tested on a number of machine architectures. This facilitates communication by enabling groups utilizing different machines to run each other's simulations. A parallellized implementation is also being developed with the objective of extending the range of problems which Genesis can address.

Using Genesis.

The user interacts with Genesis at a number of levels.

Simulation Library.

Each registered GENESIS user has access to a library containing demonstration simulations, tutorials, and high level simulation tools. These completed simulations are manipulable through a user-friendly graphical interface without any need for interaction with the GENESIS scripting language. The demonstration simulations are intended to illustrate the range diversity of simulations that can be carried out in Genesis. They also provide starting points for the design of new simulations. Tutorials are simulations specifically designed to introduce students to fundamental concepts, and classical models, in computational neurobiology. These tutorials are currently being used in a number of graduate and undergraduate training programs in computational neurobiology (Beeman and Bower, 1991). Some tutorials also serve to introduce new GENESIS users to the structure of the scripting language.

Finally, the simulation tools are complete interfaces for developing new research simulations. One of the most useful tools is "neurokit", which provides an environment for developing, editing, and testing multicompartment models of single neurons.

Prototype library.
At a slightly greater stage of sophistication, one can develop simulations using the script language directly. Such new simulation development takes advantage of the simulation component prototypes in the extensive and growing library of ion channels, cellular compartments, whole neurons, and even networks. These prototypes are genesis modules, or assemblies thereof, which have already through use in other simulations, been assigned appropriate parameters.

Implementing new prototypes.
At the next level of interaction with GENESIS, users can implement new prototypes by collecting data from the literature or their own experiments and assigning these parameters to the appropriate pre-existing computational elements.

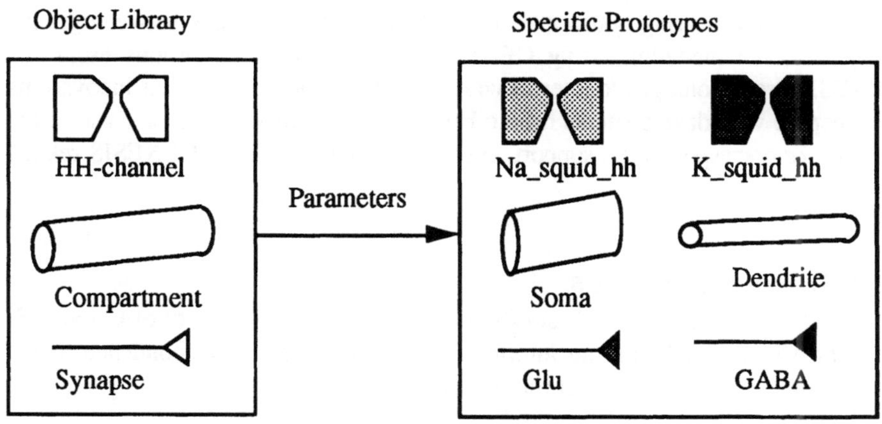

Figure 1. Implementing new prototypes.

Implementing new modules.
Finally, the most advanced level of interaction with GENESIS involves the construction of new modules to perform new computational tasks. This interaction with the simulator is the only level that requires advanced programming skills. However, the object-oriented nature of Genesis means that only a few basic rules must be observed in writing new modules, to ensure that they mesh smoothly with existing GENESIS structure.

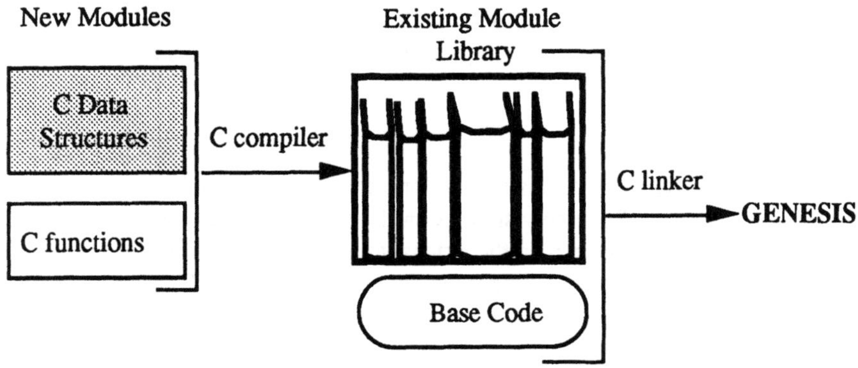

Figure 2. Implementing New Modules.

Users Group.
In order to support the continued development and use of GENESIS, we have established a active users group called Babel. The objective of the users group is to act as a simulation clearing house and to facilitate communications between groups doing neuronal modeling, using GENESIS. Accordingly, all serious users of GENESIS are encouraged to interact with the users group, to send their innovations and keep up with developments from other users. Babel makes extensive use of the internet for correspondence, support, and the distribution of new GENESIS related materials.

The structure of Genesis.

In the follow sections, we will briefly describe the overall structure of GENESIS. Readers who would like more information should address correspondence to the email address listed at the end of this article.

Base code.
All the housekeeping activities, allocation of memory and CPU resources, and maintenance of simulation state is handled by the base code. It also provides the most basic functions available in the Script Language.

Objects.

As we have stressed earlier, GENESIS is written in an object-oriented manner. The building blocks of all simulations within Genesis are objects, which we have loosely referred to as modules. In programming terms, an object is both the data structure associated with a module, and the numerical operations which that module carries out. An element is an actual instance of a Genesis object. Each element has its own memory area for storing state variables and parameters associated with it, and performs the numerical operations defined for the object.

Hierarchy.

The elements in Genesis are arranged in a UNIX directory-like hierarchy. This makes it very natural to group related elements together and provides a straightforward means for identifying each individual element.

Messages and connections.

Inter-element communications are performed in two ways. For instantaneous information transfer between elements, such as is needed for evaluating electrical properties for the integration routines, messages are used. For delayed transfer of information, as in spike propagation down an axon to a synapse, connections are used.

Simulation setup

As already alluded to, the Script Language Interface contains all the commands for setting up a simulation in genesis. A typical simulation might be set up in the following stages :

1 **Creating elements**

create compartment /soma
create hh_channel /soma/Na
create hh_channel /soma/K

2 **Assigning parameters**

set /soma Rm 1e7 // Ohms
set /soma Cm 0.01 // Farads/m^2

3 **Establishing inter-element communications**

sendmsg from /soma to /soma/Na VOLTAGE Vm
sendmsg from /soma/Na to /soma CHANNEL Ek Gk

4 **Setting up input/output**

create xform /form
create xgraph /form/graph
sendmsg from /soma to /form/graph PLOT Vm

5 **Running the simulation**

step 100

Single Neuron specification

One of the central functions of any neuronal simulator is the specification of multi-compartmental models of single neurons. This process is greatly simplified in Genesis by the use of a special set of functions and tools. The most important of these is the cell-reader function. This Genesis utility reads a neuronal specification file which contains the geometry, branching structure, and channel distribution of a particular neuron. In association with "neurokit" (see above), this provides a straightforward way for single neuron models to be made and tested with minimal knowledge of the details of GENESIS and the script language.

Numerical aspects.

The main numerical operation carried out by most objects in Genesis is numerical integration of time-varying state variables. In keeping with the object-oriented methodology, each element is responsible for updating its own variables. At the present time there are eleven integration methods available in Genesis, including Euler, Gear, Adams-Bashforth, Runge-Kutta, Backward Euler and Crank-Nicolson. All of these are intercompatible, so one can evaluate different methods using different integration schemes. The two implicit methods are based on the Hines algorithm (Hines, 1984), and are intrinsically non-object oriented and are available only to compartment and channel objects.

Input/Output.

Three interconvertible forms of IO are available in Genesis. The first is simply the script language, which can be used to explicitly set variables and have them printed out on the screen. The second in Xodus, the graphical interface. This is usually the main method for controlling and monitoring a simulation. It can also read output stored into files. The third option is file I/O. Typical file inputs are for stimulus patterns, cell parameters, and so on. File output can be used, for example to save the entire time history of a simulation. An additional I/O feature is the SAVE/RESTORE option, which can save the current state of a simulation to a file, for resumption later.

Libraries.

The development of Genesis as a multipurpose simulation tool has led to the establishment of libraries of components of the system, at different levels.

1 Code/object libraries. These are functions and objects developed for different simulations and projects. There are currently over 200 commands and over 50 different Genesis objects in the various code libraries.

2 Prototype libraries. These are simulation components such as voltage dependent channels, synapses, compartments and even complete neurons whose parameters have been worked out from experiments and the literature. They greatly simplify the task of developing new simulations.

3 Simulation libraries. These are complies simulations at a wide range of detail and complexity, providing a resource of examples and demonstrations of the use of Genesis.

Simulator Specifications.

Memory requirements

Genesis currently consists of about 45000 lines of simulator code and about 25000 lines of graphics code in Xodus. The size of the binaries are strongly machine dependent, varying from 1 to 2 Meg. The amount of additional memory needed for a particular simulation can be estimated from the size and number of modules in the simulation. Typically, elements use about 100 bytes, connections 16 and messages 20. Graphical elements use from 5 to 20 K each. In a typical simulation, messages and connections account for the bulk of the memory use.

Performance.

This varies roughly linearly with the number and type of elements in the simulation. As an example, a simulation of the olfactory bulb with 7 kinds of neuron having from 9 to 52 compartments and roughly 2 channels per compartment occupies 10 meg for a 400 cell model. On a Sun Sparc2 this simulation runs at about 1 second per time step.

Access/use of Genesis

A demonstration version of Genesis is freely available over the network by anonymous ftp from genesis.cns.caltech.edu. This includes a number of example simulations and full source for GENESIS.

Babel newsgroup.

Serious users of Genesis are encouraged to join the Babel newsgroup which provides access to the latest versions of Genesis, bugfixes, new simulations and libraries, and a forum for interacting and exchanging simulations. Interested persons should contact Chris Ploegaert (cp@smaug.cns.caltech.edu).

Acknowledgements

Genesis was devised and brought into being principally by Matt Wilson. Many people have contributed to its development, and continue to do so. To acknowledge the contributions of a few of them: Dave Bilitch and Matt Wilson : the Script Language Interface; John Uhley : Xodus; Matt Wilson, Mark Nelson, Erik de Schutter and many others : the function and module libraries; Dave Beeman, Matt Wilson, Mark Nelson, Erik De Schutter, and several generations of students at the "Methods in Computational Neuroscience" course at the Marine Biological Laboratory, Woods Hole : the library of simulations, demos and tutorials; The National Science Foundation (DIR-9017153), Sun Microsystems Inc., Digital Equipment Corporation, Intel Corporation, and the Lockheed Corporation : financial support.

References

Touretzky, D. S. 1990 Advances in neural information processing systems. Morgan Kaufman.

Koch, C. and Segev, I. 1989 Methods in Neuronal Modeling.
MIT Press.
Zornetzer, S. F., Davis, J. L., and Lau, C. 1990 An Introduction to Neural and Electronic Networks. Academic Press.
Bower, J. M. 1991 Piriform cortex and olfactory object recognition. In: J. Davis and H. Eichenbaum (Eds.) Olfaction as a Model System for Computational Neuroscience. MIT Press, Cambridge, MA. (in press).
Beeman, D. and Bower, J. M. 1991 Soc Neurosci Abstr.
Hines, M. 1984 Efficient computation of branched nerve equations. Int J. Bio-Med. Comp. 15:69-76.

CAJAL - 91:
A Biological Neural Network Simulator

Edward K. Blum and Peyvand M. Khademi
with the assistance of
Kevin Chau, Patrick Leung and Xin Wang

University of Southern California
Los Angeles, California 90089

Abstract

We describe CAJAL-91, a programming tool for simulating biological neural networks at a detailed compartment level. The underlying models of dendrites, axons, synapses and network connectivity are presented. Accuracy of numerical results is discussed.

Introduction

It is by now generally recognized that the speed and memory capacity of computers available to neuroscientists makes it computationally feasible for them to simulate biological neural networks in great detail. This can be done even on work stations of modest cost for networks of modest size. At the present rate of progress in computer hardware, it will soon be feasible to simulate quite large biological neural networks in great detail on individual work stations. For very large networks, existing supercomputers make detailed simulation feasible. Furthermore, new developments in parallel computing will make possible ever larger, more detailed and more realistic simulations.

The level of detail we have in mind is such that experimental neurobiologists can compare laboratory data against the numerical results of simulations directly or with so little transformation of the numerical results as to preclude questions of interpretation. This type of simulation can be a valuable aid to the experimentalist as well as to the theorist. A general-purpose simulator program which makes it relatively easy to set up such simulations can be a useful tool both in the biological laboratory and in the modeling laboratory. Modelling and "wet-lab" experimentation are becoming complementary interdependent research modes. The age of computational neuroscience is fast upon us.

Several groups of researchers have taken notice of this situation and developed simulators which meet the needs of neurobiologists to varying degrees. In this paper, we present the simulator called CAJAL-91, developed by the two principal authors with the valuable assistance of the other three listed. Part of this work is part of the Ph.D. thesis of the second author done at U.S.C. under the direction of the first author. Other parts are the work of the other U.S.C. students listed. The design, development and programming of a large-scale, general-purpose, neural network simulator of this type is a major undertaking in terms of man-hours of effort, close supervision and cooperation of the participants, and mastery of the biological-mathematical-computational concepts underlying the simulator. We have named our simulator after the discoverer of the neuron to emphasize its intended use: simulation of biological neural networks. The suffix "91" indicates that it is this year's version and will surely undergo modifications as new knowledge and techniques are discovered and further needs are made evident by usage. The conceptual design of CAJAL took shape in the mind of the first author during several years of teaching a graduate course on mathematical models of neurons and neural networks. The educational uses of such a simulator were evident. Its utility as a research tool would depend on the choice and proper design and implementation of programming elements to achieve generality, user-friendliness and numerical accuracy while satisfying the practical constraints of speed of execution and limitations of memory size. Our initial tests of CAJAL-91 as a research tool demonstrate that it meets these design goals.

Mathematical-Computational Models of Neurons and Neural Networks

The design of CAJAL began with a choice of model for neurons and a model for networks of neurons. The two models are closely related. The neuron model, adopted after considerable research, is essentially mathematical, based on differential equations which are derived from the bio-physics of the electro-chemical reactions known to underly neuron function. In essence, the neuron model represents the dynamics of the electric potentials due to the various ion currents which flow across most neuronal membranes and through the cell cytoplasm. The network model, likewise adopted after lengthy research, represents the synaptic interactions between neurons and requires a model of synapses. Based on the first author's research into the known physiology of synapses, a decision was made not to model the kinetic equations of chemical synapses, since kinetics of most synaptic channels and neurotransmitter release is still undergoing much laboratory study. Instead, we provide a model which allows the user to select the shape of the postsynaptic potential or current, for which there is much experimental data. Further, CAJAL-91 provides only for axo-dendritic synapses. Confronted with the question of how to specify the location of such a synapse, one is forced to make a decision on the question of how to represent the morphology of the dendritic tree of a neuron. This also is involved in the choice of the model of the electrophysiology of the dendritic structure, which comes down to a choice between a continuous cable model and a compartment model of dendritic fibres. The continuous cable model involves a system of partial differential equations (pde) with boundary conditions at points where the fibres meet. Non-uniform fibres complicate the equations. The numerical solution of these pde with boundary conditions would be very involved. This led us to choose a compartment model, which treats the problem of branch points of the dendritic tree in a natural way so that boundary conditions do not arise. Likewise, the non-uniform diameter of tapering fibres is incorporated into the compartmentalization process. This leads to a system of ordinary differential equations (ode). In fact, this is an old technique in numerical analysis for discretizing the space variables in a parabolic pde., known as the method of lines. Although not a leading method, in this case it provides an effective numerical procedure for computing neuro-electrodynamics. However, it should be remarked that the resulting ode system of a neuron compartment model can sometimes be stiff and, therefore, may require special stiff-equation solvers. We are conducting research into this question, but our simulation results on moderate-size models have so far not indicated numerical difficulties due to stiffness. In a well-designed simulator, the user should be freed from concerns about the numerical accuracy of results and spurious numerical instabilities. In the design of the CAJAL simulator, we have addresses these numerical issues. We have incorporated a well-tested ode solver with provision for stiffness, with most of its options properly preset and with many built-in error indicators. (See below.) From an electric circuit perspective our neuron compartment model can be viewed as a lumped-parameter circuit model of the type studied by Rall [Rall64]. Finally, we remark that the neuron compartment model also provides a natural way to model network connectivity. Synapses are specified as being situated in a dendritic compartment closest to the actual site. The synaptic channel is then modeled as a variable membrane conductance in that compartment. The conductance is activated by monitoring the condition of the specified presynaptic axon.

From this preliminary overview of the design of CAJAL we can proceed to a more detailed description of each of the underlying models.

The Neurons Model in CAJAL-91

We are aware that there is no typical neuron. They vary widely in morphology and physiology and the present state of knowledge of both of these aspects is incomplete. Yet, enough is known about the anatomical and electrophysiological properties of neurons in the central nervous system to enable us to set up a model that approximates in considerable detail many of the neurons which are the subject of current research. In CAJAL-91, we have set up a neuron model consisting of three submodels: a dendrite model, an axon model and a synapse model. This reflects our underlying hypothesis, based on established neuroanatomy and neurophysiology data, that most neurons have these components. We give an outline of the main features of each submodel.

The Dendrite Model

The dendrites of a neuron are modeled as a tree of compartments of a type introduced by Rall [Rall64], which we have modified to reflect our own synapse model and our network model. Anatomically, each compartment is a cylindrical section of passive membrane enclosing cell cytoplasm. Its electro-dynamical behavior is modeled as a lumped RC circuit. When setting up a simulation of a particular network, the user (prompted by a window-menu display) specifies for each neuron its membrane specific conductance (in mhos/cm^2), membrane specific capacitance (in μF/cm^2) and the cytoplasm specific conductance (in mhos/cm). The user is also prompted to specify each compartment length (μm) and diameter (μm). From these physical quantities, CAJAL-91 computes several of the lumped-parameter constants which are used to represent a compartment as an RC circuit. The other constants are specified in the synapse model. (See below.) Future versions of CAJAL will provide for active (i.e. voltage-dependent) conductances in dendritic compartments, as in the axon model (below).

The tree-structure of compartment connectivity is specified by the user (again prompted by CAJAL) by assigning a number to each compartment and giving the number of the lower compartment to which it is connected. It is suggested that the user prepare beforehand a schematic drawing of each dendritic tree with each compartment labeled by its identifying number. For trees of moderate size, this is a reasonable technique. However, for a large number of compartments it is tedious. Future editions of CAJAL will provide abbreviated ways of specifying tree geometry for large regular trees. A tree terminates usually in a trunk compartment which is connected to a soma compartment. The soma may be specified either as a spherical or cylindrical compartment. It is connected to the axon compartment. All compartment connections are through cytoplasmic conductances computed by CAJAL from the specific conductance and compartment dimensions.

Axon Models

Each neuron has a single cylindrical axon compartment, which models the actual axon hillock (or trigger zone), where action potentials are generated. In CAJAL-91, three "standard" action potential models are available: the Hodgkin-Huxley (H-H) model [HH52], the Adelman-Fitzhugh model [AF75] and the Fitzhugh-Nagumo model [Fit61], [NAY62]. The user is prompted to select one of these axon models for each neuron. The H-H model consists of the four non-linear ordinary differential equations of the space-clamped version of the H-H equations. The user may specify his own values of the parameters in these equations, in which case CAJAL will prompt him for each value, or he may choose a set of default values provided by CAJAL. In future editions, other axon models will be added (e.g. the Frankenhauser-Huxley equations, the Conner-Stevens equations, bursting equations, etc.). CAJAL does not provide a multi-compartment cable model of the axon. To model the propagation of action potential spikes along a real axon and its collaterals, CAJAL provides for delay parameters to represent the time required for a spike to travel from the trigger zone to a synaptic terminal. The delays are user-specified as parts of the synapse model. Before describing the synapse model, we shall describe the network model.

The Network Model

A network consists of a specified number, N, of "system" neurons and a specified number, M, of "input" neurons. The values of N and M are arbitrary, but of course, are subject to the practical constraints imposed by computer memory and speed. Each system neuron is assigned an identifying number (Id) between 1 and N and is user-defined by a neuron compartment model as described above. The connections between neurons may be completely general: i.e., no constraints are placed on possible network architectures. To specify a connection the user is prompted for the Id of the presynaptic neuron, the Id of the postsynaptic neuron and the compartment number of the latter in which the synapse is situated. Then the user is prompted for the parameter values of the synapse model, explained below. While this technique has the merit of complete generality, it can be quite time-consuming for large N. In the next edition of CAJAL, the specification of connections for certain uniform modular architectures will be simplified; e.g. for layered networks with uniform receptive fields.

The external inputs to a network are assumed to come from the input neurons. Connections from input neurons to system neurons are specified in the same manner as between two system neurons, the Id of an input neuron being an integer between $N+1$ and $N+M$. An input neuron is modeled simply as a voltage waveform. The user is prompted to choose either a constant voltage value (mv) or a rectangular pulse train with arbitrary frequency, width, height (mv) and start-stop times. All simulations start at time zero and end at some user-specified time T (msec).

The Synapse Model

This consists of a presynaptic model and a postsynaptic model for each synapse in a compartment. Consider synapse k in compartment j. For the presynaptic model the user is prompted to specify two parameter values for each synapse: L_{jk}, the depolarization level (mv), and D_{jk}, the duration of the depolarization (msec). A third relevant parameter, dt, the time unit for updating all synaptic state components, is specified as the parameter in the ode solver which determines the solver output time steps, t_i, for all compartment states. The user is also asked to specify a delay, d_{jk}, which is typically interpreted as the time for the action potential to propagate from hillock to synapse, but may be interpreted to additionally include the time for diffusion of neurotransmitter. Let $VS_{jk}(t)$ be the presynaptic terminal potential at time t and $VH_{jk}(t)$ the potential at the presynaptic hillock. Then $VS_{jk}(t) = VH_{jk}(t - d_{jk})$. During simulation, CAJAL-91 will record a "neurotransmitter-release event" at time t_{n+k} if $VS_{jk}(t_i) \geq L_{jk}$ for all t_i with $t_n \leq t_i \leq t_{n+k}$ in a preceding interval with $k{\cdot}dt \geq D_{jk}$. If L_{jk}, D_{jk} and dt are suitably specified, this should model the effect of depolarization of an axon terminal and the time needed for the resultant sufficient influx of calcium ions and release of neurotransmitter. The choice of dt depends on the time-constants of these presynaptic processes, in particular on the action potential rise and fall times. Since dt is also used by the ode solver to adjust integration step-size for output, the solver's maximum step-size parameter must be set less than or equal to dt. Furthermore, the synapse model is updated only at time instants dt, $2dt$, $3dt$,...,whereas the compartments may be updated at smaller time-steps by the ode-solver. Hence, dt must also be chosen sufficiently small to accurately interpolate the fastest synaptic process; e.g. action potential rise phase. A value of about 0.1 msec has been found to be suffice for most cases, but some trial-and-error tests by the user may be needed.

Once the neurotransmitter-release event has occured, it initiates the postsynaptic model at time t_{n+k}. The latter allows the user to define the waveform of the postsynaptic current by specifying the time-variation of the receptor channel's conductance. Let G_{jk} be the maximum channel conductance. The effective channel conductance is approximated by a trapezoid having a linear rise phase from 0 to 1 at rate r_{jk}, followed by a constant level of 1 for w_{jk} msec and then a linear fall phase back to 0 at rate f_{jk}. The resulting function $S_{jk}(t)$, has a trapezoidal shape (though, with appropriate parameter settings it can reduce to a rectangular or triangular shape), and is used to compute a postsynaptic current, psc_{jk} by the formula:

$$psc_{jk}(t) = S_{jk}(t) \cdot G_{jk} \cdot (V_j(t) - E_{jk}),$$

where V_j is the compartment potential and E_{jk} the synaptic reversal potential. When setting up the network model, the user is prompted for the values of L_{jk}, D_{jk}, d_{jk}, G_{jk}, E_{jk}, r_{jk}, f_{jk} and w_{jk}.

Mathematical and Numerical Aspects

The dynamical behavior of each compartment is given by an ode for membrane potential as a function of time. Thus, for compartment j, its potential $V_j(t)$ is governed by the Rall-type [Rall64] ode:

$$dV_j/dt = -\mu_j \cdot V_j/\tau_j + \varphi_j \ ,$$

where,

$$\mu_j = 1 - \sum_k S_{jk}(t) \cdot \gamma_{jk} \ ,$$

$$\gamma_{jk} = G_{jk}^{(m)} / G_j^{(m)},$$

$$\tau_j \cdot \varphi_j = I_j / G_j^{(m)} + \sum_k S_{jk} \cdot \gamma_{jk} \cdot (E_{jk} - V_0),$$

$$I_j = \sum_q G_{jq} \cdot (V_q - V_j).$$

Here, $G_j^{(m)}$ is the membrane conductance of compartment j and $G_{jk}^{(m)}$ is the maximum conductance of synapse k in compartment j. G_{jq} is the cytoplasm conductance connecting compartments j and q. Thus, I_j is the sum of all currents flowing between compartment j and all adjacent compartments q. $\tau_j = C_j / G_j^{(m)}$ is the membrane time-constant. V_0 is the resting potential. E_{jk} is the reversal potential of synapse k. The $S_{jk}(t)$ are the conductance waveforms of the synapse models.

We shall not reproduce here the axon ode's. They can be found in references: [AF75], [HH52], [Fit62], [NAY62]. They are coupled to their respective neuron dendrite ode's by a current term computed from the soma cytoplasm conductance and potential and the axon potential.

The resulting system of ode's for all neurons is solved by the ode solver IVPAG, an IMSL subroutine having options for stiff systems and step-size control. The user may set parameters to specify desired accuracy of integration, as prompted by a CAJAL menu interface. Accuracy of numerical integration procedures in simulations at a level of detail which includes action potentials and precise timing of excitatory and inhibitory synapses is of fundamental importance. We have done test simulations to verify numerical accuracy for certain choices of IVPAG parameters. However, this aspect warrants further research.

Input-Output

We have sketched the various menu-driven input devices provided by CAJAL for specifying a network to be simulated. To avoid duplication in repeated simulations where only a few parameters are to be changed, CAJAL provides for storing parameter values in files which can be accessed using an abbreviated "turbo" input in successive runs. To check input the user can display these files. A graphical display (of tree structure and network structure) will soon be available as well.

For output of voltage waveforms during a simulation experiment, CAJAL provides an on-the-fly monitor display of graphs of voltage vs. time for up to ten compartments. Likewise, after a simulation run, any voltages recorded in an output file can also be plotted on the monitor screen (in color if available). The screen can then be dumped to a printer for hard copy.

Example Simulation

An example of a small simulation done using CAJAL is given in the paper by Blum, Khademi and Thompson [BKT91] published in this volume. It is a highly simplified four-neuron network model of a cerebellar network implicated in conditioning an eye-blink response in rabbit. The reader is referred to [BKT91] for further details. Although very small-scale, that simulation successfully demonstrated the effect of disinhibition by Purkinje cells and the importance of the timing of the spikes in producing this disinhibition. This example was also part of the computer demonstration of CAJAL-91 presented at the AMNS II conference.

Programming Features

Space does not permit a detailed description of the programming features of CAJAL-91. The implementation of such a general-purpose network simulator is an arduous and far from trivial task requiring an advanced level of knowledge and skill in software development. From the user's perspective the main feature is that CAJAL-91 runs in a window environment, currently in X-windows and soon OpenWindows. A user's manual is in preparation but a user literate in Unix and X-Windows should be able to follow CAJAL's detailed menu instructions which guide the user step-by-step in constructing a particular network simulation. The computational core of CAJAL-91 is written in Fortran. This is called by an outer C program interface. Future editions will provide various mouse-accessible "buttons" to further simplify usage.

Some Conclusions

Biological neural network simulators at the level of the CAJAL-91 simulator described here hold great promise as tools for both the theoretical neuroscientist and the experimentalist. This seems clear even from our limited experience with CAJAL-91. However, as with any scientific tool, care and skill must be exercised in its use and in the interpretation of the results obtained. The accuracy of the various models should be investigated both experimentally, by comparing simulation results against laboratory data where available, and mathematically, by analyzing the underlying ode's. For a general treatment of neural models and other approaches we refer to the article by Segev, Fleshman and Burke [SFB89] in the recent book edited by Koch and Segev [KS89]. We also refer to [KS89] for other relevant articles on biological neural models and simulation and lists of references.

References

[AF75] Adelman, W.J., and Fitzhugh, R. "Solution of the Hodgkin-Huxley equation modified for potassium accumulation in a periaxonal space." In *Membranes, Ions and Impulses,* ed. Moore, J., Plenum Press, 1975.

[BKT91] Blum, E. K. and Khademi, P. M. and Thompson, R.F. "Model and simulation of a simplified cerebellar neural network for classical conditioning of the rabbit eye-blink response." Poster at AMNS II and see this volume.

[Fit61] Fitzhugh, R. "Impulses and physiological states in theoretical models of nerve membrane." *Biophysics Journal*, 1, 445-446, 1961.

[HH52] Hodgkin, A.L. and Huxley, A.F. "A quantitative description of membrane current and its application to conduction and excitation in nerve." *Journal of Physiology*, 117, 500-544, 1952.

[KS89] Koch, C. and Segev, I., editors: *Methods in Neuronal Modeling*, MIT Press, 1989.

[NAY62] Nagumo, J., Arimoto, S. and Yoshizawa S. "An active pulse transmission line simulating nerve axon." *Proc. IREE*, 50, 2061-2070, 1962.

[Rall64] Rall, W. Theoretical significance of dendritic trees for neuronal input-output relations. In: *Neural Theory and Modeling*, edited by R. Reiss, Stanford: Stanford Univ. Press, 73-97, 1964.

[SFB89] Segev, I., Fleshman, J.W. and Burke, R.E. "Compartmental models of complex neurons." In: *Methods in Neuronal Modeling*, edited by Koch, C. and Segev, I., MIT Press, 1989.

Nodus: A User Friendly Neuron Simulator for Macintosh Computers

Erik De Schutter
Division of Biology 216-76,
California Institute of Technology,
Pasadena, CA 91125, U.S.A.

Abstract

Nodus, a software package for simulation and development of compartmental models of neurons and small networks is described. Nodus is implemented on Apple Macintosh microcomputers, with the standard user interface and graphics. Two user interface issues are discussed: control of model consistency and selection of model parameters.

The Design Goals for Nodus

Modeling of neurons and neuronal networks is a powerful tool to examine nervous system function. Other chapters in this book show examples where modeling has become as important as theory, investigation and laboratory experiments.

An often heard criticism is that some models have little or no biological relevance. One can expect that more realistic models could be developed by experimental neurobiologists themselves. An impediment may have been the lack of easy to use software. If neurobiologists are going to compare computer simulations with their experimental data, they should be able to acquire and use modeling tools as any other piece of lab equipment. This means simple installation, clear instructions, no extensive computer knowledge required, an easy to use interface, etc.

The computer program Nodus was conceived as a modeling tool for scientists with a limited computer experience. The user interface resembles biological experiments instead of emphasizing the mathematical aspects. Nodus integrates most of the required steps in modeling: neuron model conception, model input, translation of the model into simulation parameters, control over experiments, iterating integration of the simulation and graphic output. These features make Nodus also useful for teaching compartmental modeling during short courses.

A microcomputer implementation was preferred because most neurobiologists have a limited access to workstations or mainframe computers. Microcomputers are now standard lab equipment and are an acceptable solution to scientists who do not regard computer simulations as their main line of work. Nodus was implemented on Apple Macintosh computers because it is easy to make graphic oriented, user friendly software on these machines.

A compartmental approach to neuronal modeling was chosen because it is intuitively easy to use and it makes accurate modeling of morphology and electrophysiology possible. Recently network functions have been added to the software. These support the simulation of small, "hard-wired" neuron circuits as for example invertebrate pattern generators.

Integration is implemented by explicit methods. Though implicit integration methods are faster, they are harder to implement. Easy expansibility of the software to include new modeling features was considered more important than raw speed.

Program Functions

This chapter describes Nodus version 3.1, which was developed for Macintosh II computers. Nodus version 2.3 is still available for use on Macintosh computers that do not use a 68020 or 68030 microprocessor. Nodus was written in Fortran (MPW Fortran, Absoft Corporation), the source code consists of about 40000 lines. The program makes full use of the Macintosh interface with movable windows, drop down menus, popup menus, data entry in dialog boxes, file specific icons, etc. An extensive program manual is available.

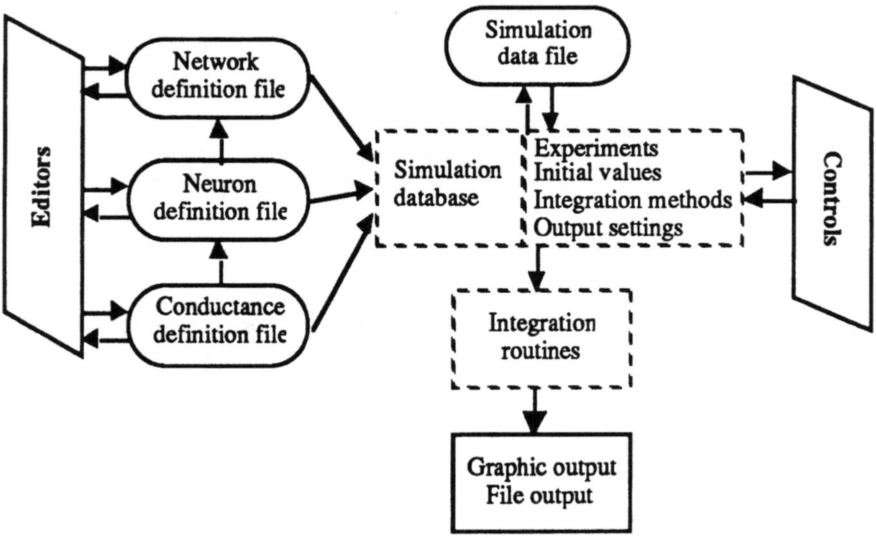

Figure 1 The internal structure of the Nodus software.

The program functions are conceptually split in two parts (Fig. 1). Database management of model definition files is performed by editor functions; integration and output of the simulation is interactively set up with simulation controls.

Models are stored in separate files: conductance definition files, neuron definition files and (optionally) network definition files. All file types have their own menu, which is enabled when the window at the front belongs to the corresponding file type. Different files specifying one model are linked together in a hierarchical structure by Nodus and automatically loaded when the top file is opened. Several conductance and neuron files can be open and edited at the same time. The dialogs which allow the user to edit model files have been constructed to look as if they represent standard physiological and morphological data.

Conductance definition files contain the Hodgkin-Huxley like equations for voltage dependent ionic conductances. These equations are saved to separate files because they are often common to several neuron models and are developed separately. Interactive plotting of conductance, time constant, (in)activation and rate factors versus membrane potential can help in the formulation of new equations. The sign convention for simulation graphics of ionic currents can be changed by the user (positive outward currents up or down).

Neuron definition files contain the compartmental model of a neuron: cable parameters, sizes and connections of all compartments (maximum 4000) and ionic and synaptic current parameters. Local variation of membrane resistance is supported. The passive, morphological structure of a neuron can be imported from different formats of text files, generated by other simulation or morphology software.

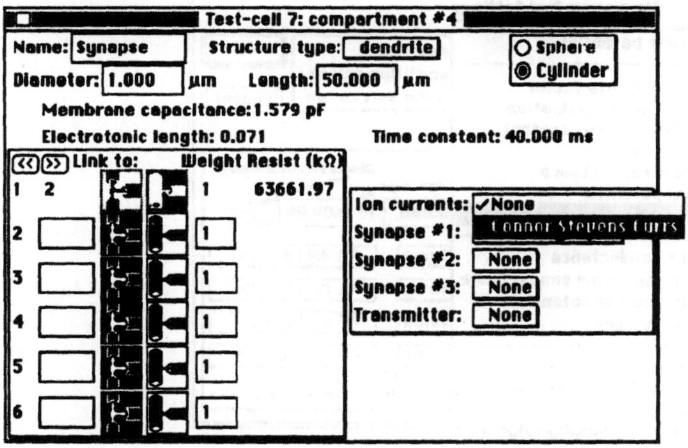

Figure 2 A passive compartment is made excitable.

The compartment structure places the nodes of the equivalent circuits at the center of the compartments. Standard symmetric connections between compartments (end to front) and asymmetric connections (center to front) are available. There are commands to split and fuse compartments; this can be done automatically to constrain the electrotonic lengths of all compartments in a model. The neuron morphology can be represented graphically in a Scholl plot.

Ionic currents and pre- and postsynaptic sites are defined in separate subdefinitions, which are stored with the neuron definition. The subdefinitions contain all the data necessary to implement the feature except for the location within the compartmental neuron model. In compartment dialog windows the user can then select with popup menus one or several subdefinitions to tie them to the compartment (Fig. 2). If no subdefinitions are selected, a compartment is passive.

Network definition files link neuron definitions to neurons in the network and specify the connections between them.

Model definition files have a good structure to define and manage neuron or network models, but are inefficient for integration of simulation equations. Nodus transforms data from selected model definition files (a single neuron or a network) into a simulation database. The simulation database structure was optimized for integration speed. Its format is invisible to the user and of no concern as all relevant simulation parameters can be accessed with selection popup menus (Fig. 3). The simulation database can be stored in a file, together with settings for integration, output, experiments, etc.

Two explicit integration methods with variable time steps are available: an accurate Fehlberg method (fifth order Runge-Kutta) and a fast forward Euler method. Accuracy of integration is controlled by a maximum relative error for the Fehlberg method or a maximum change in membrane potential per step for the Euler method. The Euler method uses tabulated values for the Hodgkin-Huxley like equations.

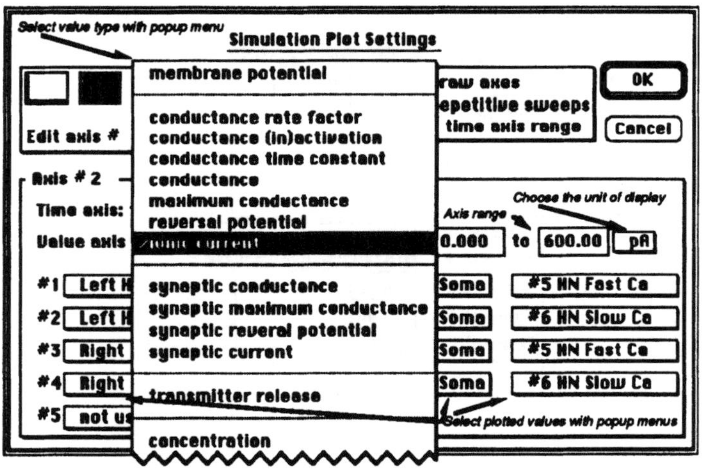

Figure 3 Selection of model parameters with popup menus.

Initial values can be essential to run successful simulations, especially if voltage dependent ionic conductances are used in the model. If conductance (in)activation factors are not matched initially, the simulated neuron may behave irregularly (the period of firing will be variable, the first action potential may look different, etc.) till a dynamic equilibrium is reached. In some cases the simulation may reach an equilibrium which has no biological meaning (for example a continuous depolarization instead of firing), caused by an initial imbalance in conductances which was too large to get corrected spontaneously. Equilibrated conductance (in)activation factors cannot be determined analytically for a firing or bursting neuron.

In Nodus the user has extensive control over the initial values of parameters. These are stored in the simulation data file, so that simulations can be initialized with equilibrated values obtained from a previous simulation. The value of any simulation parameter can be viewed and manipulated by the user.

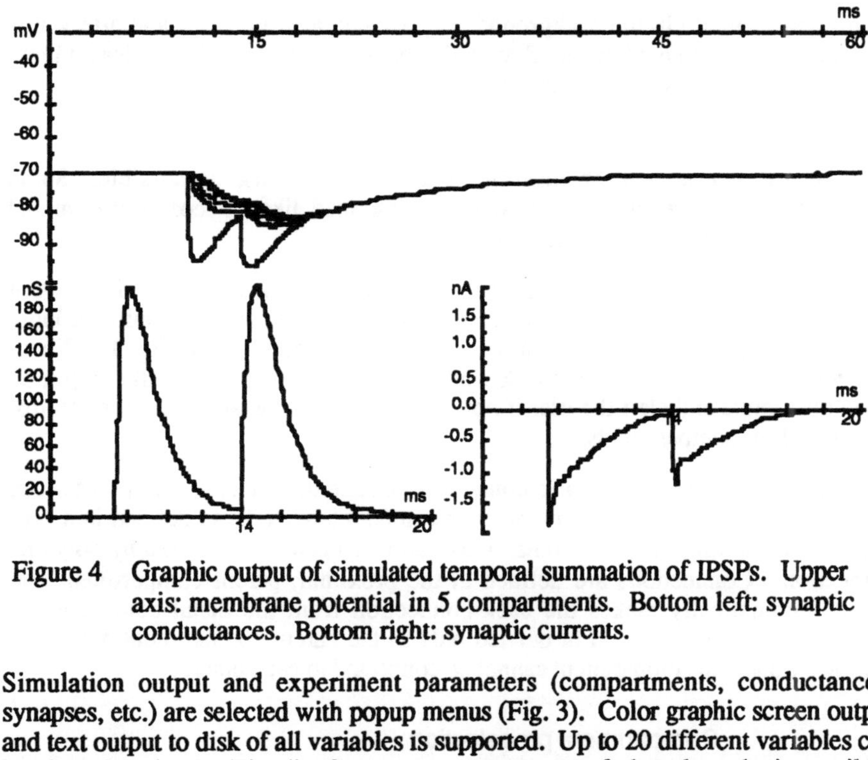

Figure 4 Graphic output of simulated temporal summation of IPSPs. Upper axis: membrane potential in 5 compartments. Bottom left: synaptic conductances. Bottom right: synaptic currents.

Simulation output and experiment parameters (compartments, conductances, synapses, etc.) are selected with popup menus (Fig. 3). Color graphic screen output and text output to disk of all variables is supported. Up to 20 different variables can be plotted on 4 axes (Fig. 4). On screen measurement of plotted results is possible.

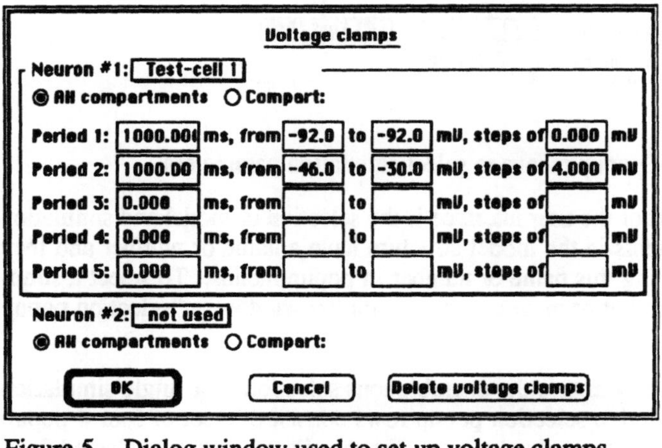

Figure 5 Dialog window used to set up voltage clamps.

Several standard electrophysiology experiments can be set up. Different currents (constant, repetitive pulses, ramps, sinus, noise) can be injected and added together in any compartment. One or two neurons can be voltage clamped (Fig. 5). Synapses can be preset to fire at specified times. Ionic currents can be blocked selectively during the simulation to simulate antagonist application.

Nodus can compute in the background, but one can also interrupt and afterwards continue prolonged simulations. Standard Macintosh printing of graphics and text is available. Graphics can be copied and pasted into other applications.

The User Interface

A lot of effort was spent on making Nodus as easy to use as possible. Some aspects of the user interface have already been described, like the separation of model database management and simulation controls.

Because Nodus may be used by inexperienced modelers it was considered important to check for model consistency. The user is constrained to use preset units for input parameters; for output the units can be selected with popup menus (Fig. 3). Whenever possible Nodus will also check wether an entered parameter falls within an acceptable range and alert the user if not. This is done for example for the electrotonic length of compartments.

There is an extensive, automatic control of consistency of connections between compartments. Compartments in the neuron model have to be to defined in a centrifugal way, starting from the soma. Compartments can be connected by only one connection. Connections are defined at the proximal, parent compartment and Nodus automatically makes the back-connection from the distal compartment. Connections can be changed or deleted only in the parent compartment definition window. Spherical compartment cannot be connected to each other.

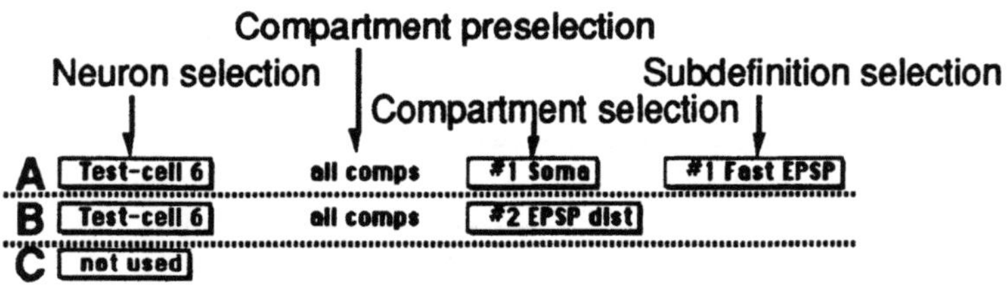

Figure 6 The use of popup menus to select simulation parameters.

Another important part of the user interface is the selection of model and simulation parameters. All elements in the model structure have a name or number and they are accessed by selecting this name or number in popup menus. To select a simulation parameter for output or in an experiment dialog window, a "selection popup row" is used (Fig. 6).

Each selection popup row makes the unambiguous selection of a single simulation parameter possible. Active selection popup rows consist of a set of 3 or 4 popup menus depending on the selected value type (Fig. 6). Parameters are selected from left to right. The neuron popup menu will show either 'not used' when nothing is selected (Fig. 6C) or the name of a neuron. The compartment popup menu will show appropriate compartments of the selected neuron, i.e. if one is selecting an ionic current only excitable compartments will be shown. For all parameters, except membrane potentials, a subdefinition selection is made with an additional popup menu.

Popup menus are a nice instrument to select parameters, but they should not be too long. Imagine selecting a compartment for membrane potential output in a 1000-compartments neuron model; a compartment popup menu showing all 1000 compartments would be hard to use! Therefore Nodus never shows more than 50 items in a popup menu. To select which subset of compartments in large models will be shown in the compartment popup menu, the compartment preselection popup menu is used. It makes selection of a subrange for the compartment menu possible.

The user can choose between three methods of compartment preselection. The default compartment preselection popup menu will show numbered subranges of compartments (#1 - #50, etc) when the model is too large. The other methods depend on how the user has labeled compartments (Fig. 2). If only a few, interesting compartments in a large neuron model have been given a name, the compartment popup menu can be set to show only named compartments and no preselection is necessary. An alternative is to preselect by structure type. The compartment preselection menu will show a list of structure types (soma, dendrite, axon, postsynaptic site, etc.) and the compartment menu will be restricted to the selected structure type.

Distribution

Nodus is available for a small fee. This includes the compiled program, example files and the manual. The source code is not available because the average Nodus user is not a programmer.

Readers interested in acquiring Nodus should contact the author. E-mail can be send to erik@smaug.cns.caltech.edu.

The mathematical methods used in Nodus and their accuracy are described in: E. De Schutter, "Computer software for development and simulation of compartmental models of neurons", Computers in Biology and Medicine, 19, 71-81, 1989.

NeMoSys: An Approach to Realistic Neural Simulation

Frank H. Eeckman[1,2], Frédéric E. Theunissen[1] and, John P. Miller[1]

[1]Dept. of Cell and Molecular Biology, University of California at Berkeley, Berkeley CA 94720

[2]O-Division, Lawrence Livermore National Lab, Livermore CA 94550

Abstract

We describe a software package that allows for efficient simulation of current flow through complex neurons. Each neuron is represented as a binary branched tree structure, where branches are constructed from linear strings of compartments. The program is set up to allow the user to simulate typical electrophysiological experimental protocols.

The modeling is done at the level of currents and voltages in individual compartments of neurons. An implicit scheme of integration which takes advantage of the branched tree structure of the neuron is used to update the voltages and currents in each neuron at each timestep. One of the major computational benefits of this method is that time scales linearly with the number of compartments used to represent the neurons. Furthermore, voltage updates are decoupled from the conductance updates, so arbitrary conductances or synaptic connections can be incorporated easily, efficiently and stably. Nemosys can also be extended to allow simulations of networks of neurons.

Design Philosophy

NeMoSys (Neural Modeling System) is a program designed by physiologists for use by physiologists. It was created specifically for the simulation of current flow through neurons consisting of one or more connected compartments. It is, however, extendable to simulations of networks of these neurons (Tromp and Eeckman, 1991). A lot of work was devoted to creating an interface that is easy to use, while at the same time powerful enough to allow one to mimic any possible experimental paradigm. The user interacts with a graphical display of a nerve cell and monitors transmembrane currents and voltages through color-coded animation. Voltage or current traces derived from "electrodes" positioned at various levels in the cell can be displayed in an oscilloscope-like window. The user interacts with these displays as he or she would interact with an experimental preparation and the associated laboratory equipment. Nemosys was originally developed by J. P. Miller and J. W. Tromp at U.C. Berkeley. The present version was prepared by the authors of this manuscript. A command-line interface was added by Michael Hines of Duke University. Nemosys is both an interactive user friendly program that can be used by novice computer users and a powerful simulator environment for the more expert programmer.

Implementation

Nemosys was written in C using Xlib to provide maximal portability between different Unix platforms. The software will run on any workstation using the Unix operating system (Unix is a trademark of AT&T) and the X11 window environment (X11 is a trademark of MIT). Nemosys has been succesfully ported to the IBM RISC/6000, the SUN SPARC 2, the SGI Personal IRIS and the SGI Indigo.

Structure

Nemosys consists of two parts. The first part is the base code that is compiled and linked to form the executable program. The base code has a semi-modular design. Users can add routines to the existing program with only minimal modifications to the existing code. The second part is an interpreter toolkit based on HOC (See Kernighan and Pike, 1984). The toolkit can be invoked directly or via the graphical interface. Users can add new functions and/or change any existing function in the original program without recompiling by using hoc. Hoc accepts commands from the keyboard or from a file and executes them line by line. Hoc commands can intercept and replace C routines before the execute. The output of the hoc toolkit can be directed to a file or to the console using the graphical interface of the main program. The interpreter adds power and flexibility to Nemosys but the drawback is slower execution, which is typical of all interpreted code.

Representation of neurons

The representation of neurons in Nemosys is an important aspect of the design. The high computational efficiency of Nemosys (see below) is due to both the integration method and the specific representation that is used. Neurons are represented as binary branched tree structures. The neurons exist in three dimensions. Various locations along the neuron arborization are used to divide the cell into "nodes". Each x, y, z position and local diameter defines one node. The area between two successive nodes is called a segment. The first node in the tree is called the root node. All other nodes are either continuation nodes (having one child, also called "right leave"), branch nodes (having two children, right and left leave), or termination nodes (having no children). Complex branching patterns are represented by a series of binary branchings. The tree is traversed in right-handed depth first mode with the root node being node number one.

For calculation purposes compartments are defined in terms of the segments. Half of the membrane surface area of each segment is associated with the node at each end, and the resistance between nodes is given by the segmental resistance. Compartments are arranged in multiply linked lists with pointers to the next and previous compartment. Each compartment has associated with it a set of flags and a set of anatomical and functional values (the position: x, y, z; the diameter: d; the electrophysiological parameters r_m, r_i, c_m, V, and I; the Hodgkin-Huxley variables: n, m, h). Each compartment also has a unique identification (id) number and can be addressed individually. At the core of this search process is a routine ("follow") that traverses the tree from the root node to the end node. Another routine ("find_closest_node") makes it possible to address each compartment directly through the graphical interface. Searching is a very fast process due to the binary tree representation that is used to define the neuron.

Input-Output (I/O)

Inputs: two kinds of input files are required.

a)The anatomy file is an ASCII file that describes the neuron by 1) a segment code ('C' for continuation, 'B' for branch, 'T' for terminal), 2) the x, y, z position of the node, and 3) the diameter at that location. The structural data files used in most of our studies are created through the use of an anatomical reconstruction, also developed in our lab. Measurements of the positions and diameters of neuronal branches are carried out semiautomatically, via a computer-controlled research microscope. The data structure is very straightforward and user-readable, however, and can be modified or directly typed in by the user.

b) The parameter file is an ASCII file with global values for all parameters (examples are R_m, C_m, R_i, Hodgkin-Huxley (HH) parameters, surface area ratios, and synaptic conductances and timecourses). A default file is provided for the first time user.

Outputs: the program generates several types of graphical and numeric output.

a) animation: pseudocolor-coded animation of transmembrane voltage in all compartments.

b) transient analysis (time domain): graphs of voltages, currents, and other parameters (such as HH parameters) at various locations.

c) bode plots (frequency domain): frequency and phase bode plots.

Calculation methods

In any simulator the accuracy and the relative efficiency (ie., computational speed) of the numerical method used for solving differential equations is a major concern. A superb review by Mascagni (1989) describes all the relevant numerical methods and provides a comparative analysis of the different methods used for different purposes. Nemosys uses a hybrid implicit-explicit integration scheme. The equations are divided into two types: voltage update equations and conductance update equations. The two types of equations will be updated separately, so that voltage changes are assumed to occur while the conductance is held constant and conductance changes are assumed to occur while the voltage is constant. The nonlinear equations are thus treated as conditionally linear equations (Mascagni, 1989).

The integration method is a first-order implicit scheme ("backward Euler"). With such a first order implicit integration method, the evaluation of the state variables at a subsequent point requires the evaluation of their derivative at that next timestep:

$$V_{t+1} = V_t + \Delta t \frac{dV_{t+1}}{dt} \qquad (1)$$

where V_{t+1} can be obtained by solving a system of N coupled algebraic equations (see below). An implicit method is used because the equations describing the voltage of a compartmental model of a neuron are said to be "stiff" (for a review on stiff equations see Press *et al.*, 1988). Stiffness in compartmental models is the result of the two different timescales that are present in the simulation: the fast timescale of intracellular events (between compartments) and the slow timescale of membrane events. In the explicit method, one would be required to follow the change in the solution on the fastest scale (microseconds in this case) to maintain stability, even though the events of interest happen on a much slower timescale (milliseconds). Use of the backward Euler integration scheme in a compartmental model involves inverting a matrix K, where

$$V_t = K\, V_{t+1}\,; \qquad (2)$$

In the case of a binary tree representation K is a sparse matrix resembling a tridiagonal matrix that can be inverted by Gaussian elimination in 2(N-1) steps, instead of the N^3 steps required to invert a full matrix. This calculation scheme was first applied to neural modeling by Michael Hines (Hines, 1984). It has been extended since (Mascagni, 1991).

The main advantage of the implicit method is that it makes it possible to use larger stepsizes while maintaining stability of integration. A second advantage of our hybrid scheme stems from the explicit part of the calculation. By decoupling the voltage and conductance updates, it is easy to incorporate any arbitrary conductances and/or synaptic events.

Interface Design

Nemosys has one central window that displays the anatomy of the cell under study. There are two menubars. One controls the viewing situation and the movements of the cursor and one is used to assign functions to the selected nodes and regions (see below). Separate buttons are used to start, stop and continue the simulation process. Colors are used extensively to display relevant aspects of the simulated active and constant parameters. The standard color code is to use a spectral colorbar (deep blue to bright red) to represent subthreshold voltages at different points. Nemosys also uses the same colorbar to display differences in R_m, different membrane infolding factors, different membrane surface area ratios, labelled regions (see below), and the location of one or more recording electrodes.

Nemosys is a cursor-driven program: all selections refer to the present position of the cursor. The interpreter also manipulates the cursor position to address various parts of the cell. A very powerful concept is the ability to define, label and assign functions to any arbitrary group of compartments ("region"). Nemosys uses a combination of cursor position, mathematical and logical operands to allow the user to mark these regions. A separate function menu is used to assign various properties to the region. Regions can be combined to form new regions. The region concept makes Nemosys behave as an object oriented program.

System Requirements

Nemosys was developed to simulate complex neurons with many compartments on small desktop workstations. The following performance measure is an indication that we were succesful in pursuing that goal. A 10 msec transient calculation using a 3,000 compartment neurons (600 of which have Hodgkin-Huxley conductances) takes about 30 seconds on a 5 MIPS (0.7 MFLOPS) Unix workstation.

Network extension

An extension of Nemosys to model networks consisting of multiple neurons was described in Tromp and Eeckman (1991).

References

Hines, M (1984) Efficient computation of branched nerve equations. *Int J. Biomed. Comput.* 15: 69-75.

Kernighan, B.W. and Pike, R. (1984) The *Unix Programming Environment*. Prentice-Hall, New Jersey.

Mascagni, M. (1989) Numerical Methods. In: *Methods in Neuronal Modelling*. Koch and Segev (eds.) MIT Press, Cambridge, MA. p 439.

Mascagni, M. (1991) A parallelizing algorithm for computing solutions to arbitrarily branched cable neuron models. *J. Neurosci. Methods*. 36: 105-114.

Press, W.H., Flannery, B.P., Teukolsky, S.A. and Vetterling, W.T. (1988). *Numerical Recipes in C*. Cambridge University Press, Cambridge. p. 592.

Tromp, J.W. and Eeckman, F. H. (1991) Efficient modeling of realistic neural networks with application to the olfactory bulb. In: *Analysis and Modeling of Neural Systems*. F.H. Eeckman (ed.) Kluwer Acad. Publ. (*In Press*).

NEURON – A Program for Simulation of Nerve Equations

Michael Hines
Dept. of Neurobiology
Duke University Medical Center
Durham, NC 27710

Introduction

Programs designed specifically to simulate nerve equations compare favorably with general purpose simulation programs in three areas. 1) The user deals directly with concepts that are familiar at the neuroscience level and is not required to translate the problem into another domain. 2) The program contains functions better suited for controlling the simulation and graphing the results of real neurophysiological problems. 3) Special methods and tricks can be used to take advantage of the structure of nerve equations to solve them much more quickly, e.g. Hines (1984) and Mascagni (1991).

However, the general domain of nerve simulation is still too large for any single program to optimally deal with every problem. In practice, programs have their genesis in the attempt to solve a much restricted class of problems — although sometimes provision is made in the underlying implementation to incrementally extend the domain of applicability. Programs do one kind of problem best but degrade, as the problem changes, both in allowing the user to maintain conceptual control and in speed of simulation.

This article describes a program, NEURON, developed in collaboration with John W. Moore, written in C, and with source code freely available to any interested person. With NEURON, nerve properties are specified, the simulation controlled, and the results graphed by writing procedural statements to an interpreter. This has the advantage of generality but the disadvantage that it is more difficult to learn than a menu driven interface. The usefulness of NEURON degrades very slowly with increased complexity of morphology and membrane mechanisms. NEURON is best suited, in terms of efficiency, for problems ranging from parts of single cells to small numbers of cells in which cable properties play a crucial role. It is best suited, in terms of conceptual control, for stylized morphologies represented as connected cable sections and where membrane channel parameters are conveniently represented as piece-wise linear functions of position within each section. Two special classes of problems for which it is well suited are those in which it is important to calculate ionic concentrations and those where one needs to compute the extracellular potential just next to the nerve membrane. It is well suited for investigating new kinds of membrane channels since they are described using a high level model description language which allows the expression of models in terms of kinetic schemes or sets of simultaneous equations.

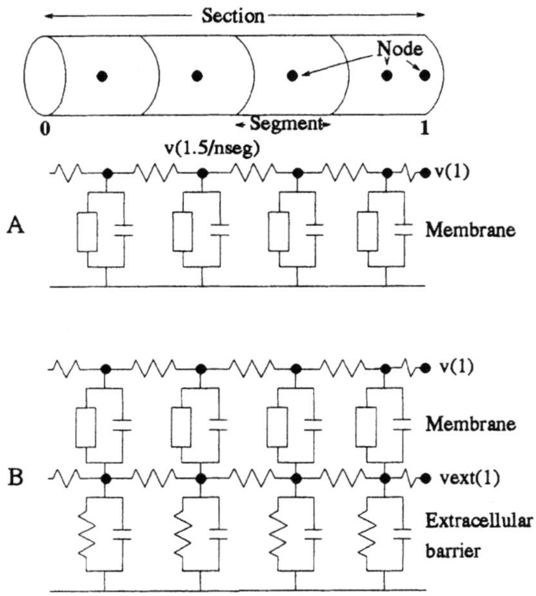

Figure 1: Section representation as electrical equivalent circuit. B shows the representation when the "extracellular" mechanism is used.

Every nerve simulation program solves for the longitudinal spread of voltage and current by approximating the cable (parabolic partial differential) equation as a series of compartments connected by resistors (Fig. 1). The compartments represent a membrane area that contains all the channels and other mechanisms that govern the total transmembrane current. The sum of all the compartment areas is the total membrane area of the whole nerve. Unfortunately, it is usually not clear at the outset how many compartments should be used. The accuracy of the approximation as well as the computation time increases as the number of compartments used to represent the cable increases. When the cable is "short", a single compartment can be made to adequately represent the entire cable. For long cables or highly branched structures, it may be necessary to use a very large number of compartments.

The problem of "conceptual control" is how best to manage all the parameters which exist within these compartments. For example, consider the membrane capacitance. The values are different in every compartment but rather than specify them all individually it is better to deal in terms of a single specific membrane capacitance that is constant over the entire cell and have the program compute the values of the individual capacitances within the compartments based on the area of the compartments.

NEURON is a program designed around the notion of continuous cable "sections" which are connected together to form any kind of branched tree structure. The value of this notion lies in the ability to specify the physical properties of a neuron without regard for the purely numerical issue of how many compartments will be used to represent each of the cable sections. This means that one can easily trade off between accuracy and speed and makes for convenient verification of the numerical correctness of the simulation.

Numerical Methods

Spatial discretization

Figure 1 showed how NEURON represents a cable section in terms of a discrete set of nodes connected by resistors. The section is divided into a number (denoted by the section parameter, **nseg**) of segments of equal length. A node (location where the internal voltage is defined) is placed at the center of each segment along with one extra node (representing 0 area) at the distal end of the section. A node at the center of a segment represents the entire area of the segment. The distal node represents 0 area and facilitates numerical computation of membrane potential when multiple branches are attached to the end of a section. NEURON takes care to calculate the value of the circuit parameters based on the position of these nodes and it can be shown that the spatial accuracy is proportional to the square of the number of segments.

Sections which have the "extracellular" membrane mechanism inserted are represented by an extra layer of of nodes representing the extracellular potential just next to the membrane (bottom of figure 1).

Time integration

NEURON uses a fully implicit (backward euler) integration method to compute the membrane voltage (cf. Hines, 1989). This method has the advantage of being numerically stable for large time steps and, in fact, a single large time step suffices to determine the steady state for passive neurons. Although the implicit method gives good qualitative results for large time steps the simulation error is proportional to the integration time step. The results of this method for two time steps can be seen in the left panel of figure 2 which shows the membrane

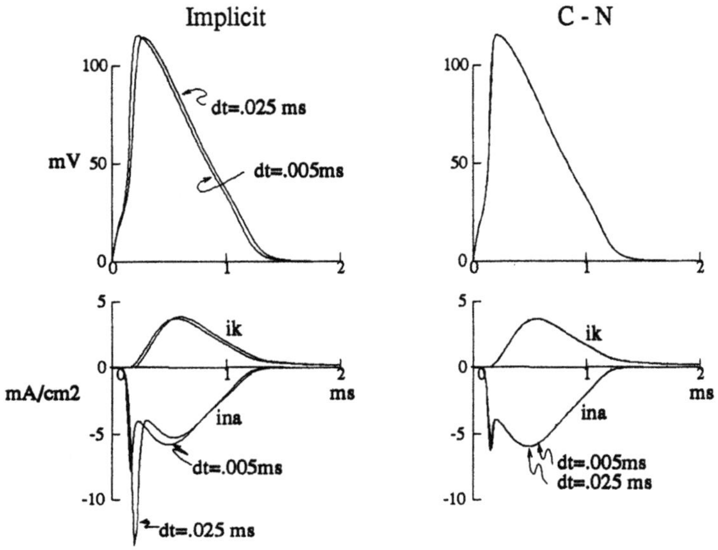

Figure 2: Frankenhaeuser - Huxley action potential: Comparison of implicit and Crank-Nicholson methods with two time steps.

voltage and ionic currents for an action potential using the Frankenhaeuser-Huxley (1964) model of the frog node. It is clear that the accurate simulation of fast changing ionic currents requires smaller time steps than that needed for the membrane potential.

A more accurate integration method (analogous to that of Crank and Nicholson, 1947) can be selected which has numerical errors proportional to the square of the time step. In fact, the simulation using the .025 ms time step in the right panel of figure 2 is more accurate than the .005 ms time step simulation with the fully implicit method. The reason the implicit method is used by default is that the C-N method doesn't work with fast voltage clamps or when the resistance coupling adjacent segments is very small. The C-N method may produce membrane potential artifacts that oscillate with large amplitude when the time step becomes too large.

Both integration methods use several efficiency tricks (Hines, 1984) that make simulations run faster. First, the geometry of the neuron is limited to a tree like structure which allows a fast direct method for solving the simultaneous equations representing charge conservation at each voltage node. Simulation time is proportional to the total number of nodes plus the total number of branches. Second, iterative solution of the nonlinear simultaneous equations is avoided by calculating channel states at the midpoint of each time step. This saves a lot of time but makes results a bit more complicated since for the C-

N method the user must be aware that voltage and concentration are most accurate at the indicated value of the time variable, `dt`, but channel states are second order correct at time, `t + dt/2`, and currents are second order correct at time, `t - dt/2`. Third, integration of channel states can be done analytically (actually second order since the voltage dependent rates use the voltage at the midpoint of the integration step) in just a single addition and multiplication operation and two table lookup operations. The tradeoff here is that the tables depend on the value of the time step and must be recomputed whenever the time step changes.

User Interface

The user interface uses broadly the same style as our previous nerve simulation program, CABLE (Hines, 1989). That is, simulation control at the user level is via a C-like interpreter called HOC (Kernighan and Pike, 1984). The built in editor for writing interpreted functions is microemacs. The integration of the nerve equations is accomplished by repeatedly calling a function which advances all the variables through a short time interval, dt.

The process of setting up and running a simulation consists in writing procedures to: 1) Define the physical properties of the neuron. This includes declaring the names of various components (continuous cable sections which may be short or long) of the cell, connecting these sections together, inserting membrane mechanisms (channels, synapses, stimulus electrodes, ionic concentrations, etc) into the sections, and setting the values of parameters for these membrane mechanisms. 2) Control the time course of the simulation. This is generally divided into at least two procedures. One initializes the membrane potential and the states of the inserted mechanisms (channel states, ionic concentrations, or extracellular potential next to the membrane). The second procedure repeatedly calls the built-in single step integration function and saves, plots, or computes functions of the desired variables at each step. In this procedure it is possible to change the value of membrane parameters during a run.

The flexibility of an interpreter makes it convenient to write procedures to handle families of related simulations, calculate new variables such as impulse propagation velocity, optimize parameters, search for thresholds, etc.

Cable Sections

The greatest problem with our previous program was that the user was responsible for knowing the correspondence between segment number and position on the nerve. All nerve properties were assigned via vector variables in which the index of the vector was the segment number. Changing the number of segments that describe the nerve was an error prone and laborious process.

This problem is overcome by the notion of a named section which can be thought of as a one-dimensional cable of length, `L`, (in μm) with a continuous position parameter, $0 \le x \le 1$, in which $x = 1$ corresponds to absolute position `L`. Most properties are functions of the position parameter and are called "range variables". Examples are the diameter in μm, **diam**, the membrane potential in mV, **v**, and, if the Hodgkin - Huxley mechanism is inserted, the maximum HH sodium conductance in mho/cm^2, **gnabar**. Specifying the values of range variables does not depend on the number of segments (parameter **nseg**) used to represent the section.

Because the same names for parameters are used in all sections it is necessary for the user to specify which section is intended. The interpreter provides several methods to determine which is the currently specified section. For example, **forall {** *statements* **}**, executes the statements within the brackets once with every section name set to the currently specified section. Special functions are provided to determine the name of the currently specified section and whether a particular membrane mechanism is inserted in that section.

The simplest use of a range variable is to assign a constant value to it as in:

```
diam=10
```

This statement places the value, 10, in every segment of the currently specified section. For values which vary along the length of a section, the assignment syntax

rangevar(*xmin* : *xmax*) = *e1* : *e2*

with $0 \le xmin \le xmax \le 1$ is used to store any piecewise linear function into the segments of the current section. The value of a range variable can appear in any expression using the syntax, "**rangevar(***x***)**", where $0 \le x \le 1$. The value is the value at the center of the segment containing point x.

Membrane Mechanisms — Model Description Language

Our previous program (CABLE) contained several built-in membrane mechanisms such as radial calcium diffusion, calcium channel, etc. However, in practice, only the Hodgkin-Huxley squid channels were enough of a standard to be used "as is" across more than one series of simulations. The other channels all required some type of modification to be useful as new situations arose. Sometimes the modifications were minor, such as changing the coordinate system for radial calcium diffusion so that there were more compartments near the membrane, but often we were forced to add an entirely new mechanism from scratch such as Frankenhaeuser-Huxley channels for Xenopus node. The problem was greatly compounded for other users of CABLE who needed to add new channels but were not familiar with the numerical issues or the detailed interface requirements.

The NEURON system overcomes this problem by allowing the incorporation of new membrane mechanisms that are defined using a high level model description language, NMODL[1] (Kohn, et. al, 1989). NMODL with NEURON is a significant improvement over CABLE with regard to adding new membrane mechanisms:

- Interface details are handled automatically. Function tables are automatically generated.
- Consistency of units is ensured.
- Mechanisms described by a kinetic scheme are written with a syntax in which the reactions are clearly apparent.
- There is often a great increase in clarity since statements are at the model level instead of the C programming level and are independent of the numerical method.

At the same time, since the model description is translated into C, the computation speed remains the same or better than a user programmed mechanism in CABLE.

Types of mechanisms

The kinds of mechanisms that can be translated by NMODL and linked into NEURON are:

- Channels in which the model consists of current-voltage relationships. This includes HH, calcium channels, etc.
- Calculation of internal and external ionic concentration changes due to currents carried by specific ions. This includes radial calcium diffusion, extracellular potassium accumulation, etc.
- Point processes which are inserted at a particular point in a section instead of being distributed throughout a section. This includes current stimuli, voltage clamps, and synapses.

More than one user defined mechanism can be simultaneously inserted into sections and NEURON will keep track of the total current for each ionic species used and the effect of that current on the membrane potential.

Types of equations

Models consisting of a mixed set of nonlinear algebraic and differential equations are written using an expression syntax of the form

[1] The model description language was originally developed at the National Biomedical Simulation Resource to specify models for simulation with their "Simulation Control Program". Only modest extensions to the syntax were necessary to allow it to translate model descriptions into a form suitable for compiling and linking with NEURON.

```
x' = f(x, y, t)
~ g(x, y) = h(x, y)
```

where the prime refers to the derivative with respect to time (multiple primes such as `x''` refer to higher derivatives) and the tilde introduces an algebraic equation. The algebraic portion of such systems of equations is solved by Newton's method and a variety of methods are available for solving the differential equations, e.g. variable step Runge-Kutta or backward euler.

Chemical reactions and kinetic schemes are expressed using a reaction style syntax illustrated with

```
~ A + 2B <-> C + D   (k1, k2)
```

Where the **k1** and **k2** are the forward and reverse rate constants. As an example of the power and convenience of the kinetic scheme approach consider a stylized model of transmitter release at the nerve terminal in response to inward calcium current. Figure 3 illustrates a model which focuses on the diffusion of calcium

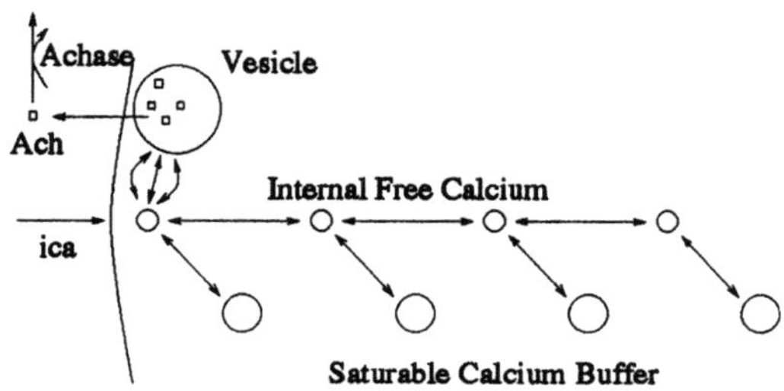

Figure 3: Calcium evoked transmitter release.

toward the interior of the terminal and its reaction with an immobile calcium buffer. At the same time, near the cell surface, three molecules of free calcium can react with a vesicle filled with neurotransmitter in such a way as to cause the vesicle to release its contents extracellularly. This transmitter is then destroyed by an enzyme.

The description of the kinetic process in the NMODL language follows. To save space the blocks which declare the parameters and variables have been left out as well as the procedure which computes factors that depend on the specific coordinate system. Ca[0] refers to the internal compartment closest to the membrane (there are N compartments indexed from 0 to N-1).

```
KINETIC calcium_evoked_release {

   : Release
 ~  Vesicle + 3Ca[0] <-> VRel   (Agen, Arev)
 ~  VRel <-> Ach (krel, 0)
 ~  Ach + Achase <-> Ach2ase    (Aasef, Aaseb)
```

```
~  Ach2ase <-> X + Achase       (Aase2, 0)

   : Buffering
   FROM i = 0 TO N-1 {
    ~  Ca[i] + Buffer[i] <-> CaBuffer[i]  (kCaB, kmCaB)
   }

   : Diffusion
   FROM i = 1 TO N-1 {
    ~  Ca[i-1] <-> Ca[i]      (Dca*f[i], Dca*f[i])
   }

   : Inward flux
~  Ca[0] <<     (ica*fac)

   COMPARTMENT i, vol[i] {Ca CaBuffer Buffer}
   COMPARTMENT    vol[0] {Vesicle VRel}
   COMPARTMENT    ext    {Ach Achase Ach2ase X}
}
```

Especially note the elegance in the way calcium diffusion is specified. The vector parameters `vol[]` and `f[]` take into account the size and surface area of adjacent compartments. More importantly, modifications to the model (such as increasing the stoichiometry of release to four calcium ions) can be easily made without introducing errors.

The above KINETIC block is translated into a C function 150 lines long which explicitly sets up the associated matrix equation — clearly the leverage is tremendous. The generated code is very efficient and the solution method is stable for all time steps (Steady state solutions are found in one step when the time step is very large; equilibrium reactions are simulated with extremely large rate constants). Furthermore the simultaneous equations are solved using a direct sparse matrix method using minimum degree ordering resulting in a tremendous decrease in computation time over the standard full matrix method.

Units

The units checking feature of NMODL has proved invaluable. Its usefulness can be illustrated by the following example, especially if the reader takes the trouble to calculate the conversion factor on her/his own.

Consider the concentration change of sodium (μM) in a sphere of radius (μm) in response to a uniform inward current density (mA/cm^2) with time measured in (ms). The equation for the concentration change (the prime refers to the derivative with respect to time, the Faraday is measured in coulombs) is written in NMODL as

```
nai' = ina/FARADAY * (4*PI*radius^2)/(4/3*PI*radius^3)
```

but is missing a conversion factor to make the units consistent.

The units checker not only generates an error message stating that the above equation is inconsistent but suggests the proper conversion factor.

References

Crank, J. and Nicholson, P.N. (1947) A practical method for numerical evaluation of solutions of partial differential equations of the heat-conduction type. *Proc. Cambridge Phil. Soc.* **43:** 50–67.

Frankenhaeuser, B. and Huxley, A.F. (1964). The action potential in the myelinated nerve fibre of Xenopus laevis as computed on the basis of voltage clamp data. *J. Physiol. (Lond.)* **171:** 302–315.

Hines, Michael. (1984). Efficient computation of branched nerve equations. *Int. J. of Biomed. Computing.* **15:** 69–76.

Hines, Michael. (1989). A program for simulation of nerve equations with branching geometries. *Int. J. of Biomed. Computing.* **24.** 55–68.

Kernighan, B.W. and Pike, R. (1984). *The Unix Programming Environment.* Prentice hall, Englewood Cliffs, New Jersey.

Kohn, M. C., Hines, M., Kootsey, J. M., and Feezor, M. D. (1989). A block organized model builder. *Proceedings of the symposium on physiological modeling at the 7th ICCM, Chicago.* In press.

Mascagni, Michael. (1991). A parallelizing algorithm for computing solutions to arbitrarily branched cable neuron models. *J. Neurosci. Methods.* **36:** 105–114.

PART II

SENSORY SYSTEMS

SECTION 3

SECTION 4

SECTION 5

SECTION 3

VISUAL SYSTEMS

A main topic in vision this year was the development of the columnar organization of the visual cortex. Our contributors presented both modeling and experimental approaches to understand the development of connectivity int he visual cortex.

Kenneth Miller examines the strengths and weaknesses of various models of the development of ocular dominance and orientation selectivity columns in the primary visual cortex. See also **Michael Lyons** and **Lionel Harrison** in the Analysis section.

Sarah Pallas et al. report the results of a series of experiments in which retinal afferents were induced to innervate the auditory thalamus. The observations suggest possible mechanisms as to how the differences in functionality arise between the different cortical areas.

Raymon Glantz and Annemarie Bartels discuss motion detection and directional selectivity in the crayfish visual system.

Josef Skrzypek and George Wu examine the receptive field properties of a vertebrate cone in the tiger salamander retina using the UCLA-SFINX simulation environment.

Philp Ulinski et al. provide a preliminary analysis of the relative contribution of both cellular and network properties to the average firing rates of turtle visual cortex cells.

Models of Activity-Dependent Neural Development

Kenneth D. Miller
Division of Biology
Caltech 216-76
Pasadena, CA 91125
e-mail: ken@cns.caltech.edu

This article is reprinted from *Seminars in the Neurosciences*, Vol. 4, No. 1 (1992), pp. 61-73, with permission of Academic Press. That issue is guest edited by K.D. Miller and is devoted to the topic "The Use of Models in the Neurosciences".

Abstract

Activity-dependent, competitive mechanisms of synaptic plasticity appear to play an important role in many processes of late neural development, where an initially rough connectivity pattern refines to a precise, mature pattern. A prominent example is the formation of ocular dominance columns in the visual cortex of many mammals. These processes may be modeled at several levels. Simple models use abstract neurons and assume synaptic modification according to a hebbian or similar correlation-based rule. These models incorporate biological constraints and attempt to predict large-scale developmental patterns from the combination of synaptic-level plasticity rules and measurable biological patterns of activation and connectivity. More detailed models attempt to incorporate various levels of biophysical realism, including membrane and channel properties and dendritic geometry. Abstract models examine the connectivity patterns that may result if biological development follows certain dynamical or other abstract rules, without concern for how such rules might be implemented at the synapse. The strengths and weaknesses of these approaches are examined through study of models for the development of ocular dominance and of orientation selectivity in the visual cortex.

Introduction

In the development of many vertebrate neural systems, an initially rough connectivity pattern refines to a precise, mature pattern through activity-dependent synaptic modification or rearrangement (reviewed in [1]). What is known about these processes is generally consistent with a hypothetical rule first proposed by Hebb [2]: synapses are strengthened if there is temporal correlation between their pre- and postsynaptic patterns of activity. The development often appears to be competitive: for a given pattern of activation, a correlated group of inputs may lose strength when competing with a more strongly activated correlated input group, yet retain or gain strength when competitors are absent ([3, 4]).

The classic example of such correlation-based, competitive development is the formation of ocular dominance columns in the mammalian visual cortex (reviewed in [5]). Visual inputs from the lateral geniculate nucleus (LGN) to the visual cortex terminate in separate stripes or patches consisting largely or entirely of terminals serving a single eye (Fig. 1). There is a regular, periodic alternation across the cortex of patches dominated by each eye. This segregated projection develops in an activity-dependent manner from a diffuse, overlapping initial projection in which inputs serving the two eyes project roughly equally throughout cortical layer 4.

Orientation columns are another striking feature of visual cortical organization ([6]; reviewed in [7]). Most cortical cells are orientation selective, responding selectively to light/dark edges over a narrow range of orientations. The preferred orientation of cortical cells varies regularly and periodically across the cortex. It has long been a popular notion that orientation columns, like ocular dominance columns, may develop by a process of activity-dependent synaptic competition. However, evidence has been lacking, because orientation selectivity generally develops before animals are born or have functioning vision [8, 9]. This does not rule out activity-dependent development, because experiments have shown that there is spontaneous neural activity, locally correlated within each eye, in the absence of vision and in the fetus [10, 11, 12], and that this activity is sufficient to guide activity-dependent ocular dominance segregation [5, 13].

The rules of synaptic modification that drive cortical development are presumably local to the environment of each synapse and its pre- and postsynaptic cells. We would like to understand these rules, yet often what can be observed is only a large-scale developmental outcome, such as ocular dominance segregation. At the same time, we would like to understand the mechanisms underlying these large-scale outcomes: which synaptic plasticity rules could produce them, what other factors are required and how can alternatives be tested? To address these questions and others, models have been devised for the development of columnar systems through local activity-dependent synaptic modification.

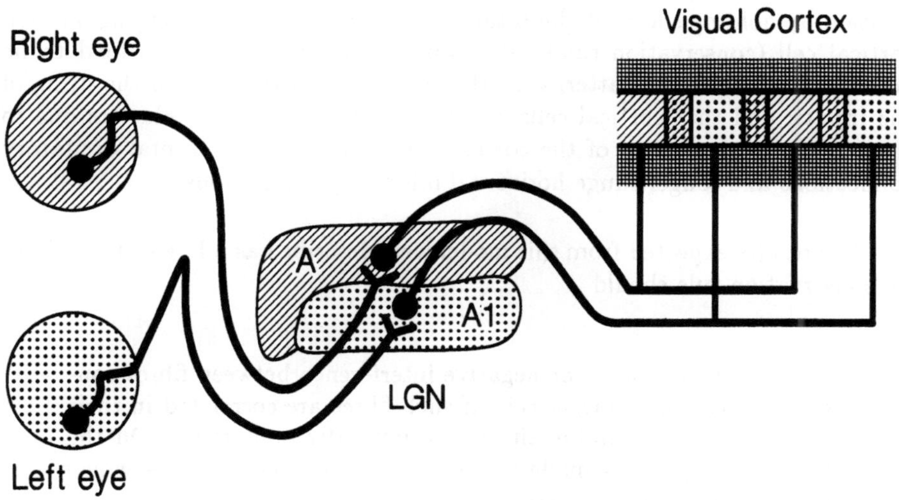

Figure 1: Elements of the Mature Visual Projection.
Retinal ganglion cells from the two eyes project to separate laminae of the lateral geniculate nucleus (LGN). LGN neurons serving the two eyes in turn project to separate patches or stripes, known as ocular dominance columns, within layer 4 of the primary visual cortex. Binocular regions receiving inputs from both LGN layers are shown at the borders between ocular dominance columns; such binocular regions exist in some species, such as cats, but not in others, such as monkeys. The cortex is shown in cross section, so that layers 1 through 3 are above and layers 5 and 6 below the layer 4 projection region. Reprinted by permission from [21]. © 1989 by the AAAS.

The von der Malsburg Model of Column Development

Von der Malsburg first formulated such a model for the development of visual cortical columns through 'self-organization' [14, 15, 16] (see also related models developed at about the same time [17, 18, 19, 20]). This model established many of the elements of current models, so is worth examining in detail. Von der Malsburg assumed that synapses of LGN inputs onto cortical neurons are modified by a hebbian rule, and that the process is competitive, so that some synapses are strengthened only at the expense of others. He enforced the competition by holding constant the total strength of synapses converging on each cortical cell (conservation rule). He assumed further that inputs tend to be activated in clusters or patterns, so that there are correlations in the firing of the inputs; and that cortical cells also tend to be activated in clusters due to the intrinsic connectivity of the cortex, e.g. short-range horizontal excitatory connections and longer-range horizontal inhibitory connections.

The results expected from this model were described as follows. The synaptic conservation rule should

> " ... (lead) to positive or negative interference between fibres connected to the same target cell. If such fibres are correlated in their activity ... their training effects are mutually reinforcing. On the other hand, synapses made on the same target cell by fibres which are anticorrelated never get reinforced.
>
> "After some time the modifiable synapses will have been rearranged so that each possible pattern of activity in the presynaptic sheet will evoke one of the possible patterns of activity in the postsynaptic sheet, thereby establishing associations between pairs of patterns. For discovering which patterns become linked together we use the crucial fact that each cell in a sheet belongs to several overlapping patterns. Under this condition the final projection has the property that overlapping stimuli will evoke overlapping responses and the responses to non-overlapping stimuli will not overlap. The geometrical interpretation of the resulting projections depends on the structure of the patterns of activity allowed." [15]

Von der Malsburg applied this model to the development of orientation columns [14], the development of ocular dominance columns [15] and, with Willshaw, to the development of topography (refinement of the retinotopic map) [16]. Computer simulations demonstrated that the models could work in each case. For orientation columns, inputs were activated in oriented patterns of all possible orientations. Individual cortical cells developed selective response to one such oriented pattern, with nearby cortical cells preferring nearby orientations. In the case of ocular dominance columns, inputs were activated in monocular patterns: each activation pattern was a localized cluster of cells from a single eye, and the two eyes were never simultaneously active. Individual cortical cells came to be driven exclusively by a single eye, and clusters of cortical cells came to be driven by the same eye. The tendency of overlapping cortical clusters to prefer the same eye resulted in a final arrangement of stripes of cortical cells preferring a single eye. "The spacing between the stripes is determined by the range of inhibition" [15], that is, by the diameter of the intrinsic clusters of cortical activity.

Thus, a synaptic conservation rule was used that forced a hebbian rule to become competitive. This led individual cells to become selective for a single correlated pattern of inputs. Combined with the idea that the cortex was activated in intrinsic clusters, this suggested an origin for cortical columns: coactivated cells in a cortical cluster would tend to become selective for similar, coactivated patterns of inputs.

A More General Framework for Analyzing Cortical Development

This model provided many of the elements of a model of cortical development, and it included a fundamental experimental prediction: the width of an ocular dominance column should be determined by measurable intracortical inhibition. But otherwise, the study of the model was largely a set of demonstrations that the ideas could work. The range of conditions under which the model *would* work, and the dependence of the model results on quantities a neurobiologist could measure, were largely unexamined. Such knowledge provides the means of testing a model or distinguishing it from alternative mechanisms.

Consider the model for ocular dominance columns: how do the results depend on the correlations in activities in the input layers? If the two eyes are sometimes coactivated, either because they are activated independently or because they are partially correlated by vision, can ocular dominance segregation occur? If inputs are correlated over a larger or smaller distance within an eye, how will this alter the outcome of development? How will the fact that inputs from the LGN have localized arborizations ("arbors") in cortex affect development? Can the size of these arbors, or the distance over which inputs are correlated, influence the width of the final patches? Can columns develop under alternative patterns of intracortical connectivity, for example, if the intracortical inhibition is weak or nonexistent during development? How would these same questions be answered under alternative models of column development?

To address questions like these, we developed a similar model of ocular dominance column development that allowed analytical characterization of the outcome of development [1, 5, 21, 22]. This analysis gives a general understanding of the framework for column formation proposed by von der Malsburg, and extends this framework to include alternative plasticity mechanisms, alternative structures of connectivity and of correlation, and alternative implementations of the postsynaptic conservation rule that ensures a competition. A detailed discussion of the mathematical and biological assumptions necessary for this analysis is in Ref. 22.

We showed that a large class of developmental models could be expressed in terms of three measurable elements (Fig. 2): a set of correlation functions, describing the correlation between the activities of input pairs as a function of their eyes of origin and their retinotopic separation; an arbor function that describes the spread of input arborizations in cortex allowed by retinotopy; and a cortical interaction function describing interactions within the cortex by which activity at one location influences the effectiveness of correlated synapses on different cortical cells at nearby locations. In the absence of such intracortical influences, the competition occurring at each cortical cell would be independent, and hence development of large-scale clustering of cortical properties, such as ocular dominance, would not be expected.

The cortical interaction function is particularly dependent on the biological mechanism proposed to underly plasticity. In a hebbian mechanism, as used in von der Malsburg's model, the cortical interaction function is determined by intracortical synaptic connections. It is positive over distances at which cortical cells tend to excite one another and negative over distances at which cortical cells tend to inhibit one another. For alternative mechanisms that involve the activity-dependent release and uptake of a trophic or modification factor, the cortical interaction function also incorporates the spread of influence across the cortex due to diffusion.

Predicting the Outcome of Development

The outcome of development under mechanisms that can be expressed within this framework can be predicted as follows (Fig. 3) [1, 5, 21]. We define the *arbor radius* as the radius of the arbor function, that is, the radius of the retinotopically allowed arborization of an input to cortex. First, cortical cells tend to develop receptive fields consisting of a subset of inputs that are as correlated as possible (Fig. 3A). The most correlated subset is determined by the correlation functions but also by the arbor function, which restricts the set of inputs that can potentially innervate a cortical cell. This subset will be from a single eye, and hence cortical cells will tend to develop monocular receptive fields, provided that (1) inputs within each eye are locally correlated, and (2) at all separations within an arbor radius, inputs are better (or no worse) correlated with inputs from their own eye than with inputs from the opposite eye (Fig. 3A, top). Second, cortical cells tend to develop receptive fields that are as correlated as possible with other cortical receptive fields at excitatory distances across the cortex, but as anticorrelated as possible with other cortical receptive fields at inhibitory distances (Fig. 3B, top left). For ocular dominance, correlated receptive fields are those representing the same eye. Thus, as suggested by von der Malsburg, cortical cells should be arranged

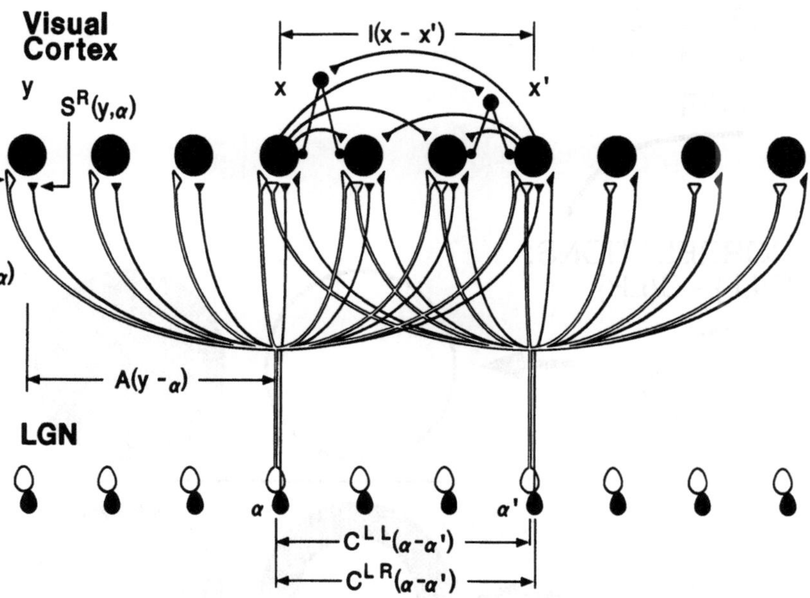

Figure 2: Elements of a Model of Cortical Development.
Left (white) and right (black) input cells innervate cells in layer 4 of the visual cortex. α and α' label positions in the input layer (LGN) and x and x' label the retinotopically corresponding positions in the output layer (cortical layer 4); y labels an additional cortical position. The layers are taken to be two-dimensional, that is x, α, etc. are two-dimensional variables. The afferent correlation functions C^{LL} and C^{LR} measure, respectively, the correlation in activity between two left-eye afferents, and the correlation in activity between a left-eye and a right-eye afferent, as a function of the retinotopic separation of the two afferents. The arbor function A measures anatomical connectivity from a geniculate location to a cortical location, as a function of the retinotopic separation between the two locations. The cortical interaction function I measures the effect of neural activity at one cortical location (x') on the development of synapses at a lateral cortical site (x); in a hebbian model, this function summarizes the effects of intracortical synaptic connections, by which activation at x' influences postsynaptic activation at x. The left-eye and right-eye synaptic strengths S^{L} and S^{R} represent the total physiological strength with which a given afferent activates a given cortical cell. These synaptic strengths are the dynamical (time-varying) variables in the model. The same framework can be applied to a competition between ON-center and OFF-center inputs from a single eye, rather than between left-eye and right-eye inputs: for example the correlation functions would then be $C^{\mathrm{ON-ON}}$, $C^{\mathrm{ON-OFF}}$, etc. Reprinted by permission from [21]. © 1989 by the AAAS.

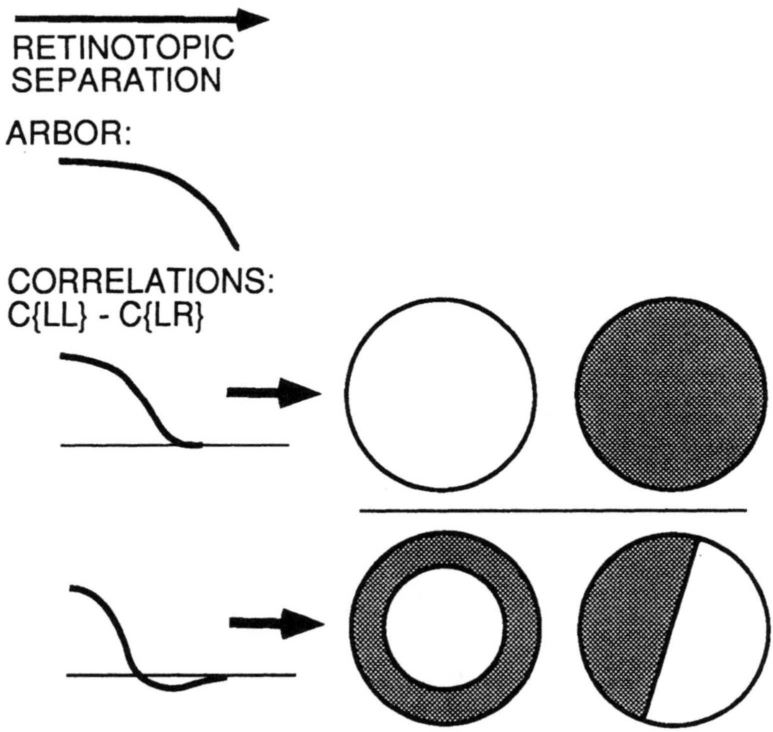

Figure 3: Determinants of the Outcome of Cortical Development as Described by the Model.
A. The correlations among input activities determine receptive field (RF) structure. Each large circle indicates the receptive field of a single cortical cell. White indicates left-eye inputs, shading indicates right-eye inputs. RF structure is determined by the degree to which two same-eye inputs with a given retinotopic separation are better correlated with one another than with an opposite-eye input at the same separation. This is described by the difference correlation function, $C^D \equiv C^{LL} - C^{LR}$, where we have assumed equality of the two eyes, $C^{LL} = C^{RR}$. TOP: If C^D is always positive within an arbor radius, ocular dominance segregation occurs: a cortical cell receptive field comes to receive only left-eye or only right-eye input. BOTTOM: If C^D were to change sign, so that at small separations same-eye inputs were best correlated, but at larger separations within the arbor radius opposite-eye inputs were best correlated, then receptive fields would develop segregated left-eye and right-eye subregions. Two alternative possible arrangements of such subregions are indicated. The correlation structure alone does not strongly distinguish between these; but the intracortical interactions lead to oriented arrangements, such as that at the lower right. The examples shown may be generalized: receptive fields tend to form with an alternation of left- and right-eye subregions whose spatial frequency maximizes the Fourier transform of C^D.

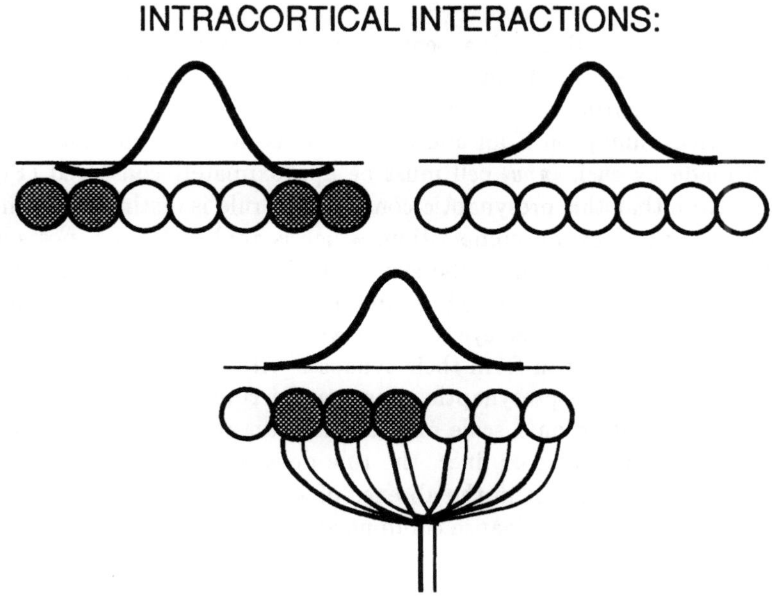

Figure 3: Determinants of the Outcome of Cortical Development as Described by the Model (Continued).
B. Intracortical interactions determine the arrangement of cortical cells. Only the case of ocular dominance segregation (A, top) is illustrated. TOP: Cells become arranged to best match the interactions. If the interactions are purely excitatory, arbitrarily large regions will become dominated by a single eye (top right). Note that interactions are assumed to be a function of distance, identical for every cell, although the figure shows them only for one cell. BOTTOM: If total synaptic strength is conserved over each presynaptic arbor, then ocular dominance columns can develop in the absence of intracortical inhibition. Initial arbors are shown stretching over six cortical cells. For each eye's inputs to retain approximately equal strength in the final arrangement, as a presynaptic conservation rule requires, eye patches can be no larger than half the arbor diameter (3 cells wide); so patches of this size will develop in the absence of inhibition.
This description of the determinants of development applies equally to a competition between ON- and OFF-center inputs from a single eye: in that case, $C^D \equiv C^{ON-ON} - C^{ON-OFF}$, white indicates ON-center inputs, shading indicates OFF-center inputs.

in patches such that the patch width of a single eye corresponds approximately to the diameter of an excitatory region in cortex. More precisely, the width of a left-eye patch plus a right-eye patch corresponds to the spatial period that maximizes the Fourier transform of the cortical interaction function.

If intracortical inhibition is absent, then large clusters of cortical cells dominated by a single eye will form (Fig. 3B, top right). It is, however, possible for periodically alternating ocular dominance columns to develop in the absence of intracortical inhibition if an additional rule is invoked: the total synaptic strength made by each *input* cell must be approximately conserved (Fig. 3B, bottom). Note that this presynaptic conservation rule is distinct from the conservation rule that ensures competition, which is applied to each postsynaptic cell. The arbor function limits the synapses of each input to an arbor radius from its 'best' cortical location. Thus, presynaptic conservation limits the width of a patch of inputs from one eye to be no wider than the arbor radius, so that inputs from both eyes can form their synapses within the diameter of any arbor. In this sense, the presynaptic conservation rule has an effect similar to intracortical inhibition on a scale of about half an arbor radius. If inhibition is present on such a scale or finer, the rule does not alter the unconstrained course of development; but in the absence of such inhibition, the rule leads to development of ocular dominance columns with a width set by the arbors.

When ocular dominance segregation does occur, our analysis also allows prediction of the relative degree of segregation. In cats, where ocular dominance segregation in cortical layer 4 is not complete, there is some binocular overlap at the borders between the patches of the two eyes. In monkeys, where there is complete segregation in layer 4, there is no overlap. To understand such alternative outcomes, it is helpful to think of the development as a competition between monocular and binocular patterns of input. Because ocular dominance segregation does tend to occur, the fastest-growing pattern of inputs to a single cortical cell is a monocular pattern (Fig. 3A, top); but how much faster does it grow than a binocular pattern consisting of inputs from one eye in half of the receptive field, and inputs from the other eye in the other half (Fig. 3A, bottom right)? If the correlations within each eye are weak and occur only over input separations that are very small compared to an arbor radius, then the binocular pattern grows nearly as well as the monocular one. In this case the binocular pattern has slightly less total correlation than the monocular one because of the boundary between left-eye and right-eye inputs, but this loss is small because the correlations are weak and do not extend very far. Alternatively, if the correlations are strong and extend throughout the arbor, then inputs in one half of the receptive field cooperate with correlated inputs in the other half and the monocular pattern greatly outgrows the binocular one. In both cases ocular dominance segregation will occur, but in the first it will occur weakly: binocular patterns will develop on some cells, particularly those at the boundaries between the patches from the two eyes. In the second case, there will be complete segregation.

The method used to ensure competition can alter the outcome of development, by altering the minimal correlation structure needed for development of ocular dominance [21, 23]. Recall that competition is modeled by conservation of total synaptic strength over each cortical cell. If this conservation is subtractive, that is, implemented by reducing each synaptic strength by the same amount, then ocular dominance can develop when the inputs from one eye are even slightly more correlated with their neighbors than with those from the other eye, as described above (Fig. 3A). If, however, this conservation is multiplicative (as was von der Malsburg's), implemented by multiplying all synaptic strength by a renormalizing constant, then the tendency to develop ocular dominance can be spoiled if the two eyes sometimes fire together, even randomly. Thus, if conservation is multiplicative, the inputs must be strictly monocular in order for ocular dominance to develop (the exact requirement depends on details of the hebbian rule but this is a reasonable intuitive characterization). Ocular dominance begins to develop in the fetus in some animals, and continues its development after birth in the presence of vision, despite the fact that the two eyes may fire together randomly in the fetus and in darkness [10, 11, 12] and are partially correlated by vision after birth. This suggests that subtractive rather than multiplicative mechanisms may be more appropriate for modeling the competitive nature of a hebbian rule in the visual cortex.

Simulations have confirmed that this framework is sufficient to account for many aspects of visual cortical development, including the development of monocular cells and their organization into periodic ocular dominance patches, the degree of monocular segregation, the restriction of afferent arbors to periodic patches of cortical innervation, and the effects of monocular deprivation including a critical period [1, 5, 21].

Application to the Development of Orientation Selectivity

In cats, the cortical cells in the layers receiving the LGN inputs are primarily 'simple cells' [6, 24]: orientation-selective cells whose receptive fields consist of oriented, spatially segregated subregions receiving exclusively ON-center or OFF-center excitatory input. The understanding we have achieved of the determinants of development raises the possibility that oriented cortical simple cells could result from a competition between ON-center and OFF-center inputs, very much as ocular dominance segregation results from a competition between left-eye and right-eye inputs [25].

The parameter regime in which ocular dominance segregation does not develop (Fig. 3A, bottom) results in receptive fields reminiscent of simple cells, with segregated subregions each receiving a different class of input. In particular, oriented receptive fields develop, as in Fig. 3A, lower right, if development occurs in the presence of intracortical interactions. For a competition between left- and right-eye inputs, the correlation structure that yields this outcome is not biologically reasonable. But this correlation structure, in which sign changes as a function of distance, is plausible for a competition between ON- and OFF-center inputs (Fig. 4).

Thus, orientation selectivity could develop through competition between ON-center and OFF-center inputs. The outcome of such competition is the formation of receptive fields like those of Fig. 3A, lower right, but with segregated subregions of ON-center and of OFF-center inputs rather than of left-eye and of right-eye inputs. Such receptive fields strongly resemble cortical simple cells. The parameter regime leading to this result (Fig. 3A, bottom) is that in which ON-center cells are best correlated with other ON-center cells at small retinotopic separations, but are best correlated with OFF-center cells at larger retinotopic separations within an arbor radius (Fig. 4). Experimental measurement will be necessary to determine whether such a correlation structure is actually present during development.

Simulations have demonstrated that competition between ON- and OFF-center inputs under these conditions leads both to the development of oriented simple cells and to their continuous arrangement in periodic orientation columns resembling those seen experimentally [25] (Fig. 5). The determinants of development follow the same general framework discovered for ocular dominance. The receptive fields consist of a subset of inputs that are as correlated as possible. Because of the change in correlations with distance, this subset consists of segregated, adjacent subfields of ON- and OFF-center inputs. For example, ON-inputs in an ON-subfield are, given this correlation structure, better correlated with OFF-inputs in an adjacent subfield than they would be with ON-inputs at that adjacent location. The intracortical interactions determine that the receptive fields of cortical cells are as correlated as possible at excitatory distances and as anticorrelated as possible at inhibitory distances. This both determines that individual receptive fields will be oriented (Fig. 3A bottom right rather than bottom left), and determines the arrangement of the orientation map. The determination of the width of the orientation columns by this mechanism is complicated, involving intracortical interactions, correlations and arbors. The fact that the width of orientation columns is determined somewhat differently than the width of ocular dominance columns is consistent with experiment: in both cats and monkeys, the periods of these two systems differ by 20-30% [26, 27]. A novel finding of the model is that the spatial phase of cortical simple cells — that is, the spatial location within the receptive field of the ON-regions and of the OFF-regions — is a critical variable in determining the orientation maps. Hence, the maps of orientation and of spatial phase must be measured simultaneously to understand the origin of the orientation map alone.

Note that these ideas have diverged from those of von der Malsburg. He

A

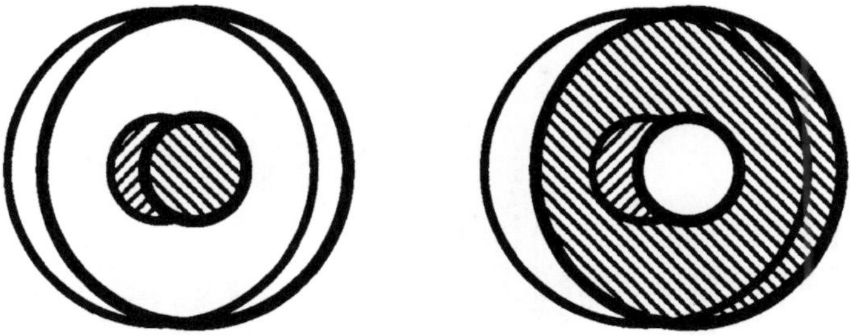

B

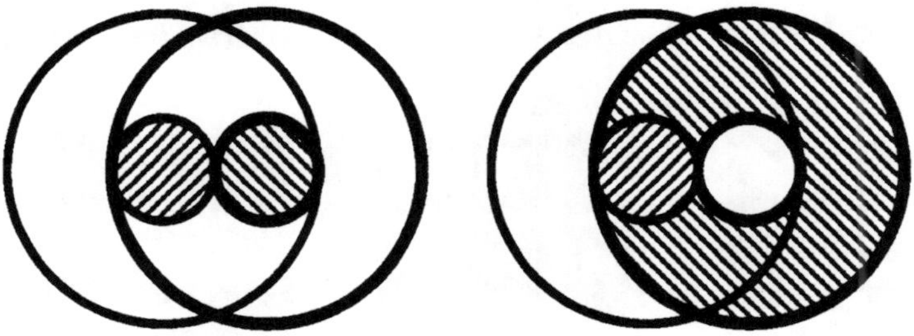

Figure 4: Plausibility of a Correlation Structure Between ON- and OFF-Center Inputs That Would Lead to Development of Orientation Selectivity.
Striped regions represent ON regions, and white regions represent OFF regions. Orientation selectivity develops in the dark [8, 9], so we consider correlations in spontaneous activity, in the absence of vision, due to common input from photoreceptors [10]. A: Correlations at small retinotopic separations, where receptive field (RF) centers overlap. Left: Two ON-center RFs have overlapping ON-centers and overlapping OFF-surrounds, and hence would be expected to frequently receive common input and thus to be well correlated. Right: For an ON-center and an OFF-center RF at similar separations, ON-center overlaps OFF-center, and ON-surround overlaps OFF-surround, so anticorrelation is expected. Such correlations and anticorrelations at separations where centers overlap have been observed in dark activity of retinal ganglion cells [10]. B. Correlations at larger retinotopic separations, where RF center overlaps RF surround. Left: Two ON-center RFs. The ON-center of each RF overlaps the OFF-surround of the other, and so the two would be expected to rarely receive common input, and thus to be poorly correlated or anticorrelated. Right: An ON-center and an OFF-center cell. The ON-center of one RF overlaps the ON-surround of the other, and similarly OFF-center overlaps OFF-surround, so better correlation might be expected than at left. While such correlations have not been observed, measurements have only been made in the retina; the LGN has stronger RF surrounds compared to the retina, so this effect might originate in or become enhanced in the LGN.

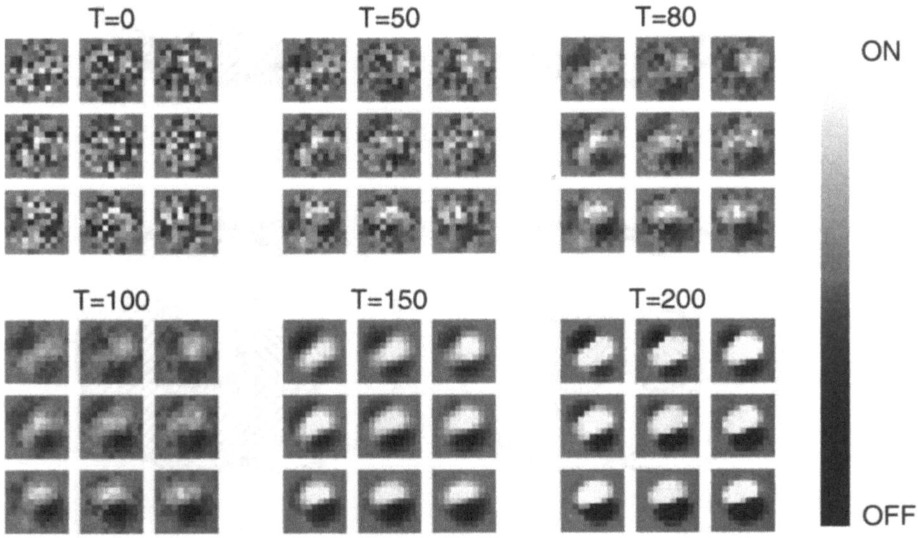

Figure 5: Development of Orientation Selectivity Through Competition Between ON- and OFF-center Inputs.
Results from simulation of development in a 31 by 31 array of cortical cells. A. Development of orientation selectivity in receptive fields of a 3 by 3 group of cortical cells. Difference between ON-center and OFF-center synaptic strength at each point in each receptive field is shown, at six timesteps ranging from a randomly assigned initial condition (T=0) to the final state.

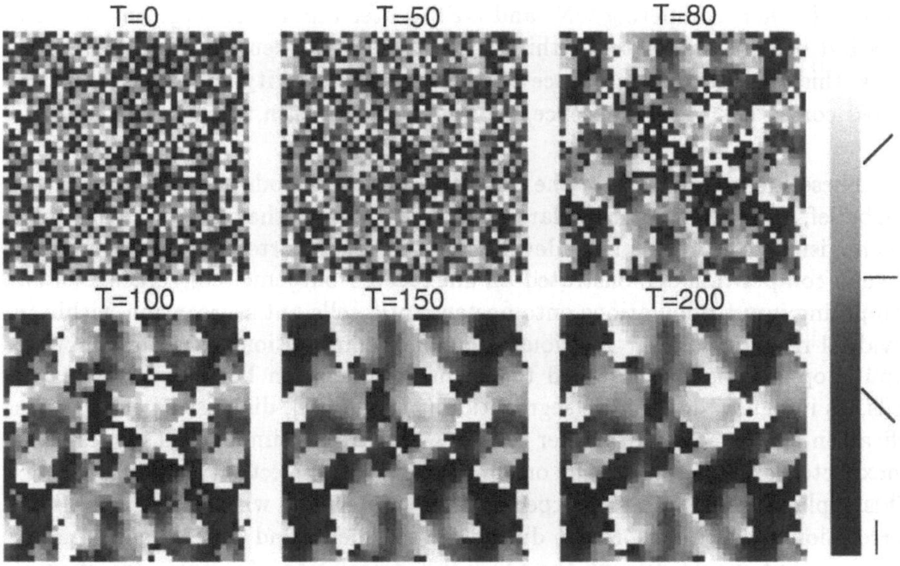

Figure 5: Development of Orientation Selectivity Through Competition Between ON- and OFF-center Inputs (continued).
B. Time development of map of preferred orientation from the same simulation. Each pixel shows the "best" orientation of one cortical cell. Orientation selectivity (preference for best orientation) is initially extremely slight, as seen in A. Note that, because orientation is a circular variable, black and white extremes both represent vertical. If one grid interval is regarded to represent about 100 μm, the final map shows reasonable resemblence to maps measured in cats. This scale is biologically plausible, corresponding for this simulation to arbor diameters of about a millimeter, purely excitatory intracortical interactions over about 200 μm, and a change in sign in ON/OFF correlations (Fig. 3A bottom) at a retinotopic separation corresponding to about half an arbor radius.

conceived of orientation selectivity arising through a process by which competing oriented patterns of inputs become associated with different clusters of coactivated cortical cells. In the present model, there are no oriented patterns of inputs. Rather, competing ON- and OFF-center inputs converge onto cortical cells but become segregated within receptive fields. Given intracortical interactions, this leads to the emergence of orientation selectivity and its organization across cortex despite the absence of oriented patterns in the inputs.

These ideas demonstrate the potential power of modeling. We began with the belief, based on work in ocular dominance columns, that a hebbian or similar mechanism is operating in the development of visual cortex. A general analysis of such competition demonstrated an unexpected outcome: convergence of two competing input populations onto postsynaptic cells but segregation within individual receptive fields. Previously, hebbian competition between competing input populations was assumed to produce segregation between postsynaptic cells, as in ocular dominance segregation (but see [28], discussed below). Identification of this novel parameter regime with cortical simple cells suggested an unexpected explanation for the origin of orientation selectivity in visual cortex. This explanation is testable: experiment can determine whether the postulated correlation structure is present during development; and experimental manipulations that would disrupt the ON/OFF correlation structure, for example forcing all inputs to fire in synchrony, should prevent the development of orientation selectivity if the hypothesis is correct.

Strengths and Limitations of This Approach

This framework allows understanding of the results of development based on the spatial arrangements of correlations, arbors, and intracortical interactions. This is the same level of detail as that studied in many experiments on column formation [5]: measurement or perturbation may be made of the correlations between competing sets of inputs, the arborizations of those inputs in the cortex, or the types and interconnections of cells in cortex, as well as of the biochemical mechanisms underlying plasticity. The model thus systematically connects experimentally measurable and perturbable quantities with expected developmental outcome at similar levels of detail. It also demonstrates that formulation of the problem at this level is sufficient to account for a wide range of observed phenomena, and it makes a number of novel predictions, for example, that the development of orientation selectivity may depend on the ON/OFF correlation structure.

Our analysis focuses on predictions that are independent of the details of biological nonlinearities, because the nature of these is largely unknown. We accomplish this by concentrating on the initial development of a pattern of the difference between two similar input projections; this difference is initially small, allowing linearization of the equations [22]. We restrict our analysis to elements that develop in the early, linear regime of the model and thus are very robust to implementation details. These include the column width, and the very facts that cortical cells become monocular or orientation-selective. We do not address questions, like the detailed layout of ocular dominance columns (i.e. long straight stripes vs. irregular patches), that depend on nonlinearities and hence may vary with details of model implementation. The results must also be robust to noise, meaning that they develop from essentially any randomly chosen initial condition. Similarly, we respect biological constraints such as the exclusively excitatory nature of the LGN input to cortex.

This framework does not address the details of neuronal structure or the temporal structure of activation and plasticity. The conjunctions of activation that lead to hebbian plasticity depend in a complex way on the temporal and spatial distributions of input activities, on interactions through the dendritic structure of the postsynaptic cell, and on the biophysical and learning mechanisms at the synapse. Yet in our model this is expressed simply through a correlation function that summarizes the ability of inputs to cooperate in achieving hebbian plasticity on any postsynaptic cell, and the postsynaptic cell is collapsed to a point without dendritic structure. In a sense we are begining with the answer: we know that the competition ultimately leads an entire cell to become dominated by a single pattern of inputs. Thus, competition is ultimately integrated over the cell as a whole, so we ignore finer levels of competition.

Our framework also ignores cell- and cell-type specific connectivity. Given the complexity and specificity of cortical connections, why should the intracortical interactions be described simply as a function of the horizontal separation of two cortical sites? There is at present no compelling answer for this question based on realistic models of cortical circuitry, although simplified circuits with this property can be constructed [14, 17, 29], and there are bits of evidence for the real cortex [5]. Again, we begin with the solution: without an interaction function that depends on separation, it is difficult to account for such regular spatial properties as periodic columns. Should the cortical interaction function and correlation functions be independent of one another? It may be that the average distance over which intracortical interactions are excitatory depends on the average size of a coactivated set of inputs, and not simply on intrinsic circuitry in cortex. By beginning with the separate role of each of these functions, we reach an understanding that will serve us even if it turns out that one function partly determines the other.

The Role of Realistic Models

Realistic modeling may help bridge the gap between biophysically realistic neuronal properties and the level at which our modeling begins. If the plasticity rule is coupled to local postsynaptic voltage, why should the competition on a single cortical cell result in the left eye or right eye winning the entire cell, rather than in an independent competition for each dendrite? The answer will ultimately be determined by detailed single-cell modeling [30], which is just beginning to be used to explore synaptic plasticity [31, 32, 33]. Using realistic but linear membrane biophysics, realistic synaptic conductances, and a hebbian plasticity rule that depends on local postsynaptic voltage, it has been found that there are conditions in which one or a small number of patterns of synaptic inputs 'takes over' an anatomically detailed hippocampal pyramidal cell [33]. The parameters that give this result are still under investigation, but are difficult to characterize due to the complexity of the model. Should intracortical interactions be described simply as a function of intracortical separation? This could be investigated by constructing biophysically and anatomically realistic models of activation in cortical networks, using methods like those described in [34], in order to determine the cortical interactions that emerge on a time scale appropriate to affect plasticity in a correlated group of inputs.

Such piece-by-piece use of biophysical modeling to evaluate or alter key assumptions in a simplified model should be contrasted with an attempt to directly attack a developmental problem using a biophysically detailed simulation. The latter approach faces two major hurdles. First, it must necessarily use a very simplified version of the biophysics, for practical reasons. It is at present computationally impossible to simulate detailed model networks like those, say, of Traub and Miles (this issue [34]) for the hours or days of real time that are the minimum needed to study development. Second, despite such simplification a biophysically detailed model of plasticity necessarily has a large number of parameters that are not constrained by experimental knowledge. Some means of managing this complexity must be found.

Inclusion of biophysical detail does not necessarily render a model more realistic (see related discussion in [35]). For example, one recent model of development in somatosensory cortex [29] used a plasticity rule involving hypothetical receptors with complicated kinetics and 14 arbitrary parameters, and rules for cortical cell activation requiring specification of 10-15 additional parameters. But, under the conditions studied, the plasticity rule reduced to a hebbian rule in which plasticity occurs only when the postsynaptic cell is depolarized, and the pattern of cortical activation was simply short-distance excitation and longer-distance inhibition [5]. Thus, our analysis could be applied, so that the results could be understood from the correlations in the input patterns used, the arbors, and the intracortical interactions as specified by the range of excitation and inhibition [5]. Interesting developmental results were obtained, but understanding was obscured rather than enhanced by using biologically unconstrained complexity to implement a very simple model. In particular, because the authors did not understand what determined the width of the cortical patches that developed in their model, they did not realize that certain aspects of their results could be ruled out by existing physiological observations [5].

In sum, models must be informed and constrained by biological knowledge at their particular level of realism or abstraction. An increase in detail does not imply an increase in realism; unrealistic or unconstrained details may obscure rather than enhance understanding. Genuine study of the problem of plasticity at deeper levels of biological realism will extend and transform our knowledge, but this will likely need to be guided by, and will in turn guide, the understanding we gain at a more simplified level of description.

Abstract Models of Activity-Dependent Development

Models of development may also be formulated at an abstract level. These models explore dynamical or optimization principles that may lead to patterns like those observed in development. The elements of these models cannot readily be identified with biologically measurable quantities; but the simplified setting may expose principles that can carry over to a biological setting.

One set of models characterizes the Fourier transforms of biological patterns, either directly [36] or by positing developmental rules that are simple in Fourier space [37, 38]. Swindale proposed that ocular dominance at one cortical site influences the growth of ocular dominance at another cortical site according to a simple function of distance, and that these influences add linearly; the biological nature of the influence was not specified [37]. Such a rule selects periodic patterns of ocular dominance whose spatial period corresponds to the peak of the Fourier transform of the influence function. A similar rule, proposing that orientations influence one another's growth according to a function of distance [38], selects a periodic pattern of preferred orientations in the same way.

The strength of these models is that they accurately describe observed maps of orientation and ocular dominance in a very simple way. Furthermore, the simplicity of the models allows interesting reasoning about issues such as monocular deprivation [37] or the spacing of orientation singularities [38]. The problems with these models are that they cannot obviously be distinguished from more complex models that would yield a similar final distribution of power in Fourier space, and they do not propose a biological mechanism by which the posited influences might be exerted.

In the case of the ocular dominance model, there is a simple bridge to a biologically identifiable model. In the limit in which each eye fires as a unit, so that correlations are completely specified by eye of origin without need to specify retinotopic position, the model we proposed for ocular dominance reduces to one in terms of an influence function. The influence function is then specified in terms of biologically identifiable factors: correlations, arbors, cortical interactions, and presynaptic conservation rules if any. However, in the case of orientation selectivity our model does not provide such a connection: in our model, the interaction between two cells of the same orientation can vary from maximally positive to maximally negative depending on the spatial phases of the two receptive fields, so an influence function cannot be derived that depends on orientation difference alone.

Another set of models posit simplified versions of hebbian rules. Linsker used linear hebbian rules with synapses that could be either positive or negative, and a constraint that fixed the total percentage of positive and of negative synaptic strength in a receptive field [28] (analyzed in [22, 39, 40]). Variation of those percentages and of the input correlations revealed circumstances in which interesting receptive field structures arise. Correlations that oscillate led to spatially structured receptive fields with segregated regions of positive and negative synaptic input. Symmetry could spontaneously break, so that circularly symmetric inputs and arbors led to oriented receptive fields. However, the orientation-selective cells developed only when the constraint was finely tuned, and then only in the nonlinear regime of development [40]. Most of the results similarly depended on the constraint and its tuning, on the ability of synapses to become either positive or negative (or more generally, on the assumption of complete indistinguishability of positive and negative synapses [22]), and on the particular choice of saturating synaptic nonlinearities [22, 39, 40]. Thus, this model demonstrated previously unknown dynamical outcomes that could arise from hebbian rules, but it did not demonstrate biologically plausible or robust ways to achieve those outcomes.

Other simplified Hebbian models use the self-organizing feature maps of Kohonen [41] or similar algorithms [42] to model the development of ocular dominance and orientation columns [42, 43, 44]. The Kohonen mappings are an abstraction of a hebbian rule. The major abstractions are that, for any given input pattern, the output of cortical cells is determined without intracortical interactions; and reinforcement of active inputs then occurs only on the most activated cortical cell and its near neighbors, a representation of localized clusters in cortical activation. Multiplicative constraints conserve the sum of squares of the synaptic weights over each cortical cell. These mappings are amenable to analysis and have a number of interesting features. They lead to continuous maps in which all inputs gain equal representation in cortex if activated equally often, nearby inputs develop nearby cortical representations, and a constant distance across the cortex corresponds to a roughly constant distance in "input space". If the input space has more than the cortex's two dimensions — for example, the five-dimensional space of ocular dominance, preferred orientation, orientation selectivity, and 2D retinotopic position — this means that, when one feature is changing rapidly across cortex, the others will be changing slowly. The algorithms produce realistic maps of orientation [42, 43] and of ocular dominance [41, 44], and reproduce many of the experimentally observed relationships between orientation and ocular dominance columns [44].

These mapping algorithms have several problems as biological models: they do not predict the column widths in any straightforward way from biologically identifiable parameters; as implemented thus far, they cannot break symmetry to develop oriented responses from non-oriented input patterns; and more generally, the biological interpretations of the various abstractions used are not clear. But these models demonstrate simple rules that can account for complicated aspects of cortical maps, and their formal closeness to hebbian rules suggests that they may serve as a guide for more biologically based models.

The abstract models demonstrate that differing dynamical mechanisms may converge on similar results resembling those seen biologically. Conversely, very different biological mechanisms may hide similar underlying dynamics. Furthermore some aspects of the results of a model may vary with details of implementation and thus not be firmly tied to the mechanism being studied. Thus, resemblance of model and biological results should not be viewed as a validation of a proposed model; rather it is a demonstration that a proposed mechanism can be involved, under some circumstances, in achieving results like those found biologically.

Conclusion

What makes a useful model of neural development? One important contribution of modeling is to demonstrate that proposed biological mechanisms can be sufficient to account for experimental results. The von der Malsburg model is a classic example. But such demonstrations alone do not provide tools to experimentally distinguish one mechanism from another. To draw such distinctions, the connection between measurable biological quantities and developmental outcomes must be established.

Perhaps the most important task for the future of developmental modeling is to deepen the connection between theory and experiment. Experimentally, this requires detailed and difficult measurements or experimental perturbations of the correlations among inputs and the intracortical connectivity existing during development. Simultaneous measurement of the maps of spatial phase and orientation of mature simple cells will provide important information for the understanding of orientation column development.

Theoretically, the number of open problems is enormous. How will inclusion of additional plasticity mechanisms, such as sprouting and retraction of synapses or plasticity of intracortical connections, alter the analytical understanding thus far achieved? What precisely determines the width of orientation columns in the model presented here? Can the relationship between ocular dominance and orientation columns be understood from developmental rules in a testable way? The existing framework may be extended to a three-dimensional cortex and to more complex models of intracortical connectivity. It may also be applied to other developmental phenomena including the development of lamination in the LGN [13, 45], the formation of visual maps in experimentally altered auditory cortex [46], and the mapping of visual and auditory maps in the optic tectum [47]. For each system the goal is to develop testable predictions as to the patterns of activity and connectivity that could or could not lead to the results observed given a proposed mechanism of plasticity. Incorporation of deeper levels of biophysical realism will extend, deepen, and perhaps fundamentally alter the framework presented here. An important goal for the future will be to understand the computational and functional significance of developmental rules.

References

[1] Miller, K.D. (1990). "Correlation-based mechanisms of neural development", in *Neuroscience and Connectionist Theory* (Gluck MA, Rumelhart DE, eds), pp. 267–353. Lawrence Erlbaum Associates, Hillsboro NJ.

[2] Hebb, D.O. (1949). *The Organization of Behavior.* John Wiley and Sons, NY.

[3] Guillery, R.W. (1972). "Binocular competition in the control of geniculate cell growth", *J. Comp. Neurol.* **146**, 407–420.

[4] Wiesel, T.N., and D.H. Hubel (1965). "Comparison of the effects of unilateral and bilateral eye closure on cortical unit responses in kittens", *J. Neurophysiol.*, **28**, 1029–1040.

[5] Miller, K.D., and M.P. Stryker (1990). "Ocular dominance column formation: mechanisms and models", in *Connectionist Modeling and Brain Function: The Developing Interface* (Hanson SJ, Olson CR, eds), pp. 255–350. MIT Press/Bradford, Cambridge, MA.

[6] Hubel, D.H., and T.N. Wiesel (1962). "Receptive fields, binocular interaction and functional architecture in the cat's visual cortex", *J. Physiol.* **160**, 106–154.

[7] LeVay, S., and S.B. Nelson (1991). "Columnar organization of the visual cortex", in *The Neural Basis of Visual Function* (Leventhal A, ed), pp. 266–315. MacMillan Press, London.

[8] Sherk, H., and M.P. Stryker (1976). "Quantitative study of cortical orientation selectivity in visually inexperienced kitten", *J. Neurophysiol.* **39**, 63–70.

[9] Wiesel, T.N., and D.H. Hubel (1974). "Ordered arrangement of orientation columns in monkeys lacking visual experience", *J. Comp. Neur.* **158**, 307–318.

[10] Mastronarde, D.N. (1989). "Correlated firing of retinal ganglion cells", *Trends Neurosci.* **12**, 75–80

[11] Maffei, L., and L. Galli-Resta (1990). "Correlation in the discharges of neighboring rat retinal ganglion cells during prenatal life", *Proc. Natl. Acad. Sci. USA* **87**, 2861–2864.

[12] Meister, M., R.O.L. Wong, D.A. Baylor, and C.J. Shatz (1991). "Synchronous bursts of action potentials in ganglion cells of the developing mammalian retina", *Science* **252**, 939–943.

[13] Shatz, C.J., and M.P. Stryker (1988). "Prenatal tetrodotoxin infusion blocks segregation of retinogeniculate afferents", *Science* **242**, 87–89.

[14] von der Malsburg, C. (1973). "Self-organization of orientation sensitive cells in the striate cortex", *Kybernetik* **14**, 85–100.

[15] von der Malsburg, C., and D.J. Willshaw (1976). "A mechanism for producing continuous neural mappings: ocularity dominance stripes and ordered retino-tectal projections", *Exp. Brain Res.* **Supp. 1**, 463–469.

[16] Willshaw, D.J., and C. von der Malsburg (1976). "How patterned neural connections can be set up by self-organization", *Proc. R. Soc. Lond. B.* **194**, 431–445.

[17] Wilson, H.R., and J.D. Cowan (1973). "A mathematical theory of the functional dynamics of cortical and thalamic nervous tissue", *Kybernetik* **13**, 55–80.

[18] Perez, R., L. Glass, and R. Schlaer (1975). "Development of specificity in the cat visual cortex", *J. Math. Biol.* **1**, 275–288.

[19] Nass, M.M., and L.N. Cooper (1975). "A theory for the development of feature detecting cells in visual cortex", *Biol. Cybern.* **19**, 1–18.

[20] Grossberg, S. (1976). "On the development of feature detectors in the visual cortex with applications to learning and reaction-diffusion systems", *Biol. Cybern.* **21**, 145–159.

[21] Miller, K.D., J.B. Keller, and M.P. Stryker (1989). "Ocular dominance column development: analysis and simulation", *Science* **245**, 605–615.

[22] Miller, K.D. (1990). "Derivation of linear hebbian equations from a non-linear hebbian model of synaptic plasticity", *Neur. Comp.* **2**, 319–331.

[23] Miller, K.D., and D.J.C. MacKay (1992). "The role of constraints in Hebbian learning", in preparation.

[24] Bullier, J., and G.H. Henry (1979). "Laminar distribution of first-order neurons and afferent terminals in cat striate cortex", *J. Neurophysiol.* **42**, 1271–1281.

[25] Miller, K.D. (1992). "Development of orientation columns via competition between ON- and OFF-center inputs. *Neuroreport* **3**, 73–76.

[26] Hubel, D.H., T.N. Wiesel TN, and M.P. Stryker (1978). "Anatomical demonstration of orientation columns in macaque monkey", *J. Comp. Neurol.* **177**, 361–379.

[27] Lowel S., H.J. Bischof, B. Leutenecker, and W. Singer (1988). "Topographic relations between ocular dominance and orientation columns in the cat striate cortex", *Exp. Brain Res.* **71**, 33–46.

[28] Linsker, R. (1986). "From basic network principles to neural architecture" (series), *Proc. Nat. Acad. Sci. USA* **83**, 7508–7512, 8390–8394, 8779–8783.

[29] Pearson, J.C., L.H. Finkel, and G.M. Edelman (1987). "Plasticity in the organization of adult cerebral cortical maps: a computer simulation based on neuronal group selection", *J. Neurosci.* **7**, 4209–4223.

[30] Lytton, W.W., and J.C. Wathey (1992). "Realistic single-neuron modeling", *Seminars in the Neurosciences* **4**, 15-25.

[31] Zador A., C. Koch, and T.H. Brown (1990). "Biophysical model of a Hebbian synapse", *Proc. Nat. Acad. Sci. USA* **87**, 6718–6722.

[32] Holmes, W.R., and W.B. Levy (1990). "Insights into associative long-term potentiation from computational models of NMDA receptor-mediated calcium influx and intracellular calcium concentration changes", *J. Neurophysiol.* **63**, 1148–1168.

[33] Brown, T.H., A.M. Zador, Z.F. Mainen, and B.J. Claiborne (1991). "Hebbian modifications in hippocampal neurons", in *Long term potentiation: A debate of current issues* (Davis J, Baudry M, eds), pp 357–389. MIT Press, Cambridge MA.

[34] Traub, R.D., and R. Miles (1992). "Modeling hippocampal circuitry using data from whole cell patch clamp and dual intracellular recordings in vitro", *Seminars in the Neurosciences* **4**, 27-36.

[35] Sigvardt, K.A., and T.L. Williams (1992). "Models of central pattern generators as oscillators: the lamprey locomotor CPG", *Seminars in the Neurosciences* **4**, 37-46.

[36] Rojer, A.S., and E.L. Schwartz (1990). "Cat and monkey cortical columnar patterns modeled by bandpass-filtered 2D white noise", *Biol. Cybern.* **62**, 381–391.

[37] Swindale, N.V. (1980). "A model for the formation of ocular dominance stripes", *Proc. R. Soc. Lond.* **B208**, 243–264.

[38] Swindale, N.V. (1982). "A model for the formation of orientation columns", *Proc. R. Soc. Lond.* **B215**, 211–230.

[39] MacKay, D.J.C., and K.D. Miller (1990). "Analysis of Linsker's simulations of Hebbian rules", *Neur. Comp.* **2**, 169–182.

[40] MacKay, D.J.C., and K.D. Miller (1990). "Analysis of Linsker's application of Hebbian rules to linear networks", *Network* **1**, 257–298.

[41] Kohonen, T. (1989). *Self-organization and associative memory*, 3rd ed. Berlin, New York: Springer-Verlag.

[42] Durbin, R., and G. Mitchison (1990). "A dimension reduction framework for understanding cortical maps", *Nature* **343**, 644–647.

[43] Obermayer, K., H. Ritter, and K. Schulten (1990). "A principle for the formation of the spatial structure of cortical feature maps", *Proc. Natl. Acad. Sci. USA* **87**, 8345–8349.

[44] Obermayer, K., K. Schulten, and G. Blasdel(1991). "A neural network model for the formation and for the spatial structure of retinotopic maps, orientation and ocular dominance columns", Theoretical Biophysics Technical Report TB-91-09, The Beckman Institute, University of Illinois Urbana-Champaign.

[45] Hahm, J., R.B. Langdon, and M. Sur (1991). "Disruption of retinogeniculate afferent segregation by antagonists to NMDA receptors", *Nature* **351**, 568–570.

[46] Roe, A.W., S.L. Pallas, J.O. Hahm, and M. Sur (1990). "A map of visual space induced in primary auditory cortex", *Science* **250**, 818–820.

[47] Knudsen, E.I., and M.S. Brainard (1991). "Visual instruction of the neural map of auditory space in the developing optic tectum", *Science* **253**, 85–87.

Visual Inputs and Information Processing in Sensory Cortex: An *in vivo* Developmental Study

Sarah L. Pallas†, Laurie S. Carman, and Mriganka Sur
Department of Brain and Cognitive Sciences,
Massachusetts Institute of Technology, E25-618, Cambridge, MA 02139
†Present Address: Division of Neuroscience, Baylor College of Medicine, One Baylor Plaza, Houston, TX 77030

Introduction

The cerebral cortex of mammals can be subdivided into many different areas based on differences in internal structure, functional properties, and behavioral role. We are interested in the way these differences come about developmentally, and in how they are represented in terms of the actual cortical neural networks. One intriguing possibility is that the identity of a cortical area is determined by the type of input that it receives during its early development. To address this question, we have utilized a preparation in which we can change the modality of the input that a cortical area receives, before the stage in the animal's life when that area has established its final identity. Specifically, we have induced visual inputs to innervate the auditory pathway, and subsequently examined the anatomical, physiological, and behavioral consequences of this rewiring of the auditory cortex.

We can change the modality of the input to AI from auditory to visual without changing the identity of the fibers (the thalamocortical pathway from MGN to AI)

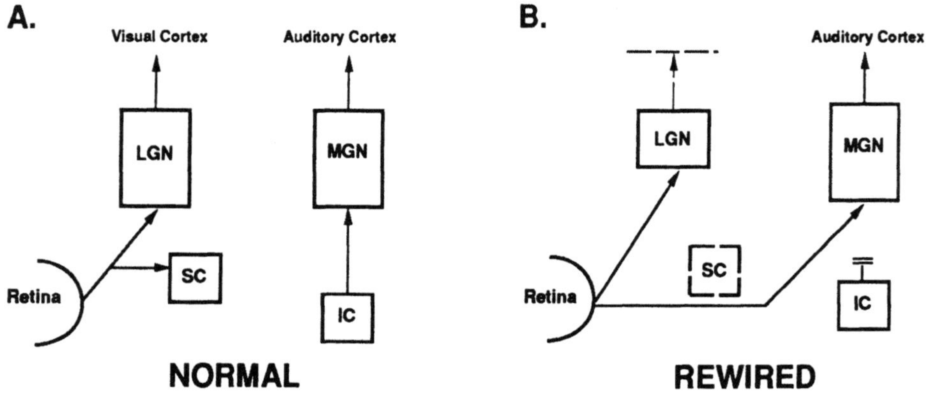

Figure 1. Diversion of retinal axons into auditory thalamus early in development. The normal targets of the retinal axons (shown in A) are eliminated or reduced by direct ablation of the superior colliculus (SC) and indirect reduction of the size of the lateral geniculate nucleus (LGN) by visual cortical ablation (B). Deafferentation of the medial geniculate nucleus (MGN) by section of the afferents from the inferior colliculus enables retinal fibers to innervate the MGN. The MGN to auditory cortex projection then carries visual rather than auditory information.

carrying the information. This provides us with a relatively straightforward system in which to study the effects of input modality and activity on cortical development. The way in which this modality switch is achieved is described in Figure 1. Normally, the retina of the ferret (*Mustela putorius furo*) projects topographically to two main targets in the brain, the superior colliculus (SC) in the midbrain, and the lateral geniculate nucleus (LGN) in the thalamus. The LGN projects directly to the primary visual cortex (V1), again in a topographic fashion. The primary auditory cortex (AI), on the other hand, gets topographic input from the medial geniculate nucleus (MGN) of the thalamus, which in turn gets its input from the inferior colliculus (IC) in the midbrain. The cochlea projects via intermediate stations to the IC. If the normal targets of the retina are removed early in development, and alternative space is made in the MGN by removing the projection from IC, the retina will innervate the MGN, thus giving the auditory cortex access to visual information

(Sur et al., 1988). Because the lesions are made unilaterally, each animal can to some extent serve as its own control.

In ferrets "rewired" in this way at birth, we have addressed several related questions in relation to the effect of visual inputs on the auditory pathway:

1) How is the connectivity pattern of AI affected by the modality switch?

2) Does the auditory cortex contain a map of visual space, and if so, how is this mapping achieved?

3) Does the auditory cortex respond to visual stimuli, and if so, does it respond in a way similar to normal visual cortex, or does it respond in some unique way?

4) What is the identity of the retinal fibers that invade the MGN, and how do they compare with fibers that normally innervate the MGN?

5) If the auditory cortex can respond to visual stimuli, can it then mediate visual behavior?

Results

Connectivity Patterns of AI in Lesioned Ferrets

Our overall aim is to see whether anomalous visual input changes the developmental program of auditory cortex. To this end, we can look for changes in the internal or external connectivity of AI. AI in normal ferrets, as in cats (Aitken et al., 1984), is reciprocally connected with the auditory thalamus (MGN and associated structures) as well as with other auditory cortical areas. In similar fashion, visual cortex is connected with other visual cortical and subcortical areas. Our connectivity studies in the rewired ferrets (Pallas et al., 1988, 1990) show that, with the exception of a novel connection between AI and the lateral posterior nucleus of the thalamus, this wiring pattern is not substantially changed as a result of the early lesions. Thus, although AI gets visual input during its postnatal development, it is not induced to form connections with other visual structures as a result.

AI in normal animals contains a one-dimensional map of frequency, with high frequencies represented medially and low frequencies laterally. When looking at the results of our connectivity studies in more detail, we found that, in the rewired ferrets as well, the thalamocortical and corticothalamic projections between MGN and AI are restricted in one plane (the tonotopic plane in normal animals) and highly overlapped along the isofrequency plane. This situation is similar to the normal connectivity pattern between IC and MGN, but different from the projections between the retina and the MGN in rewired ferrets (see below). This pattern would suggest that, even if there is a two-dimensional mapping of the retina in the MGN of rewired ferrets, this map is lost as a result of the one-dimensional nature of the thalamocortical projection. In the next section, we will explain the results of our mapping studies, which show that AI does contain a two-dimensional visual map in spite of the overlapped input from MGN.

Formation of Topographic Sensory Representations

Normal primary visual cortex in ferrets and in all other mammals studied to date contains a two-dimensional, point-to point mapping of the retina. In contrast, primary auditory cortex contains a one-dimensional, slab-to-slab mapping of the cochlea. What happens when these two mapping systems are overlaid, as in our rewired ferrets? Our receptive field mapping studies (Roe et al., 1990a) have revealed the presence of a two-dimensional visual map in AI of the rewired ferrets, with azimuth mapped mediolaterally along what is normally the tonotopic axis, and elevation mapped along the rostrocaudal, isofrequency axis.

How does this map arise? Mapping studies by Roe et al. (1991) in our lab have demonstrated that the MGN in rewired ferrets does contain a rough two-dimensional visual map. This raises the possibility that the visual map in AI results simply from a translation of the MGN map onto cortex. However, as discussed above, our anatomical results show that the MGN-AI projection is one-dimensional in the rewired ferrets. This suggests that the elevational map is recreated within AI.

This idea is supported by some important details of the visual map in AI (Roe et al., 1990a). We find that the precision of the azimuthal mapping is much higher than that of the elevational mapping. In addition, the polarity of the visual map is consistently the same along the tonotopic axis (with central azimuths represented medially and peripheral azimuths laterally), but variable along the isofrequency axis (the elevational dimension).

Our hypothesis is that the azimuthal map is a natural result of the projection between MGN and AI, and is thus hard-wired, but that the elevational map, because it lies along the isofrequency dimension of AI, must be physiologically created out of the overlapped inputs from MGN. This could occur via an alteration in the strengths of either inhibitory or excitatory circuitry within AI, such that some of the anatomically observed, overlapped inputs are not expressed functionally. An activity-dependent stabilization process, such as that operating in the formation of ocular dominance columns in primary visual cortex of the cat (Hubel and Wiesel, 1977; LeVay et al., 1978; Stryker and Harris, 1986; see Shatz, 1990, for review) could be the mechanism responsible for this modification of cortical circuitry. Since this process is dynamic, it can occur at relatively late stages of development or even in the adult, as shown in primate somatosensory cortex (Recanzone et al., 1990; Pons et al., 1991; see Merzenich et al., 1988, for review) or cat visual cortex (Kaas et al., 1990; Gilbert et al., 1990). It is difficult to understand from an engineering perspective, however, why the brain would go to the trouble of creating a two-dimensional visual map in MGN only to discard it in the projection between MGN and AI. But in examining this apparent paradox from a developmental perspective, one sees that the thalamocortical projection is probably formed before our rewiring lesions (Jackson et al., 1989; McConnell et al., 1989; Ghosh et al., 1990; see Pallas, 1990 and Sur et al., 1990, for review), and thus the cortex is simply making the best of a confusing situation by resynthesizing the map *post hoc*.

Formation of Stimulus Processing Networks

The response properties of cells in sensory cortex are not a simple reflection of their input, but are the result of complex processing circuitry within cortex which synthesizes and transforms the information contained in the inputs. For example, cells in primary visual cortex (or V1) prefer bars of light moving at a particular orientation, direction, and velocity, while most cells in the LGN prefer flashing spots of light (Hubel and Wiesel, 1962). At the time that we make our neonatal lesions, ferret cortex is in a very immature state, with most of the cortical cells still migrating into position (Jackson et al., 1989). Thus we are in a position to maximally affect the intracortical circuitry and the resulting response properties of primary auditory cortex.

To determine the effect of visual input during development on stimulus processing circuitry, we recorded from single cells in both V1 of normal animals and in AI of rewired animals in response to visual stimulation. We measured the response of the cells to the orientation, direction, velocity, and length of the stimulus. We also examined ocular dominance– the differential response of the cells to stimulation of either eye– and the fine structure of the receptive fields (simple or complex). Our objective was to find out whether the AI cells performed any of the same transformations on their inputs as cells in normal V1.

We found that the responses of cells in AI of the rewired ferrets were remarkably similar to the responses of cells in normal V1 (Roe et al., 1990b). Not only were cells in AI selective for orientation, length, direction, and velocity of a visual stimulus, but the degree and distribution of selectivity were comparable. In addition, we found cells of both simple and complex receptive field types as well as some binocular cells (although our lesions are unilateral, there is a small projection from the ipsilateral eye to the MGN in the lesioned animals). The proportions of various cell types were about the same as in normal V1. Thus AI in rewired animals is capable of processing visual input, and indeed must contain at least some of the same circuitry as normal V1.

What these experiments do not tell us, however, is whether in normal animals, AI and V1 are fundamentally different in their stimulus processing networks. Does the early redirecting of visual information into AI result in a change in its intrinsic processing circuitry, or do all primary sensory cortices process inputs in a similar way, with response properties only appearing different due to different modalities of representation? We have argued most strongly in favor of the first possibility, that inputs influence the development of processing circuitry. We believe this interpretation is supported by the fact that the rewired AI has a two-dimensional representation of the retina in place of a one-dimensional representation of the cochlea. Cells in AI come to represent visual stimulus characteristics such as orientation selectivity, length tuning, and simple/complex receptive field types, which have no obvious correlate in the normal auditory cortex. Perhaps the same mechanism which promotes the formation of a two-dimensional visual map also directs the formation of the intrinsic cortical circuitry responsible for receptive field properties.

Identity of Retinal Afferents to MGN

The ferret retina contains three types of retinal ganglion cells, 'X', 'Y' and 'W', which differ in morphology and physiology (Vitek et al., 1985; Roe et al., 1989). Because of these differences, it is of interest to know which of the ganglion cell types are responsible for the retino-MGN projection. Using several morphological and physiological characteristics, we have suggested that the ganglion cell axons innervating MGN in our rewired ferrets are of the 'W' type. Specifically, we find that the axons have slow conduction velocities and sluggish responses (Sur et al., 1988), and that they have small diameter axons, small, sparse axonal arbors, few boutons, and low bouton density (Pallas et al., 1989) compared to X and Y axons. We are now investigating the dendritic morphology of these retinal ganglion cells to see how they compare with the normal cell types in the retina.

We have also compared the retinal ganglion cell axons in MGN with the axons of IC cells (Pallas et al., 1991), which provide the major input to MGN in normal animals. Although the two types have similar axon diameters and bouton morphology, the axons from the IC have much larger arbors, with more boutons, than the retinal ganglion cell axons. In particular, the IC axons are quite extensive along the rostrocaudal plane. This is to be expected, due to the one-dimensional map of frequency in the medio-lateral plane of MGN. The isorepresentational (isofrequency) plane runs roughly rostrocaudally and dorsoventrally (Andersen et al., 1980; Middlebrooks and Zook, 1983; Imig & Morel, 1985;). The fact that the retinal arbors are more restricted provides a substrate for a two-dimensional map of visual space in MGN of the lesioned ferrets. This is consistent with the physiological results that were discussed above.

Behavioral Output of Rewired Auditory Cortex

A critical test of our paradigm is to see whether the auditory cortex in rewired animals can mediate behavior, and whether that behavior is auditory or visual. To this end, we have been conducting behavioral experiments with both rewired ferrets and normal controls. Because our lesions are unilateral, it is possible to test either the lesioned side or the normal side of the brain independently by presenting stimuli to the ipsilateral or contralateral monocular visual field. As a further control, we ablate the remaining visual structures in the thalamus to ensure that the only visual pathway to the cortex in the rewired hemisphere is through the MGN.

Rewired ferrets are trained to remain stationary and to fixate straight ahead so that we can in fact present stimuli to only one part of their visual field at a time. They are then asked to make one behavioral response for any light stimulus and a different response for any sound stimulus. They are trained with light stimuli presented to the normal visual field, then tested with light stimuli in the rewired field. In the test condition, they are rewarded for either response to avoid preferentially training one response over the other. Our results to date indicate that rewired ferrets perceive the test light stimulus to be visual. We have confirmed that the response to test stimuli is mediated by the rewired auditory cortex by ablating AI and retesting; following AI lesions, performance decreases to chance levels.

In another series of experiments, we are taking the analysis of visual perception by AI a step further, and asking the ferrets more detailed questions about what they see with their rewired auditory pathway. For example, the lesioned ferrets are asked to choose between two equally luminant stimuli, one of which is a grating. By varying the location and spatial frequency of the stimulus, we can test the spatial acuity of the normal visual and rewired auditory pathways. The same method can be used to test contrast sensitivity.

These experiments are still in their early stages, but they already promise to yield some very interesting results, and enable us to ask questions about the nature of perception itself. Some of the more intriguing questions we can now ask are: how does the acuity and contrast sensitivity of the auditory pathway compare with the normal visual pathway? Does the behavior reflect its W-cell input? How do the ferrets label the retinal stimulus through the auditory pathway as visual if the auditory cortex is not overtly connected to other visual structures? Does the formation of this percept require experience?

Summary and Conclusions

We have shown that diversion of retinal afferents into auditory thalamus early in development confers visual responsiveness on auditory cortex. Not only is there a topographic map of visual space in AI (and in MGN), but individual cells there respond to visual stimuli in much the same way as cells in normal visual cortex do. This demonstrates that auditory cortex is capable of the same transformations in stimulus representation as visual cortex. One way for this to occur is if visual activity during development induces circuits in AI that are typical of normal V1. It is also possible that all primary sensory cortices contain the same general processing circuitry, and that this allows AI to process its anomalous visual input appropriately. One kind of experiment to distinguish between these two possibilities, now being performed in our laboratory, involves the study of intrinsic circuitry in AI of rewired and normal ferrets in the isolated cortical slice preparation.

The manipulation to induce visual inputs to the auditory pathway does not, however, lead to substantial changes in the extrinsic connectivity pattern of AI. Our hypothesis is that the connectivity pattern of a sensory cortical area is specified to a large extent quite early in development, whereas topography, neuronal response properties, and even the behavioral role of the cortical area are specified later (Pallas, 1990; Sur et al., 1990). Visual inputs can thus effectively respecify these features of auditory cortex in our rewired ferrets, turning it in essence into a visual area.

References

Aitken, L., D.R.F. Irvine and W.R. Webster (1984) Central neural mechanisms of hearing. In: *Handbook of Physiology: The Nervous System III*, Amer. Physiol. Soc., Bethesda, pp. 675-737.

Andersen, R.A., P.L. Knight and M.M. Merzenich (1980) The thalamocortical and corticothalamic connections of AI, AII, and the anterior auditory field (AAF) in the cat: Evidence for two largely segregated systems of connections. *J. Comp. Neurol.* 194: 663-701.

Ghosh, A., A. Antonini, S.K. McConnell and C.J. Shatz (1990) Requirement for subplate neurons in the formation of thalamocortical connections. *Nature* 347: 179-181.

Gilbert, C.D., J.A. Hirsch and T.N. Wiesel (1990) Lateral interactions in cerebral cortex. *Cold Spring Harbor Symp. Quant. Biol.* 55: 663-677.

Hubel, D. and T.N. Wiesel (1962) Receptive fields, binocular interaction and functional architecture in the cat visual cortex. *J. Physiol. (Lond.)* 160: 106-154.

Hubel, D. and T.N. Wiesel (1977) Plasticity of ocular dominance columns in the monkey striate cortex. *Phil. Trans. R. Soc. Lond. B* 278: 377-409.

Imig, T.J. and A. Morel (1985) Tonotopic organization in ventral nucleus of medial geniculate nucleus in the cat. J. Neurophysiol. *53*:309-340.

Jackson, C.A., J.D. Peduzzi and T.L. Hickey (1989) Visual cortex development in the ferret. I. Genesis and migration of visual cortical neurons. *J. Neurosci.* 9: 1242-1253.

Kaas, J.H., L.A. Krubitzer, Y.M. Chino, A.L. Langston, E.H. Polley and N. Blair (1990) Reorganization of retinotopic cortical maps in adult mammals after lesions of the retina. *Science* 248: 229-231.

LeVay, S., M.P. Stryker and C.J. Shatz (1978) Ocular dominance columns and their development in layer IV of the cat's visual cortex: a quantitative study. *J. Comp. Neurol.* 179: 223-244.

McConnell, S.K., A. Ghosh and C.J. Shatz (1989) Subplate neurons pioneer the first axon pathway from the cerebral cortex. *Science* 245: 978-982.

Merzenich, M.M., G. Recanzone, W.M. Jenkins, T.T. Allard and R.J. Nudo (1988) Cortical representational plasticity. In: *Dahlem Workshop on Neurobiology of Neocortex*, P. Rakic and W. Singer, eds., John Wiley & Sons, NY, pp. 41-67.

Middlebrooks, J.C. and J.M. Zook (1983) Intrinsic organization of the cat's medial geniculate body identified by projections to binaural response-specific bands in the primary auditory cortex. *J. Neurosci.* 3: 203-224.

Pallas, S.L. (1990) Cross-modal plasticity in sensory cortex: Visual responses in primary auditory cortex in ferrets with induced retinal projections to the medial geniculate nucleus. In: *The Neocortex: Ontogeny and Phylogeny*, B.L. Finlay, G. Innocenti and H. Scheich, eds., NATO Advanced Research Workshop, Plenum, NY, pp. 205-218.

Pallas, S.L., J. Hahm and M. Sur (1989) Retinal axon arbors in a novel target: Morphology of ganglion cells induced to arborize in the medial geniculate nucleus of ferrets. *Soc. Neurosci. Abstr.* 15: 495.

Pallas, S.L., J. Hahm and M. Sur (1991) Axonal arborizations in the medial geniculate nucleus of ferrets: Comparison of arbors from the inferior colliculus with induced inputs from retinal ganglion cells. *Soc. Neurosci. Abstr.* 17: 304.

Pallas, S.L., A.W. Roe and M. Sur (1988) Retinal projections induced into auditory thalamus in ferrets: Changes in inputs and outputs of primary auditory cortex. *Soc. Neurosci. Abstr.* 14: 460.

Pallas, S.L., A.W. Roe and M. Sur (1990) Visual projections induced into the auditory pathway of ferrets. I. Novel inputs to primary auditory cortex (AI) from the LP/Pulvinar complex and the topography of the MGN-AI projection. *J. Comp. Neurol.* 298: 50-68.

Pons, T.P., P.E. Garraghty, A.K. Ommaya, J.H. Kaas, E. Taub and M. Mishkin (1991) Massive cortical reorganization after sensory deafferentation in adult macaques. *Science* 252: 1857-1860.

Recanzone, G.H., T.T. Allard, W.M. Jenkins and M.M. Merzenich (1990) Receptive–field changes induced by peripheral nerve stimulation in SI of adult cats. *J. Neurophysiol.* 63: 1213-1225.

Roe, A.W., P.E. Garraghty and M. Sur (1989) Terminal arbors of single ON-center and OFF-center retinal ganglion cell axons within the ferret's lateral geniculate nucleus. *J. Comp. Neurol.* 288: 208-242.

Roe, A.W., J. Hahm and M. Sur (1991) Experimentally induced establishment of visual topography in auditory thalamus. *Soc. Neurosci. Abstr.* 17: 898.

Roe, A.W., S.L. Pallas, J. Hahm and M. Sur (1990a) A map of visual space induced in primary auditory cortex. *Science* 250: 818-820.

Roe, A.W., S.L. Pallas, Y.H. Kwon and M. Sur (1990b) Visually innervated auditory cortex in ferrets generates visual response properties similar to those in normal visual cortex. *Soc. Neurosci. Abstr.* 16: 984.

Shatz, C.J. (1990) Impulse activity and the patterning of connections during CNS development. *Neuron* 5: 745-756.

Stryker, M.P. and W. Harris (1986) Binocular impulse blockade prevents the formation of ocular dominance columns in cat visual cortex. *J. Neurosci.* 6: 2117-2133.

Sur, M., P.E. Garraghty and A.W. Roe (1988) Experimentally induced visual projections into auditory thalamus and cortex. *Science* 242: 1437-1441.

Sur, M., S.L. Pallas and A.W. Roe (1990) Cross-modal plasticity in cortical development: Differentiation and specification of sensory neocortex. *Trends in Neurosciences* 13: 227-233.

Vitek, D.J., J.D. Schall and A.G. Leventhal (1985) Morphology, central projections, and dendritic field orientation of retinal ganglion cells in the ferret. J. Comp. Neurol. *241*: 1-11.

Motion Detection and Directional Selectivity in the Crayfish Visual System

Raymon M. Glantz and Annemarie Bartels
Department of Biochemistry and Cell Biology,
Rice University, Houston TX 77251

Abstract

The detection of movement by visual systems begins with the time varying signals of the photoreceptor array and critically depends upon a neuronal integrative step that compares the output of at least two sampling points of the receptor array. The receptors are approximately linear for continuously modulated illumination. Their response to a moving point of light may be calculated from the convolution of a dynamic transfer function (with constant τ*) and the spatial receptive field profile (a Gaussian with acceptance angle* ϕ*). The ratio* ϕ/τ *determines the relative velocity sensitivity. The tangential cells of the distal medulla externa (second optic neuropile) are the most peripheral elements in the visual pathway exhibiting directional selectivity (DS). Certain receptive field properties and antagonistic synaptic inputs from cholinergic and GABAergic pathways provide a basis for simulating DS. Our simulation is based upon an asymmetric lateral inhibition model that requires a delay* τ *in the inhibitory pathway and a separation* Δp *between sampling stations. For a single bar with a Gaussian intensity profile DS is maximal when the velocity is* $\Delta p/\tau$*. For sinusoidal gratings a correct DS requires that the spatial wavelength exceed* $2 \times \Delta p$ *and that the contrast frequency,* $CF < 0.5/\tau$*.*

Motion detection is a general property of all visual systems. Even a single detector can signal movement induced, variations in light intensity (e.g., in barnacles). But a single receptor cannot distinguish movement from flicker, and it cannot separate variations in target velocity from variations in the target's contrast or spatial intensity gradient. Furthermore, a single receptor cannot discriminate the direction of movement. All of these distinctions require the comparison of signals arising at a minimum of two spatially separated sampling points and are thus properties of neuronal networks. Because photoreceptors have finite temporal frequency bandwidths, acceptance angles and contrast sensitivity, the detectable range of velocities and spatial frequencies accessible to a neuronal mechanism of motion detection are limited by the spatiotemporal properties of the receptor array. These considerations also suggest that in systems optimized for motion detection there should be a close relationship between the spatiotemporal properties of the receptor array and those of the associated neuron network. Undersampling may lead to evolutionary extinction and excessive oversampling is energetically and developmentally expensive.

Directionally selective (DS) motion detection has been characterized in many vertebrates and in the optomotor behavior of numerous arthropods; most notably in flies (Reichardt, 1961; Buchner, 1984) and crabs (Horridge and Shepherd, 1966; Nalbach, 1989). In many systems the behavioral data supports a correlational model of DS. Such models require that the output of one sampling channel be multiplied by the delayed output of a second channel (Poggio and Reichardt, 1973; Buchner, 1984; Egelhaaf *et al.*, 1989). Despite considerable effort, however (Grzywacz and Koch, 1987; Franceschini *et al.*, 1989; Srinivasan and Bernard, 1976; Gilbert *et al.*, 1990), the cellular mechanisms of multiplication are unknown. In rabbit retina (Barlow and Levick, 1965; Wyatt and Daw, 1975) the signal pathway of DS ganglion cells appears to be subjected to an asymmetric form of lateral inhibition that suppresses the response when the stimulus moves in the null direction. The inhibitory mechanism depends upon the neurotransmitter GABA (Ariel and Adolph, 1985; Caldwell *et al.*, 1978) but the source of asymmetrical inhibition in vertebrate retina and the basis of the synchronization of the excitatory and inhibitory signals is also unknown.

Here I will describe recent efforts to experimentally characterize and simulate motion detection in the first three cellular layers (receptors, lamina neurons and medullary neurons) of the crayfish visual system. There are two good reasons for focusing our efforts on crayfish: i) Directional selectivity is well described in crayfish optomotor behavior (Okada and Yamaguchi, 1985; Hisada and Higuchi, 1973) and optomotor neurons (Wiersma and Oberjat, 1968; Wiersma and Hirsh, 1974); and ii) There is a reasonably good description of the neural elements of the optomotor pathway and the functional connectivity between these elements (Wiersma and Oberjat, 1968; Kirk *et al.*, 1982; Wang-Bennett and Glantz, 1987a, b; Glantz and Nudelman, 1984; Pfeiffer and Glantz, 1989).

The Photoreceptor Movement Response

Motion detection begins with the time varying distribution of excitation in the photoreceptor array. The functional unit of arthropod compound eyes is the ommatidium. Each ommatidium contains a single corneal facet and a collection of retinular cells (receptors) which vary in their absorption spectra and optimal e-vector preference. In the crayfish Procambarus, there are eight receptors per ommatidium. About 2500 ommatidia form an oblate hemisphere which subtends about 180° of visual space.

Several characteristics of photoreceptors lend themselves toward an analytical description of movement sensitivity. For illumination modulated around a fixed mean intensity, the receptor response is approximately linear (Fuortes and Hodgkin, 1964; Brodie *et al.*, 1978; Naka *et al.*, 1987; Glantz, 1991) and the receptive field can be approximated by a point spread function (i.e., a Gaussian sensitivity profile) (Glantz, 1991). Thus the receptor response to a point of light (at fixed intensity) and moving at a constant angular velocity can be derived from the output of two concatenated filters. For a Gaussian receptive field (Fig. 1A, top, left) with acceptance angle ϕ and target angular velocity, s, the effective stimulus is:

$$I(t) = I \exp[(-\ln 2)\,(2st/\phi)^2] \qquad (1)$$

where t is time and I is the point intensity. The time varying signal (E(t) is derived from the impulse response (Fig. 1A, top, right) (transfer function) of a low-pass filter based upon first order linear cascade of n stages and time constant (τ):

$$E(t) = A[\tau(n-1)!]^{-1}\,(t/\tau)^{n-1}\,e^{-t/\tau} \qquad (2)$$

In the crayfish n is 3 to 5, and τ can be estimated from the time to peak (t_p) of the impulse response, $t_p = (n-1)\tau$. A is a constant, proportional to the contrast sensitivity.

Since the spatial and temporal filters are linear, the receptor response (V(t)), to a point moving at any velocity can be calculated from the convolution of the spatial and temporal transfer functions (Fig. 1A):

$$V(t) = I(t) * E(t) \qquad (3)$$

An interesting feature of this convolution is that the peak value of V(t), for a target moving at any velocity, depends on the ratio ϕ/τ, (which has dimensions of an angular velocity). In crayfish ϕ/τ is about 200°/s and the velocity at 1/2 maximal response is about 100°/s. For a given number (n) of cascaded stages (in the dynamic filter), the function relating peak response amplitude to target angular velocity is identical, for all ϕ and τ, if the ratio ϕ/τ is constant (Fig. 1B).

The above characteristic has important implications for the velocity sensitivity of photoreceptors in different states of visual adaptation. A common property of both vertebrate and arthropod receptors (Fuortes and Hodgkin, 1964; Naka *et al.*, 1987; Glantz, 1991) is that light adaptation is associated with a two to three fold increase in the receptor response rate (or a 50 to 70% decline in τ). In addition, many arthropods (particularly those with superposition eyes) exhibit a pupil-like mechanism (e.g., migratory shielding pigments) and during light adaptation the pigment migrations reduce the receptor's acceptance angle by 50 to 70%. As a result the ratio ϕ/τ changes much less than ϕ or τ taken alone. Thus the velocity sensitivity of most arthropod photoreceptors may be independent or only slightly dependent upon the ambient light intensity.

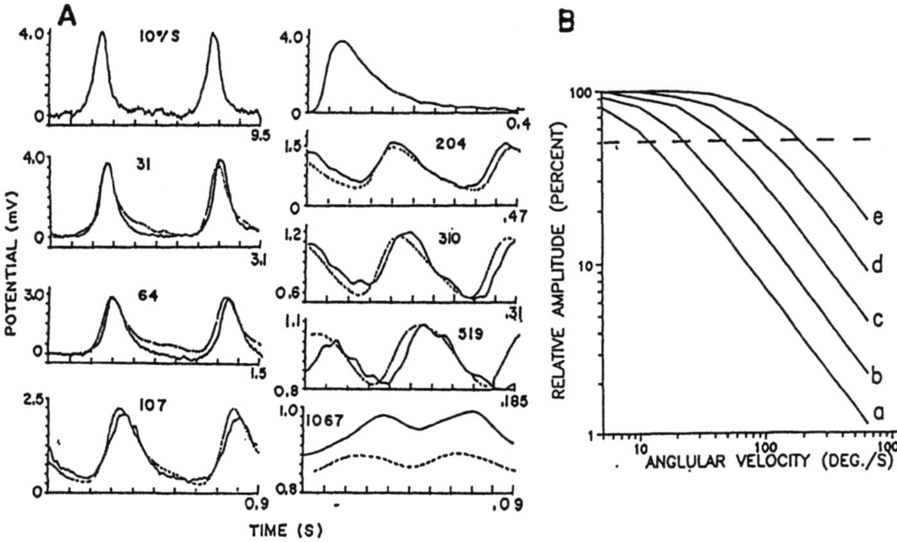

Figure 1. A. Receptor responses to the oscillating motion of a white bar at the velocity indicated in each frame. The length of the horizontal axis indicates the time required for the bar to make a forward and backward traverse of 48°. Each response is the average of 10 to 100 successive responses. The broken traces are produced by convolution of the impulse response (upper right) and the sensitivity profile indicated at the upper left for seven velocities. B. The calculated and measured velocity sensitivities for various time constants and acceptance angles. Curves are derived from Equation (3). The separate curves a-e represent successive doublings of the ratio ϕ/τ from 25°/s to 400°/s. The dashed line indicates the half-maximal response. Modified from Glantz (1991).

Directional Selectivity in Tangential Neurons

A central question for students of motion detection is the nature of the synaptic and cellular mechanisms that give rise to directional selectivity. A substantial literature spanning 30 years documents efforts to extract the nature of elementary movement detectors from the responses of higher order neurons (Egelhaaf *et al.*, 1989; Franceschini *et al.*, 1989). A more direct approach is to begin by determining the earliest stage in the visual pathway in which DS first appears.

Our results on DS should be viewed in the context of the organization of the crayfish visual pathway (Fig. 2). The basic structure is a retinotopic columnar projection system composed of small field nonspiking interneurons. Retinular cell (photoreceptor) axons project to lamina monopolar neurons (LM). The five monopolar cells in each lamina cartridge are innervated by receptors from a single ommatidium and in turn the LMs provide the principle retinotopic projection from

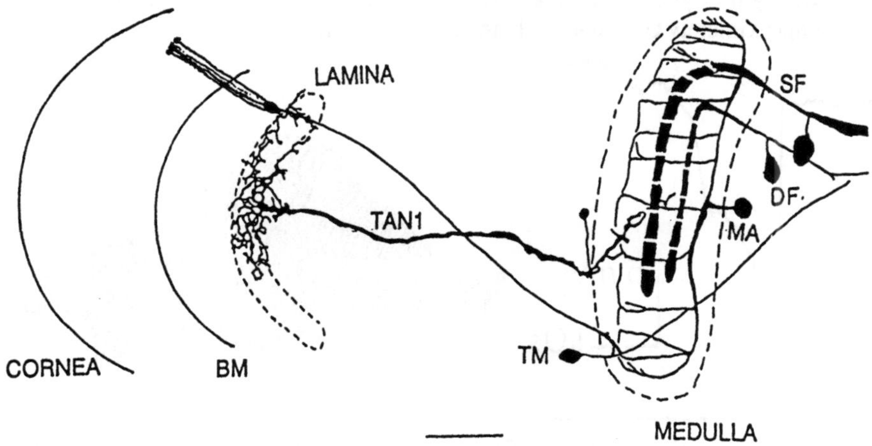

Figure 2. Morphological representation of crayfish lamina and medullary cells. Abbreviations represent: Basement membrane (BM); medullary amacrine cell (MA); tangential cell (TAN1); transmedullary neuron (TM); dimming fiber (DF) and sustaining fiber (SF). Scale is 100 µm. The SF and DF dendrites are planar structures viewed from their dorsal edge and striped for clarity. The TM neuron is drawn without projections in the medulla. Our results suggest that Tan1, MA, SF and DF and all innervated by TM neurons. From modified Pfeiffer-Linn and Glantz (1991a).

the lamina to the medulla externa (the second optic neuropile). The LM's response to a step increase in illumination is a rapid transient hyperpolarization followed by a rapid partial repolarization. LMs have narrow receptive fields; 2-3° in light adapted preparations and are subject to strong lateral inhibitory influences from neighboring ommatidia. Off-axis illumination elicits a slow depolarization of LMs. Immunocytochemical and pharmacological studies indicate that some LM axons are GABAergic (Pfeiffer-Linn and Glantz, 1989) and others are glutamatergic (Pfeiffer-Linn and Glantz, 1991b). The LM response to the movement of a sinusoidal grating appears to be nondirectional.

The LMs terminate in the distal half of the medulla externa where they retinotopically innervate columnar transmedullary neurons (medullary monopolar neurons). Transmedullary neurons (6 per medullary column) have depolarizing light responses and narrow receptive fields. Some are cholinergic, others are GABAergic and still others appear to be glutamatergic. The transmedullary neurons continue the retinotopic projection and within the medulla they innervate multicolumnar neurons such as tangential cells and sustaining fibers (Fig. 2).

The tangential cells (Tan1) are the most distal elements of the visual pathway to exhibit reproducible and strong directional selectivity (Fig. 4). The Tan1 dendrites (Fig. 2) lie in the distal medulla externa where they typically cross several medullary columns. A thick (~10 μm) axon projects centrifugally to the lamina. The Tan1 response to a step increase of light is a transient hyperpolarization (Fig. 3A) which subsequently decays by 50% to a sustained plateau. The Tan1 receptive field (Fig. 3B) is approximately Gaussian with an acceptance angle of 20 to 30°.

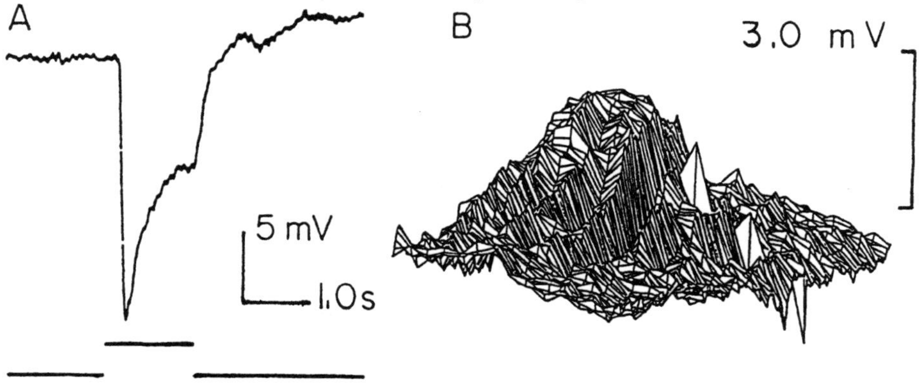

Figure 3. Functional properties of Tan1. A. Hyperpolarizing response to an 1.0 sec flash of light. B. Two-dimensional Tan1 response profile to a scanning spot (diameter is 3°) over a 100° x 100° visual field. The responses are inverted (x-1) for clarity.

Stationary sinusoidal modulation of a point source of light elicits a complex response in Tan1. The rising phase of the sinewave produces a response not unlike the response to a step (transient hyperpolarization and partial decay). The negative sine phase elicits a smaller and slower depolarization. The Tan1 response to a drifting sinusoidal grating is a membrane potential oscillation at the stimulus contrast frequency (Fig. 4A, B). The peak to peak amplitude of the oscillatory response varies with the direction of drift, the spatial wavelength (λ) of the grating and the contrast frequency (temporal frequency at any one point on the retina equals drift velocity/λ). There is a consistent preference (Fig. 4C) for vertically (±45°) drifting gratings with preferred directions (in the various Tan1s) about equally distributed between upward and downward motions. Presumably another class of neurons (as yet unidentified) encode horizontal motions.

Acetylcholine (ACh) released by transmedullary neurons elicits an outward Cl^- current in Tan1 with the same reversal potential as the light response (Pfeiffer and Glantz, 1989) and ACh meets all of the criteria for the transmitter of the light evoked hyperpolarization. GABA (released by a second set of transmedullary neurons) depolarizes Tan1 by inhibiting the ACh-elicited outward Cl^- current (Pfeiffer-Linn and Glantz, 1989). GABA's action is delayed relative to that of ACh by about 100 to 200 ms. The GABA antagonist bicuculline blocks the delayed repolarizing phase of the Tan1 light-step response. Thus the Tan1 response to step increase in illumination is shaped by a relatively rapid cholinergic action and a delayed antagonistic GABA action. In many tangential cells, scanning the receptive

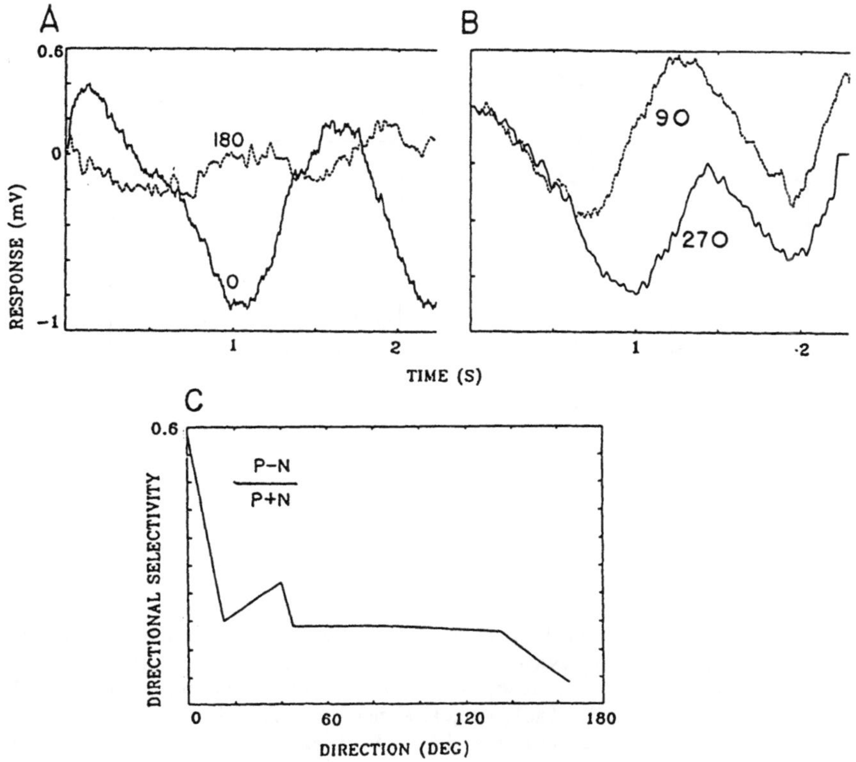

Figure 4. A. Responses to 1.5 cycles of a vertically drifting sinewave grating. O is upward, 180 is downward. Each trace is the average of 12 cycles. B. Responses to horizontally drifting gratings. 90 is forward (rostral) and 270 is backward (caudal). C. Directional selectivity (DS) calculated from the peak-peak amplitudes of responses as in A and B. P and N are the magnitudes of the responses in the preferred and null directions, respectively.

field with a point of light (<2° diameter) at velocities above 5°/s elicits oscillatory responses suggestive of distinct loci which predominantly elicit cholinergic or GABAergic responses. The sites eliciting the antagonistic responses are separated by 5-6°.

Simulation of Directional Selectivity

One way to fashion a directionally selective motion detector from the above synaptic interactions is to consider the effects of a spatial asymmetry in the receptive fields of the GABAergic and cholinergic inputs to Tan1. One possible scheme is represented in Fig. 5A and B.

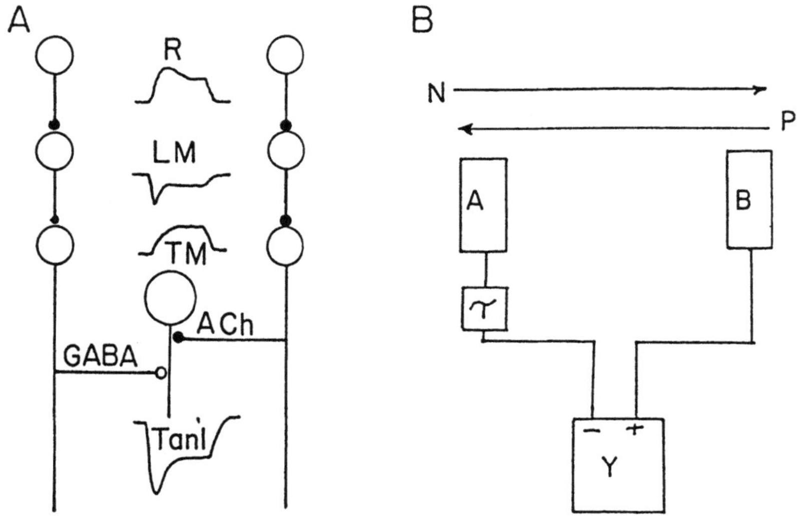

Figure 5. A. A proposed neuronal circuit for asymmetric lateral inhibition. Small filled circles indicate sign inverting synapses and small open circles are sign conserving synapses. Abbreviations: R, receptors; LM, lamina monopolar cells; TM, transmedullary monopolar cells; Tan1, tangential neurons. The characteristic response to a pulse of light is shown adjacent to each label. B. Diagrammatic representation of an asymmetric lateral inhibitory model for directional selectivity. The model is presented in its most reduced and essential form. A and B are parallel visual pathways (receptors plus interneurons) which converge on motion detector Y. A elicits a delayed inhibition of Y and B is displaced from A by a sampling angle Δp and excites Y without a delay. The null direction N emerges from that fact that at an appropriate target velocity (s) the signals generated by A and B will coincide at Y. A is delayed because of τ, B is delayed by the time required for the stimulus to traverse Δp, i.e., by $\Delta p/s$. Coincidence maximizes the cancelling effects of the B-A antagonistic interaction. In the preferred direction, P, both τ and $\Delta p/s$ serve to separate the arrivals of the A and B signals at Y and thus minimize the antagonistic interaction and maximize the peak-peak output of Y.

The GABAergic pathway (channel A) is subjected to a delay (τ) (due to the slower time course of GABA action) and operates through an antagonistic (sign conserving) synapse. The cholinergic pathway is displaced from A by a sampling interval Δp and is represented as an excitatory (sign inverting) synapse. If we assume that ACh and GABA interact in a linear antagonistic fashion then the output of the summing stage (Y) is:

$$Y(t) = B(t) - A(t-\tau) \tag{4}$$

We evaluated the model for several stimulus parameters; λ and CF for sinusoidal gratings; velocity and angular width for single bars. In these simulations, τ represents a simple delay in the sign conserving channel. The responses in the preferred (P) and null (N) directions were evaluated from the peak-peak amplitude of Y(t). The directional selectivity (DS) is defined as (P-N)/(P+N) (Fig. 6). For a drifting sinusoidal grating the peak-peak amplitude of Y(t) is sensitive to the direction of drift. For τ=200 ms and spatial wavelength $\lambda = 5(\Delta p)$, DS is modes but

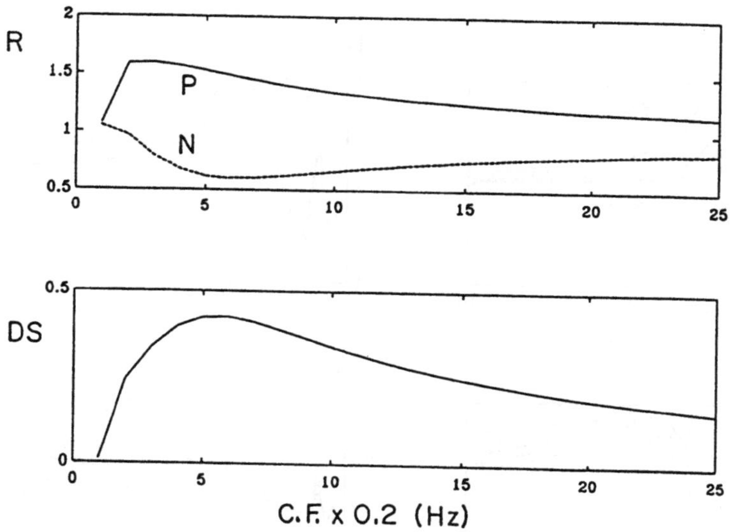

Figure 6. The response (R) and directional selectivity (DS) of the model in Fig. 5B to sinewave gratings (λ=5Δp) drifting in the preferred and null directions at contrast frequencies of 0.2 to 5.0 Hz, for τ = 150 ms and $\Delta p = 6°$.

positive for contrast frequencies above 0.2 Hz and the DS has a peak of about 0.5 at CF = 1.0 Hz (Fig. 6). Identical results are obtained if the detectors are given Gaussian sensitivity profiles with acceptance angles of 2° to 20°. Fig. 3 presents the results of a parametric study of the model for spatial wavelengths of 2 to 60° and contrast frequencies of 0.24 to 4.8 Hz. As a consequence of the Nyquist sampling limit the minimal value of λ for a correct DS is >2Δp. The range of CF for correct DS is also bounded by the requirement that CF < 0.5/τ. It is possible that these ranges can be extended by parallel detectors exhibiting variations in Δp between 2° (minimal receptor acceptance angle) and any larger number that can be accommodated by the integrative mechanisms in the lamina and medulla. Directional selectivity is also obtained for the responses to single drifting bar. A positive DS is obtained if the bar has a Gaussian intensity profile along its width, with half widths of Δp or larger and for velocities of 2 to 100 °/s. The DS is maximal (1.0) for a velocity equal to $\Delta p/\tau$ (30°/s for $\Delta p = 6°$ and τ = 0.2 s). The peak-peak amplitude of the response to a single bar is independent of the direction of contrast (black on white vs. white on black). This follows from the fact that the leading and trailing edges of a Gaussian bar produce responses of equal magnitude, but opposite sign. Changing the direction of contrast only alters the order of occurrence of the positive and negative phases of the response. Since this is a linear model with dynamic properties determined by Δp and τ, it is not surprising that the operating range for velocity depends upon $\Delta p/\tau$.

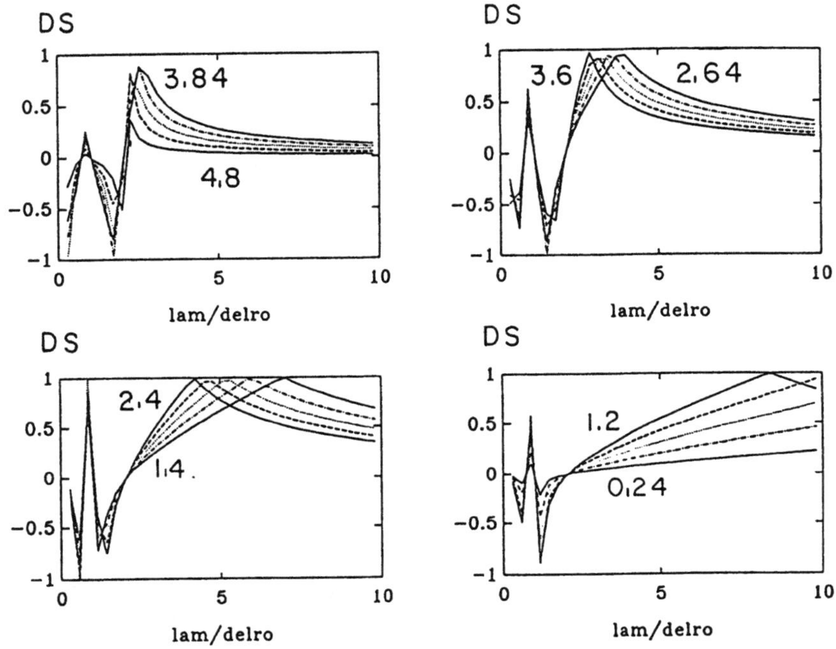

Figure 7. The directional selectivity of an asymmetric inhibition model (with $\tau = 100$ ms and $\Delta p = 6°$) for sinusoidal gratings of spatial period (lam/delro) of 1 to 60° and contrast frequencies of 0.24 to 4.8 Hz. The model produces a correct DS for $\lambda > 2\Delta p$ and for contrast frequencies in which $\tau \times CF < 0.5$.

When the results of these simulations are compared to those based upon a correlation model (Buchner, 1984; Egelhaaf *et al.*, 1989) quantitative differences are apparent with respect to the expected behavior *vis a vis* variations in λ, CF, velocity and contrast. It is more remarkable, however, that qualitatively, the two models are quite similar. Both produce approximately linear relationships with velocity over a defined velocity range and both classes of models have optimal CF and λ related to τ and Δp.

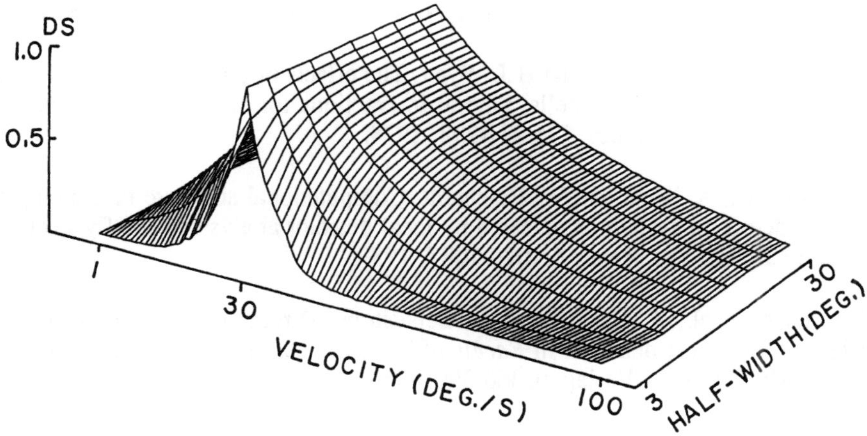

Figure 8. The directional selectivity of the response of an asymmetric inhibition model to the motion of a single bar and as a function of the velocity and width of the bar. The bar has a Gaussian intensity profile along its width. The DS is independent of the direction of contrast of the bar. Model parameters; τ is 200 ms, Δp is 6°.

We wish to emphasize that the purpose of our efforts to date has been: a) to determine the most distal site in the visual pathway that exhibits directionally selective responses to motion; and b) to determine by a systems level simulation if the types of synaptic interaction found in such cells (asymmetric lateral inhibition) are sufficient to support directional selectivity. We have not attempted so far to fit the simulation to the detailed behavior of Tan1 nor have we included any of several physiologically demonstrable nonlinearities in the model. These are the tasks before us.

Acknowledgment. These studies are supported by the National Science Foundation, Grant No. BNS-9021216 and a predoctoral fellowship to A.B. from the W.M. Keck Center for Computational Biology.

References

Ariel, M. and A.R. Adolph, Neurotransmitter inputs to directionally selective turtle retinal ganglion cells. *J. Neurophysiol.* 54:1123-1143, 1985.

Barlow, H.B. and W.R. Levick, The mechanism of directionally selective units in rabbit's retina. *J. Physiol.* 178:477-504, 1965.

Brodie, S.E., B.W. Knight and F. Ratliff, The spatiotemporal transfer function of the *Limulus* lateral eye. *J. Gen. Physiol.* 72:167-202, 1978.

Buchner, E., Behavioral analysis of spatial vision in insects. In *Photoreception and*

Vision in Invertebrates. M.A. Ali, ed. New York: Plenum, p. 561-621, 1984.

Caldwell, J.H., N.W. Daw and H.J. Wyatt, The effects of picrotoxin and strychnine on rabbit retinal ganglion cells: Lateral interactions for cells with more complex receptive fields. *J. Physiol.* 276:277-298, 1978.

Egelhaaf, M., A. Borst and W. Reichardt, Computational structure of a biological motion-detection system as revealed by local detector analysis in the fly's nervous system. *J. Opt. Soc. Amer.* 6:1070-1087, 1989.

Franceschini, N., A. Riehle and A. LeNestour, Directionally selective motion detection by insect neurons. In *Facets of Vision.* D.G. Stavenge and R.C Hardie, eds. Berlin: Springer-Verlag, p. 360-390, 1989.

Fuortes, M.G.F. and A.L. Hodgkin, Changes in time and sensitivity in the ommatidia of *Limulus*. *J. Physiol.* 172:239-263, 1964.

Gilbert, C., D.K. Penisten and R.D. DeVoe, Discrimination of visual motion from flicker by identified neurons in the medulla of the flesh fly *Sarcophaga bullata*. *J. Comp. Physiol.*, submitted, 1990.

Glantz, R.M. and H.B. Nudelman, Linear integration of convergent visual inputs in an oculomotor reflex pathway. *J. Neurophysiol.* 52:1213-1225, 1984.

Glantz, R.M., Motion detection and adaptation in crayfish photoreceptors: A spatiotemporal analysis of linear movement sensitivity. *J. Gen. Physiol.* 97:777-797, 1991.

Grzywacz, N.M. and C. Koch, Functional properties of models for direction selectivity in the retina. *Synapse* 1:417-434, 1987.

Hisada, M. and T. Higuchi (1973) Basic response pattern and classification of oculomotor nerve in the crayfish, *Procambarus clarkii*. *J. Fac. Sci. Hokkaido Univ. Ser. VI* 2001 18:481-494, 1987.

Horridge, G.A. and P.R.B. Shepheard, Perception of movement by the crab. *Nature* 209:267-269, 1966.

Kirk, M., B. Waldrop and R.M. Glantz, The crayfish sustaining fibers: Morphological representation of visual receptive fields in the second optic neuropil. *J. Comp. Physiol.* 146:175-179, 1982.

Naka, K.-I., M.A. Itoh, and R.L. Chappell, Dynamics of turtle cones. *J. Gen. Physiol.* 89:321-337, 1987.

Nalbach, H.-O., Three temporal frequency channels constitute the dynamics of the optokinetic system of the crab, *Carcinus maenus*. *Biol. Cybern.* 61:59-70, 1989.

Okada, Y. and T. Yamaguchi, Eyestalk movements in the crayfish, *Procambarus clarkii*. *Comp. Biochem. Physiol.* 81A:157-164, 1985.

Pfeiffer, C. and R.M. Glantz, Cholinergic synapses and the organization of contrast detection in crayfish optic lobe. *J. Neurosci.* 9:1872-1882, 1989.

Pfeiffer-Linn, C. and R.M. Glantz, Acetylcholine and GABA mediate opposing actions on neuronal chloride channels in crayfish. *Science* 245:1245-1251, 1989.

Pfeiffer-Linn, C. and R.M. Glantz, GABA mediated inhibition of visual interneurons in the crayfish medulla. *J. Comp. Physiol. A* 168:373-381, 1991a.

Pfeiffer-Linn, C. and R.M. Glantz, An invertebrate NMDA receptor. *Synapse*, in press, 1991b.

Poggio, T. and W. Reichardt, Considerations on models of movement detection. *Kybernetik* 13:223-227, 1973.

Srinivasan, M.V. and G.D. Bernard, A proposed mechanism for multiplication of neural signals. *Biol. Cybern.* 21:227-236, 1976.

Wang-Bennett, L. and R.M. Glantz, Functional organization of the crayfish lamina ganglionaris. I. Nonspiking monopolar cells. *J. Comp. Physiol.* 161:131-145, 1987a.

Wang-Bennett, L. and R.M. Glantz, Functional organization of the crayfish lamina ganglionaris. II. Large field spiking and nonspiking cells. *J. Comp. Physiol.* 161:147-160, 1987b.

Wiersma, C.A.G. and T. Oberjat, The selective responsiveness of various crayfish oculomotor fibers to sensory stimuli. *Comp. Biochem. Physiol.* 26:1-16, 1968.

Wiersma, C.A.G. and R. Hirsh, Memory evoked optomotor responses in crustaceans. *J. Neurobiol.* 5:213-230, 1974.

Wyatt, H.J. and N.W. Daw, Directionally sensitive ganglion cells in the rabbit retina. Specificity for stimulus direction, size and speed. *J. Neurophysiol.* 38:613-626, 1975.

Neither DoG nor LoG fits the receptive field of the vertebrate cone

Josef Skrzypek and George Wu
Machine Perception Laboratory
University of California
Los Angeles, CA 90024
email: skrzypek@cs.ucla.edu

Abstract

For a given light intensity distribution I = f(x,y) *representing a scene, the visual system seems to encode the point information in terms of intensity* I *and in terms of the second derivative of intensity (Mach 1868). The question of where in the visual systems and in what form this encoding first takes place and what are the responsible synaptic pathways and mechanisms remains unclear. In this paper we focus on physiological mechanism underlying receptive field of a cone in the vertebrate retina and we test the data with analytical model using UCLA SFINX [1] simulation environment.*

Introduction

The idea that the retinal output contains second order, spatial difference of the input such as Laplacian of intensity was first introduced by Mach in 1868. Kuffler (1953) extended this hypothesis by finding ganglion cells that encode spatial difference of intensities through antagonistically organized, concentric receptive fields. Ratliff (1965) related Mach's spatial derivative with the retinal mechanism of lateral inhibition. At the same time Rodieck and Stone (1965) reported that cats ganglion cells with the antagonistic, center-surround receptive fields (RF) can be modeled as a difference of two concentric Gaussian sensitivity profiles; difference of Gaussians - DoG. Marr and Hildreth (1980) were first to propose Laplacian of a Gaussian (LoG) as a generic analytical model for isotropic RF. They argued that Gaussian is useful in reducing noise and the Laplacian can accurately localize intensity discontinuities; LoG filter, can be modeled by a DoG operator when the diameter of center Gaussian is approximately one third of the "surround" Gaussian.

In the past twenty years, the concentric organization of RF's into antagonistic zones was reported in bipolar cells [2], in horizontal cells (HC's) [3,4] and in cones [5,4,6]. It is now well established that antagonism begins with cones [4,5,7] but the quantitative comparison of DoG or LoG formalism with actual physiological data has never been performed. We report here such tests using intracellular data from the cone and HC's in the eyecup preparation of the Tiger salamander retina. The data from physiological experiments is used to build a simulation of the outer retina that is equivalent to 128 by 128 array of cones.

Several new findings emerged requiring modification to DoG model of receptive field as applied to retinal cones. The results suggest a more accurate approximation of the cone RF as compared to DoG is the difference between central Gaussian G_c and the sum of displaced "surround" Gaussians G_s. The G_c represents the RF of a cone and it reflects the coupling between neighboring receptors. The displaced G_s represent RF's of HC's that are not concentric with the receptive field of the cone; the Difference of offset Gaussian (DooG) model [8]. In fact, our data seems to suggest that a HC that gets direct synaptic input from the cone concentric with its perikaryon, have very little or no feedback effect on the cone; the feedback effect comes mainly from HC's that have their dendritic fields (DF's) displaced laterally with respect to the cone and is mediated by the more eccentric dendritic processes.

Materials and Methods

Preparation and the general experimental procedure were described previously in detail [3,4]. Briefly, all experiments were performed using eyecup preparation from the Tiger salamander. Intracellular recordings were made with single or double barrel micropipettes filled with 4M potassium acetate. For staining experiments, one of the two barrels was filled with 3% solution of Lucifer Yellow dissolved in 1M LiCl. Cones were identified on the basis of physiological criteria [2,9,10] and later confirmed by staining with Lucifer yellow [11]. Light stimuli were generated by specially constructed double-beam photostimulator which could deliver concentric, circular and annular stimuli of white or monochromatic light. The flux density of the unattenuated light on the retina was about 15 mW/cm^2.

IDENTIFYING RETINAL CELLS Several criteria were used to distinguish cones from other cells in the OPL. First of all, the depth of recording in the retina [9] at which the cone was penetrated was correlated with histological evidence, as corresponding to the layer of cone inner segments. It also appears that at this depth, the only other cell bodies belong to rods [12]. After recording form a cone, further advancement of electrode sometimes resulted in penetration of a rod. On the other hand, after recording from bipolar cell, further advancement of electrode always produced penetration of a horizontal cell followed by rods or at times cones. This sequence of penetrations concomitant with characteristic light responses was a very useful heuristic in differentiating cones from bipolars. One exception are cone pedicles, which could be penetrated and mistaken for bipolar cells, especially, if other impalements were to follow. Hence, it is important to emphasize that no single test alone could be used to differentiate cones from bipolar cells and only a sequence of tests followed by staining could give reasonable indication of the penetrated cell type. The identification of single cones was also based on spectral response curves. For the red cones, the only ones considered in this study, their peak in wavelength spectrum, fit well with absorption spectrum for red pigment [13]. The spectrum response curve for HC or bipolar cells were often flat or even peaked in the short wavelengths [6,4,3]. Cones as compared to horizontal cells or bipolars, had smallest diameter for RF center $< 0.075mm$ [14]. The average for bipolars was about $0.15mm$ and the smallest horizontal cells were found to be about $0.3mm$. Under dark adapted conditions cones were also distinguished by the fastest time from dark potential to the peak of the light response ($50ms$) as compared with bipolars ($100ms$) or horizontal cells ($> 150ms$). One additional criterion was that the I-R curves for cones usually span 3 - 4 log units (LU) of intensity between 5% and 95% of the saturating membrane potential value. On the other hand the I-R curves in bipolars were found to span on the average 1 LU (1.3 LU in mudpuppy, [15]). These values represent averages derived from all intracellular recordings in 37 cones, 84 bipolar cells, more than 1000 horizontal cells, and more than 100 rods.

In some experiments the attempt was made to stain the physiologically identified cones and other cells with Lucipher yellow. A good staining was considered when the response of a cone did not deteriorate by more than 20%. Often, the center/surround antagonism would disappear or decrease significantly so that we could not complete the measurements of dynamic changes in I-R curves, but we could still record an unchanged center response. In most of these cases we attempted to fill the cone with the stain. The stained cells were viewed immediately after injections, under epifluorescent microscope, without going through conventional procedures of fixating and embedding tissue [3]. While the stained HC and bipolar cells always showed extensive arborization of dendritic processes, protruding from the soma, this was not the case for cones which appeared as symmetrical spheres. In all 37 cones were identified by a combination of these criteria and partial results were obtained in most cases, including 19 attempts at staining. In six of these cells the experiments were completed but only in three of the six the staining was also successful. Out of 117 attempts to stain HC's, 31 resembled axon terminals without clear perikaryon and were associated with very large RF's. The remaining 86 had classical appearance of the cell body with numerous fine dendrites and often stained in clusters of between two and six. Many of these HC's had their RF centers between 0.3*mm* and 1*mm* and could be made to display center/surround antagonism.

Results

FEEDBACK TO CONES Although the concentric organization of the cone RF into antagonistic zones of center and surround has been firmly established in lower vertebrates [5,4,16,15,10] the analytical model of this organization is still lacking. Fig.1. shows series of light responses recorded intracellularly from the red cone. White spot of light (0.15 mm at 3LU) evoked, near saturating, peak hyperpolarization of about 20 mV measured from the resting membrane potential in the dark of V_m = - 40 mV; there is little change during the plateau portion of the response. However, the response waveshape changed drastically when the diameter of the stimulus was increased to 1 mm. Although the peak response and the onset time-course remained unchanged, the plateau portion of the response showed 30% depolarization from -61 mV to -54 mv as compared to peak response. The time course of this depolarization was much slower than the onset of the hyperpolarizing response, culminating with a clear transient undershoot when the stimulus was turned off. This slow onset of depolarization reflects integrated delay caused by peripheral signal traveling from increasingly eccentric areas of the retina. This result agrees well with similar observation in turtle and fish retinas and it demonstrates that in the intact retina the waveshape of the cone's light response is a function of the stimulus diameter; illumination of peripheral area of RF antagonizes the effect of the stimulus in the center of RF.

Fig.2. shows the cone light response plotted as a function of the stimulus diameter. The peak V_m of the response (open circles) increased to a maximum value at the spot diameter of about 0.1 mm. However, during the plateau, measured one second after the peak response, the V_m (filled circles) decreased drastically from 9 mV at 0.1 mm down to 3 mV at 2 mm, the maximum diameter stimulus covering the whole eyecup preparation.

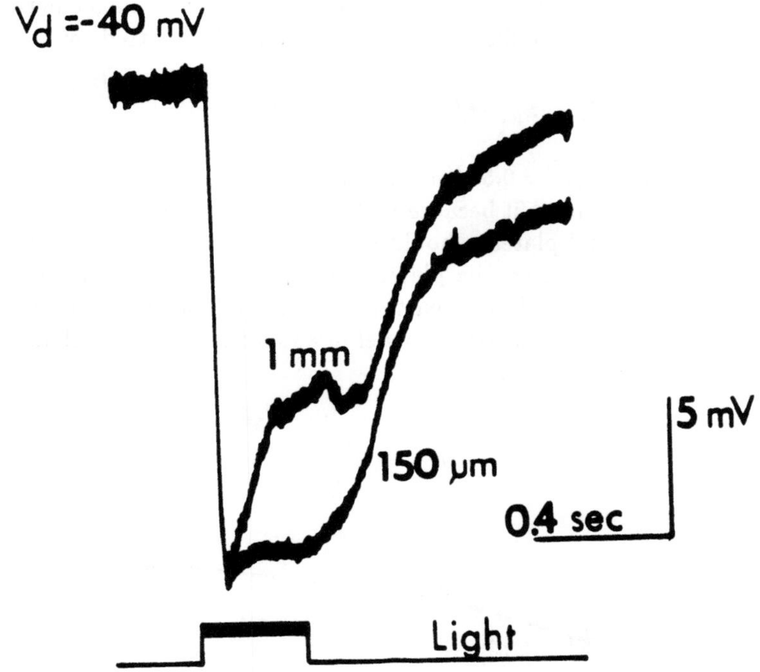

Figure 1: Series of light responses recorded intracellularly from the red cone. Two traces are overlaid, representing different diameter of the stimulating spot. 0.15 mm spot of white light elicits a peak hyperpolarization of 20mV which lasts as long as the stimulus. Large 1.0 mm spot of light causes the same peak response immediately followed by large depolarization.

The plot of the peak response can be fitted by integrating a Gaussian curve

$$G_c = A_c * exp(-r/r_c)^2$$

where r_c (0.05 mm) is the radius of the cone RF center, r is the distance from the long axis of the cone and the A_c is the gain coefficient for the center of RF, assumed to be one. The result in Fig. 2 argue against modeling cone RF as a LoG function. Laplacian of G_c with 0.1 mm diameter predicts that the antagonistic surround will disapear at 0.15 mm from the origin of RF. The data in Fig. 2. shows strong antagonism even when the radius equals 1 mm. In addition LoG function specifies the amplitude of the center to be fixed at exactly twice the amplitude of the antagonistic surround which can not be fitted against the plateau curve in Fig. 2.

A better model is the traditional DoG function [17]. The balance between G_c and G_s is described by the filled circles curve that can be fitted by

$$DoG = G_c - G_s = A_c * exp(-r/r_c)^2 - A_s * exp(-r/r_s)^2$$

where $r_c = 0.025$ mm, $r_s = 0.5$ mm, $A_c = 1$ and $A_s = 0.037$. This analytical model gives only approximate fit because it shows that at the radius value of 50 um, V_m for both peak and plateau should be equal which does not agree with data; at 0.1 mm spot diameter, the plateau response is only 3/4 of the peak response. This difference could be in part due to light adaptation process that is not being captured in the analytical model in addition to having unbalanced center and surround contributions.

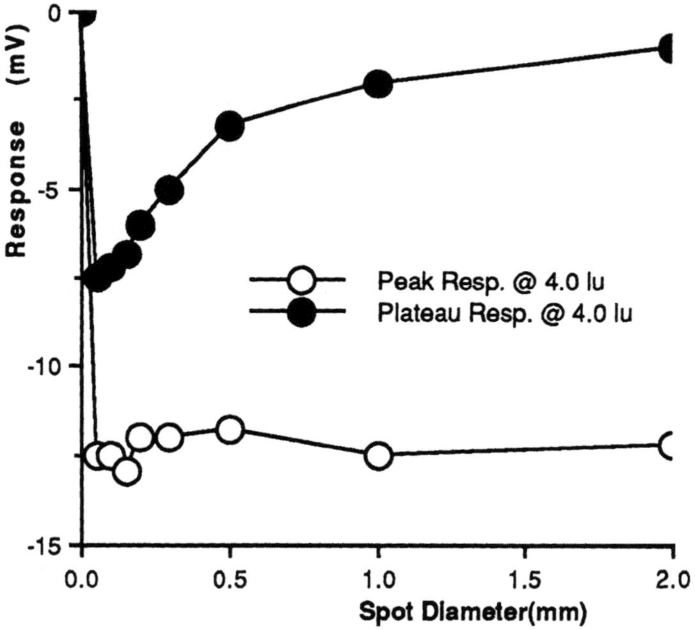

Figure 2: Membrane potential of the red cone plotted as a functions of the stimulus diameter. The peak response (open circles) remains constant with spot diamters larger than 0.1mm. The plateau response (filled circles), measured 1sec after the peak hyperpolarization, shows drastic reduction in amplitude with increasing stimulus diameter.

CENTER - SURROUND ANTAGONISM IN HORIZONTAL CELLS The DoG model assumes that the G_s is continuous and concentric with G_c of HC's and can extend from 0.3 mm to 2 mm [4,3]. This can not be reconciled with the observation that HC's display center surround antagonism [16,4]. Fig. 3. shows a series of intracellular recordings from the "narrow" RF HC in response to center spot of light (0.5 mm) and the concentric annular illumination (ID = 1 mm and OD = 2 mm).

The plot in Fig. 4. shows that the peak response of a different, narrow field HC that also displays center surround antagonism, reaches saturation level at stimulus diameter of about 0.6 mm. This is the extent of the RF center. The intensity of the annulus was fixed at 3.2 LU and when presented alone it caused the hyperpolarizing response, probably because of light scatter from the annulus to the center. Apparently the amount of scatter is sufficient only to excite rods input to the HC; at the stimulus offset, the V_m of this HC returns slowly to the dark resting potential (-31 mV) with a time course characteristic for rods. In the presence of center spot of light with increasing intensity, the hyperpolarizing "surround" response reverses in polarity and becomes depolarizing. Here, the overall time course of the center response still retains the slow recovery to dark V_m level suggesting predominance of rod input [14]. Further increase of the center intensity, up to 3.5 LU, depresses the depolarizing surround response to zero, concomitant with a change in the time course of the center - light elicited response at the offset, indicating cone input. The depolarizing surround response can be restored to the previous level by increasing center spot intensity up to 2.5 LU. But further increases eliminate the surround response again. The changes in center/surround antagonism with increasing center intensity have been observed in many "narrow" RF HC's. They are related to rod-cone differences in synaptic transmission to HC and will be addressed elsewhere (Skrzypek, 1991 in prep.). The final decrease in the mechanism of surround depolarization at center intensities of 2 LU was suggested to result from decreasing feedback conductance associated with the reversal potential of -65mV; the closer the center elicited hyperpolarization brings the cone's Vm to -65 mV the smaller the effect of any changes in feedback conductance on V_m [4,6].

The results in Fig.3. suggest that antagonistic effect of surround measured previously in cones can be also seen in some HC's. This has been previously explained as a natural consequence of HC being postsynaptic to cones and reflecting their activities. The antagonism observed in HC can be seen at different levels of stimulus intensity without any effect on the size of RF center or the strength of the center/surround antagonism (Fig. 4.).

The results in Fig 3. and Fig. 4. argue against a simple DoG function as a

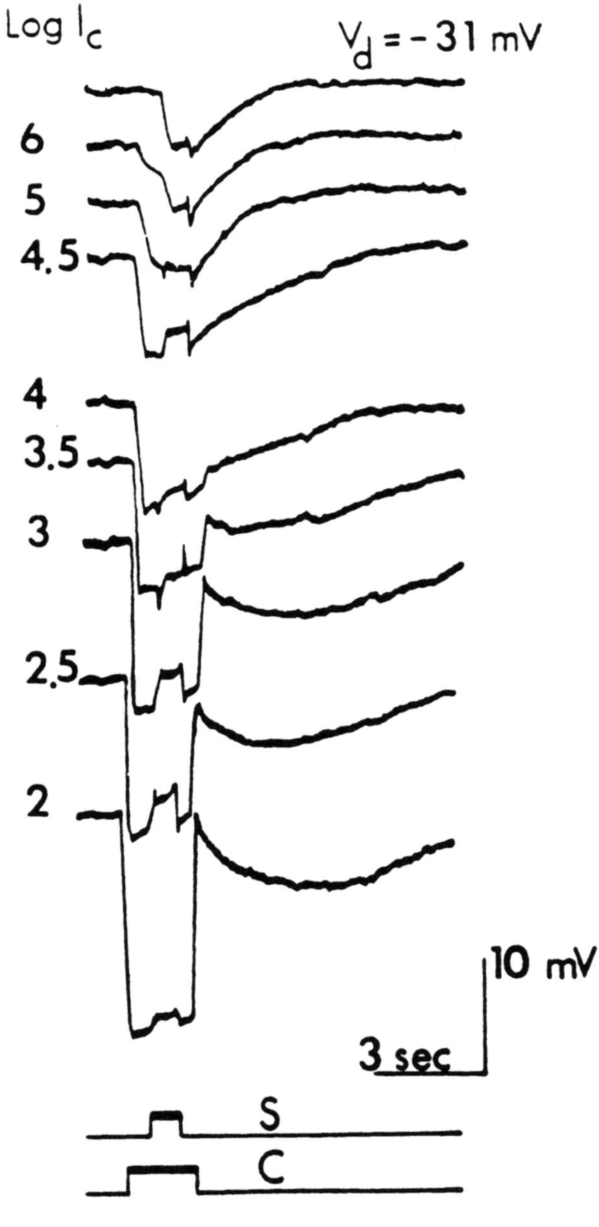

Figure 3: Series of intracellular recordings from the "narrow" receptive field HC in response to center (C) spot of light (0.5mm) and the concentric annular (S) stimulus (ID = 1mm, OD = 2mm). The intensity of the annulus was fixed at -3.2 Log units, while the intensity of the center spot was incremented as indicated by numbers to the left of each trace.

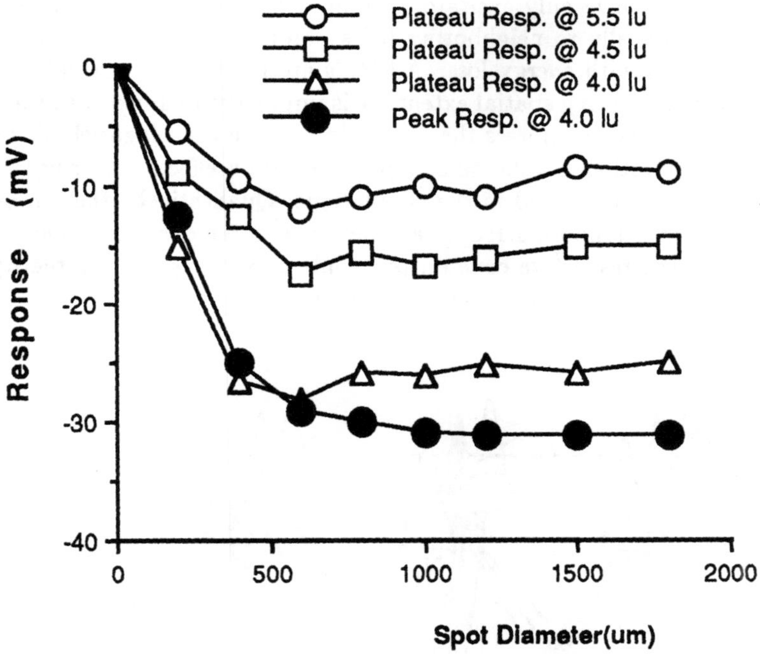

Figure 4: Membrane potential of the "narrow" RF HC plotted as a function of the stimulating spot diameter at various intensities. Plateau response (open symbols) at all intensities reaches maximum at about .6mm and declines with further increases of the spot diamter.

model for the cone RF. DoG requires that the response measured from the HC in the center of RF be maximally hyperpolarizing which is inconsistent with the observation that HC's display depolarizing surround responses to annular illumination; clearly HC's dont feedback to cones concentric with their dendritic fields which drive these HC's directly.

Simulations

MODEL OF THE OPL The computational model consists of two layers representing cone pedicles and the narrow field HC's. Each pedicle 1) is driven without any attenuation by the cone outer segment, 2) makes electricall contacts to neighboring pedicles and 3) receives negative feedback inputs from neighboring HC's. HC's receive only excitatory inputs from neighboring pedicles and are coupled electrically to neighboring HC's. Model of each cell incorporates "leaks" associated with energy loss. All RF's are modeled as normalized 2-D gaussian kernel of varying spatial extent. It is implicitly assumed that the cone outer segments merely compress the scene illuminance via a logarithmic transform. This assumption is valid since light adaptation effects are insignificant in the illuminance range used in the simulations. Qualitatively, the inhibitory connections from neighboring HC to a cone serve to remove *locally* redundant components in the respective cone output while preserving signal integrity. A schematic view of the model is shown in Fig. 5.

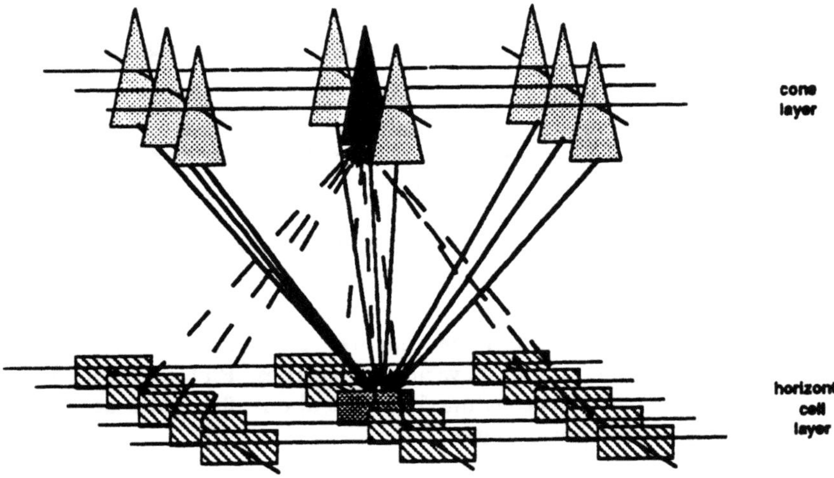

Figure 5: Simplified diagram of the OPL showing connectivity pattern between horizontal cells and cones.

The cone output is a rectifyed (*Output Rectification Function = ORF*) sum of the three *Input Conductance Functions* (ICF's) minus the leakage terms, one fixed (K_c) and the other dependent on the cell's output (L_c). The three ICF's represent input transfer functions to the cone and they differ only in parameters.

$$\begin{aligned} O_c(t) = ORF(&ICF(G_{l-c_1}, G_{l-c_2}, S_{l-c}) \\ &+ ICF(G_{c-c_1}, G_{c-c_2}, S_{c-c}) \\ &- ICF(G_{hc-c_1}, G_{hc-c_2}, S_{hc-c}) \\ &- L_c O_c(t-1) - K_c) \end{aligned}$$

where: $ORF(I) = 1 for I > 1; = 0 for I < 0; otherwise = I$

HC's receive only two excitatory inputs from neighboring cones and HC's that are expressed as two unique ICF functions. Each ICF operates on a gaussian weighted sum of inputs over a 2-D RF. The output of a HC is determined in a similar manner as the cone output.

$$\begin{aligned} O_{hc}(t) = ORF(&ICF(G_{c-hc_1}, G_{c-hc_2}, S_{c-hc}) \\ &+ ICF(G_{hc-hc_1}, G_{hc-hc_2}, S_{hc-hc}) \\ &- L_{hc}O_{hc}(t-1) - K_{hc}) \end{aligned}$$

The ICF function is really a lumped model of the actual synaptic conductance characteristics that might vary from synapse to synapse; $ICF(G_1, G_2, I) = G_2 \tanh(G_1 I)$. However, this lumped model is adequate since the goal is to identify the *robust* physiological characteristics of the OPL. A more serious limitation is the fact that all signal delays in the model are equal and constant. This limitation was introduced intentionally to simplify the model and facilitate the analysis of the steady-state behavior.

There are two important aspects of the model, its network structures and its receptive field functions. Structurally, the model consists of one 128x128 cone photoreceptor layer and one horizontal cell layer. In some experiments, the cone-horizontal cell coverage ratio was 9 which is close to convergence number found in monkeys fovea [18]. Our model allows to set the cone RF center to be anywhere from 128x128 to zero coupling with neighboring cells. For the simulation shown in Fig.6. and in Fig.7., the cone RF center was set to 49 (7x7) pedicles; pedicle diameter is assumed to be 5 μm

The feedback comes from 9 (3x3) HC's and each HC was coupled to 49 (7x7) neighboring cell. The model functionality is determined by its RF's characteristics. The respective inputs to any cell are weighted and summed and then normalized to produce the final response. The normalization of response is accomplished via a Sigmoidal squashing function. Although the convergence values used in our simulation are smaller than real connectivity pattern found in mamalian retinae (50 cones to 1 HC [19], they are sufficient to to evaluate the plausibility of approximating the cone RF with selected analytical model. In the present simulation study, all RF's are Gaussians except the cone-HC and HC-cone RF's which are inverted Gaussians such that the input at the center of the RF has the least weight. The utilization of inverted Gaussian RF's is deduced from the physiological response of cones and HC's presented in previous sections and available anatomical data on HC dendrites[19]

SIMULATIONS RESULTS The OPL model described above was intended to resolve whether the LoG, DoG or other kernels can adequately model cone

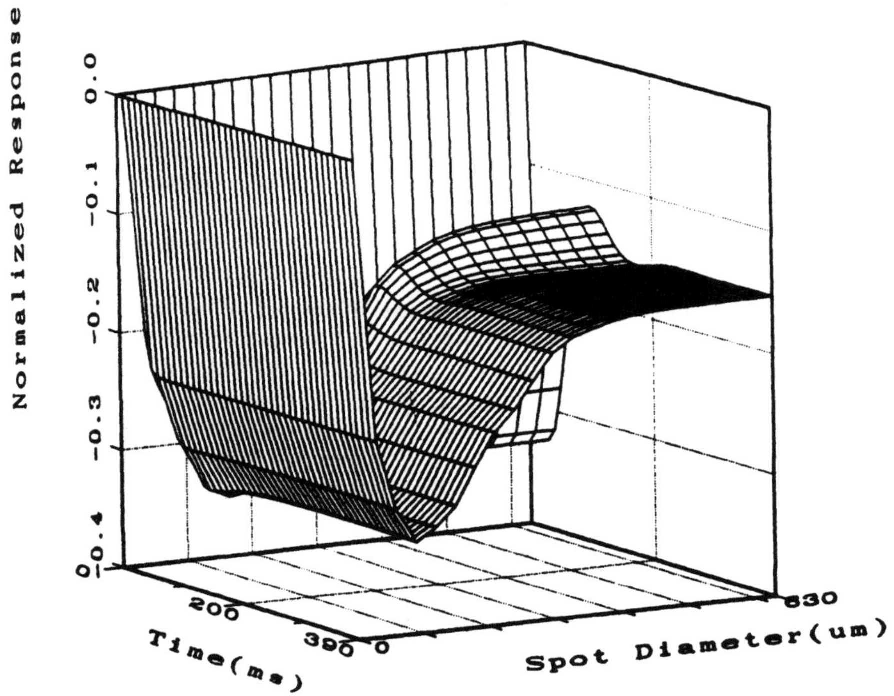

Figure 6: Membrane potential of a cone in the simulated model of the OPL, during the light response ploted in 3D as a function of time and spot diameter.

RF's. The experiments conducted on the computational model were analogous to equivalent experiments designed to gather physiological data. A fixed intensity light stimulus of variable diameter was centered on the dark-adapted, cone layer. After the onset of the stimulus, the response of the illuminated cone and concentric (central) HC was recorded for 400 ms until the steady-state response was reached. Fig. 6 shows a 3D plot of the simulated cone response as it changes with increasing stimulus diameter over the 400 ms period. The antagonistic surround is only 1.6 times larger than the agonistic center; experiments with larger G_c/G_s ratios show more pronounced antagonism in HC's but are more difficult to stabilize. The connectivity pattern between HC's and cones, used to generate this result was to have maximal feedback to cones from HC's that are laterally displaced by the equivalent of one cone RF center. The plot shows that the response reaches peak amplitude in both temporal and spatial domains, after which it settles to a lower plateau level. Here we are concerned only with spatial antagonism; decreasing hyperpolarizing response with increasing spot radius. The pronounced depolarizing transient that grows with increasing stimulus diameter is not an accidental by-product of the simulation;

a comparable, short-lived oscillatory behavior has been observed in some cones (Skrzypek, 1991 in preparation).

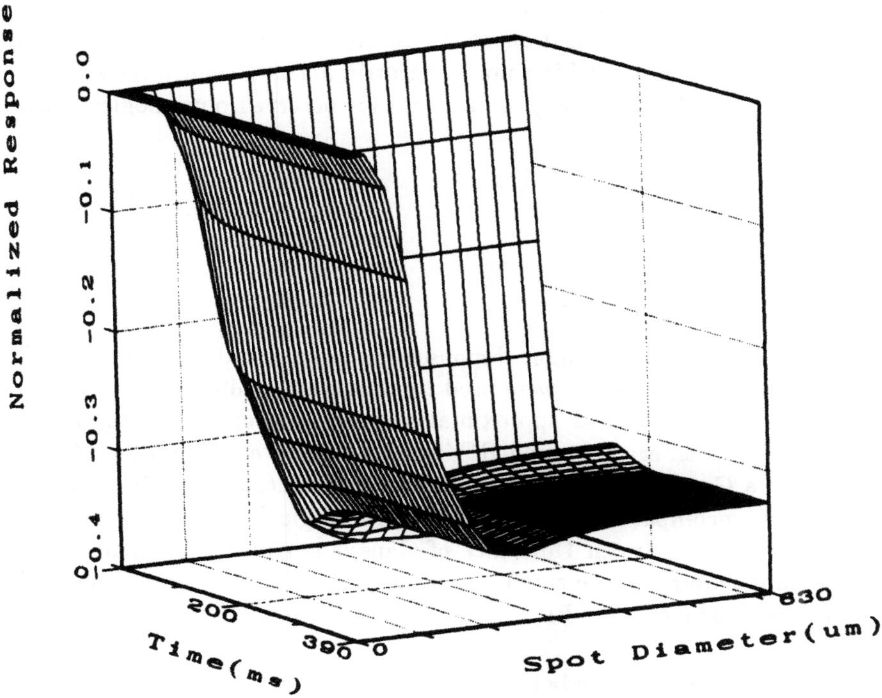

Figure 7: 3D plot of the simulated narrow RF HC that displays weak center-surround antagonism.

Fig.7. shows the simulation result observed in the central HC that displays a weak center-surround antagonistic effect. This antagonism could not be generated by modeling cone RF as DoG function where G_c and G_s are concentric with each other. The antagonism was visible only when the cone RF profile was modeled as a difference of G_c and eight G_s(one concentric and the rest non-concentric). This is essentially the DooG model [8] with a small addition. The standard DooG model does not make use of inverted Gaussians and a robust characteristic of the model is that all RF's based on it demonstrates center dominance over the periphery. The incorporation of inverted Gaussians allows for RF's with periphery dominance over the center. Simulations using center-dominant DooG and periphery-dominant DooG were conducted and both produced results that resemble physiological data. However, simulations using center-dominant DooG often produced oscillatory response profile in the spatial domain that is not consistent with data. On the other hand periphery-dominant DooG based simulation results are in superior correspondence with data as shown in Fig. 6 and Fig.7. Nevertheless, in the DooG (center- or periphery-dominant) model, the cone response is attenuated by spatially offset HC's; with increasing stimulus diameter, the response of peripheral HC's grows and increasingly attenuates the response of the center cone, more than it contributes to hyperpolarization of the central HC. Consequently, with all cells having only Gaussian RF's a non-concentric *network* connectivity pattern between cones and HC's (DooG) can account for observed physiological data.

Conclusions

The intracellularly recorded light responses from cones and narrow field HC's in the tiger salamander retina were used to evaluate five different filters (DoG, LoG, DooG, DoE, Gabor) as approximation to the concentric arrangement of the antagonistic, center - surround RF in cones. Three of these filters, Gabor, Laplacian of a Gaussian and the difference of exponentials were dismissed because of clear incompatibility with known physiological and anatomical data. The discrimination between Difference of Gaussians and the Difference of offset Gaussians required more careful analysis of data and the test against the connectionist model of the OPL.

Analytically, the balance between center and antagonistic surround (Fig.2) could be approximated with DoG function. This model however can not explain why HC's also display center/surround antagonism. Anatomical evidence suggests that HC's contact cones through invaginating processes that are lateral to the ribbon in cone pedicle; there is no clear distinction between feedback and feedforward direction of synaptic transmission. How could then HC, respond to an annulus with membrane depolarization, reflecting activities of a cone, that is concentric with HC's dendritic field, while at the same time being hyperpolarized by neighboring HC's responding directly to the annulus? One possibility is that there are two types of HC's which are not coupled electrically, and which provide separate feedback pathways like the chromaticity HC's in the fish [20]. Two morphological types of HC were reported in Ambystoma Tigrinum, one axonless and the other having both cell body and the axon terminal (AT) [21]. Of course any conclusion based on Golgi stain must be accepted with caution since the HC axon's are at times only a fraction of a micron in diameter and could be hard to impregnate. In fact in rhesus monkey there is only one morphological type of HC segregated into functionally isolated cell body and axon terminal (AT) [18]. It is possible that the narrow RF HC's which display clear center/surround antagonism are the cell bodies while the AT with their very large RF provide the feedback pathway. However, in the monkey and in the cat both AT and the cell body are functionally segregated; AT's contact only rods while the cell body connects only to cones [19,18].

What synaptic arrangement would allow the same HC to receive input from cones and also to feed back to some of them? Such a scheme is possible if we imagine that the strength of feedback from one HC to cones within its dendritic tree is minimal in the center of a perikaryon and increases with eccentricity of synaptic contacts. In the rhesus monkey retina [19] some peripheral HC's that contact up to 50 cones showed no contact with pedicles concentric with HC's dendritic tree. When the contacts were seen, cones nearest the HC perikaryon made more contacts with this HC (six) than the cones at the periphery of HC's dendritic field (one). Thus less than 10% of contacts per cone belong to one HC that is concentric with a given cone; most of the contacts come from HC's whose cell bodies are laterally displaced with respect to the cone. It is conceivable that groups of terminals near the perikaroyn are involved in feedforward transmission from cones to HC while the single terminals located on eccentric dendrites participate in feedback transmission. This direction of the signal flow is favored also by the impedance matching considerations, between relatively large perikaryon of the HC and high resistance of the peripheral dendritic terminals. The effect of feedback synapses being located on the peripheral terminals of HC's can be approximated with a DooG function.

References

[1] J. Skrzypek. Neural specification of a general purpose vision system. In *Proceedings of ACNN'90, Australian Conference on Neural Networks*, Sydney, Australia, January 1990.

[2] F.S. Werblin and J.E. Dowling. Organization of the Retina of the Mudpuppy, *Necturus maculosus*. II. Intracellular recording. *Journal of Neurophysiology*, 32:339–355, 1969.

[3] J. Skrzypek. Electrical Coupling Between Horizontal Cell Bodies in the Tiger Salamander Retina. *Vision Research*, 24(4):701–711, 1984.

[4] J. Skrzypek and F.S. Werblin. Lateral Interactions in Absence of Feedback to Cones. *Journal of Neurophysiology*, 49(3):1007–1016, 1983.

[5] D. A. Baylor, M. G. Fuortes, and P. O'Bryan. Receptive fields of cones in the retina of the turtle. *J. Physiol.*, 214:265–294, 1971.

[6] J. Skrzypek. Light sensitivity in cones is affected by the feedback from horizontal cells. In *Proceedings of Analysis and Modeling of Neural Sys-*

tems, NASA Conference on Neural Networks, UCB Berkeley, California, July 1990.

[7] D.A. Burkhardt. Responses and receptive field organizations of cones in perch retinas. *J. Neurophysiol*, 40:53–62, 1977.

[8] R. Young. The gaussian derivative model for spatial vision: retinal mechanisms. *Spatial vision*, 2:273–293, 1987.

[9] A.. Kaneko. Physiological and morphological identification of horizontal, bipolar and amacrine cells in goldfish retina. *J. Physiol*, 207:623–633, 1970.

[10] Lasansky A. Synaptic action mediating cone responses to annular illumination in the retina of the larval tiger salamander.,. *J.Physiol. Lond.*, 310:206–214, 1981.

[11] W. Stewart. Functional connections between cells as revealed by dye-coupling with a highly fluorescent napthalimide tracer. *Cell*, 14:741–759, 1978.

[12] A Lasansky. Contacts between receptors and electrophysiologically identified neurons in the retina of the larval tiger salamander. *J.Physiol.*, 285:531–542, 1978.

[13] Werblin F.S. Attwell D. and Wilson M. The properties of single cones isolated from the tiger salamander retina. *J.Physiol*, 328:259–283, 1982.

[14] L.M. Marshall and F.S. Werblin. Synaptic inputs to the horizontal cells in the retina of the larval tiger salamander. *J. Physiol.*, 279:321–346, 1978.

[15] F.S. Werblin. Integrative pathways in local circuits between slow potential cells in the retina. *The Neurosciences*, 279:193–210, 1978.

[16] J. Skrzypek and F.S. Werblin. All Horizontal Cells have Center-surround Antagonistic Receptive Fields. In *ARVO Abstracts*, page 1, 1978.

[17] R. W. Rodieck. Quantitative Analysis of Cat Retinal Ganglion Cell Response to Visual Stimuli. *Vision Research*, 5:583–601, 1965.

[18] H. Kolb. Organization of the outer plexiform layer of the primate retina: electron microscopy of golgi impregnated cells. *Phil. Trans. Roy. Soc. Lond. B*, 258:261–283, 1970.

[19] H. Kolb. The connections between horizontal cells and photoreceptors in the retina of the cat. *J. Comp. Neurol.*, 155:1–14, 1974.

[20] W.K. Stell. Functional Polarization of Horizontal Cell Dendrites in Goldfish Retina. *Invest. Ophthal. Visual Sci.*, 15:895–907, 1976.

[21] Lasansky A. and Vallerga S. Horizontal cell responses in the retina of the larval tiger salamander. *J.Physiol*, 251:145–165, 1975.

Cellular and Network Determinants of Visual Motion Properties in Cortical Neurons: Studies with an *In Vitro* Preparation of Visual Cortex

Philip S. Ulinski, Linda Larson-Prior and N. Traverse Slater
Department of Organismal Biology and Anatomy, University of Chicago, 1025 E. 57th Street, Chicago, Il 60637 and Department of Physiology, Northwestern University, 303 E. Chicago Ave., Chicago, Il 60611

Introduction

The cerebral cortex of fresh water turtles (*Pseudemys scripta* and *Chrysemys picta*) contains a visual area (review: Ulinski, 1990) that receives direct input from the dorsal lateral geniculate complex (Hall and Ebner, 1970; Heller and Ulinski, 1987). The projection of the retina to the geniculate complex is bilateral and topographically organized (Ulinski and Nautiyal, 1987), but the subsequent projection of the geniculate to the visual cortex is not topographically organized. Instead, there is a convergence of all points of visual space onto each cortical cell as a result of the spatial organization of the geniculocortical and intracortical projections (Mulligan and Ulinski, 1990; Cosans and Ulinski, 1990). The anatomy of the geniculocortical pathway is consistent with extracellular recordings in alert, paralyzed animals presented with visual stimuli (Mazurskaya, 1974). Each cortical cell in such experiments responds to moving stimuli at all points in binocular visual space. They also respond well to two stimuli presented in succession at disjunct points in visual space in an "apparent motion" paradigm and have temporal tuning curves with two facilitatory peaks, one quite constant peak at 50 - 75 msec and a second, more variable peak occuring between 250 and 600 msec. These neuronal properties suggest that visual cortex plays a role in visually guided orienting behaviors in turtles (Ulinski et al., 1991a).

We are interested in understanding how the firing properties of these cells are determined by the circuitry of the cortex and other parts of the nervous system. Neuronal firing properties can be determined, in principal, by the biophysical properties of individual cells, by the neural networks present in the cortex or other parts of the brain, or by some interaction of neural circuits with the biophysical properties of individual cells. Assessment of the relative contributions of cellular and network factors to firing properties can be quite difficult in, for example, mammalian cortex because of the problems inherent in relating detailed studies of cellular properties (obtained from *in vitro* slice preparations) to data on receptive field properties (obtained from *in vivo* preparations that preserve the neural circuitry of the cortex).

Our work has been aided by an ability to maintain the turtle's brain intact in an *in vitro* preparation (Hounsgaard and Nicholson, 1990) that allows the investigator to record intracellularly from cortical neurons while activating specific elements of the circuitry (Larson-Prior et al.,1991). Alternatively, it is possible to present natural visual stimuli to the retina while recording intracellularly from cortical cells (Kriegstein, 1987). Using an *in vitro* preparation consisting of the isolated visual cortex and geniculate afferents, we have determined that the temporal tuning properties of cortical cells demonstrated by Mazurskaya (1974) *in vivo* are also found in cells studied in an isolated, *in vitro* preparation of visual cortex (Ulinski et al., 1991b). This suggests that the firing properties of cortical neurons are determined by some combination of cellular biophysical and cortical network properties. In this paper, we provide a preliminary analysis of the relative contribution of cellular and network properties to the average firing properties of cortical cells. Emphasis is placed upon outlining the general approach, leaving a full account of the data to more lengthy publications.

Results

Biophysical Properties of Cortical Neurons

Cellular properties of cortical neurons were studied in an *in vitro* prepartion of turtle dorsal cortex that preserves the visual cortex and a segment of the geniculocortical pathway (Larson-Prior et al., 1991). Individual cells were studied by recording intracellularly in current clamp and subjecting them to a sequence of hyperpolarizing and depolarizing current pulses (1 sec in length, ranging from - 1.0 to + 1.0 nA in amplitude). This procedure yields a detailed picture of the voltage-current relationships that characterize the cell as well as information on the relationship between its firing rate and injected current.

The voltage changes resulting from current pulses were measured at 10 msec intervals from 0 to 1,000 msec and used to prepare a three dimensional plot of the voltage change, V(I,t), that occurs in the cell at a specific time, t, as a function of a change in current, I. These plots typically show both outward and inward rectification, the magnitudes of which vary as a function of time. The apparent conductance of the cell, G(V,t), as a function of voltage and time was calculated by numerically computing the derivative $\partial V / \partial I = R(I,t)$, changing the coordinate system from (I,t) to (V, t) and then plotting G(V,t) = 1 / R(V,t).

Higher amplitude current pulses resulted in voltage transients that exceeded the firing threshold of the cell and produced action potentials. These records were used to prepare plots of the instantaneous firing rates, F(I,t) of the cells (measured as the reciprocal of interspike interval) as a function of time and the amplitude of the input current pulse. The relationship between firing rate, amplitude and current is a complex one that has not yet been completely analyzed. As a first approximation, however, the equation

$$(1) \qquad F(I,t) = a(t) + b(t)\ I$$

where a(t) and b(t) are functions that describe the time dependence of the firing rate provides a reasonable fit to the data.

Network Properties of Cortical Neurons

The effects of excitatory and inhibitory circuits impinging upon individual cortical neurons were studied by sequential, electrical activation of geniculocortical fibers carrying information from two disjunct regions of visual space (Ulinski et al., 1991,b). The stimulus paradigm was designed to simulate the pattern of activation that occurs when a visual stimulus moves past the turtle along the horizontal meridian of visual space. Electrical stimulation of the geniculate afferents evokes complex patterns of EPSPs and IPSPs in cortical neurons (Larson-Prior et al., 1991). The response of a cortical neuron to sequential activation was characterized by calculating a function, P(t), which gives the probability that the cell will fire in response to the second stimulus in the sequence where the interval between the first and second stimuli is t msec. There was substantial variation in the form of the P(t) functions between cells, but all could be represented as a linear combination of three functions of the form t Exp (- at):

$$P(t) = 1 + \sum_{i=1}^{3} N_i \, t \, \mathrm{Exp}\,(-a_i t). \qquad (2)$$

The P(t) functions provide a characterization of the response of the cortical cell to a particular stimulus paradigm, and presumably reflect the excitatory and inhibitory circuits that impinge upon the given cell being studied.

Interaction of Biophysical and Network Properties of Cortical Neurons

Ultimately, we hope to use our biophysical characterizations of individual neurons in visual cortex and P(t) curves obtained for the same cells to predict their instanteous firing rates in response to specific stimulus paradigms. This necessitates calculating the total current that flows through the cell at time t in response to a synaptically induced voltage change. This current is given by the product of the conductance function and and voltage transient:

$$I(t) = G(V,t)\, V(t)\,. \qquad (3)$$

The instantaneous firing rate of the cell is determined by plotting I(t) as a function of t, which shows action potentials as rapid current fluctuations.

So far, we have restricted our attention to an estimate of the average firing rates of cortical cells in the apparent motion paradigm. Since Mazurskaya's data (Mazurskaya, 1974) on firing rates in cortical cells *in vivo* are expressed as average firing rates as a function of interstimulus interval, we have attempted to calculate the average firing rate of cortical cells using the biophysical parameters of the cell and its P(t) curve. Cells were studied with the membrane voltage held at the cell's resting membrane potential so the recordings showed both the underlying postsynaptic potentials and the action potentials resulting from the stimulus paradigm. G(V,t) was measured at time intervals during which the cell produced action potentials as the slope of the calculated current-voltage curve for values of membrane potential immediately below the cell's firing threshold. Then, equation (1) becomes

$$F_{T(t)} = a + b\,G\,V_o\,P_T(t) \tag{4}$$

where V_o is the amplitude of the postsynaptic potential produced in the cell in response to the first in the pair of stimulus pulses, and $P_T(t)$ is the time average of the P(t) curve over a time bin of T = 50 msec duration centered on time t. Figure 1 shows the predicted average firing rate, $F_T(t)$ of a cell (open circles) and the P(t) curve for the cell (closed circles). The cell fired action potentials at interstimulus intervals of 25 and 50 msec and almost reached threshold at 200 msec.

Discussion

Our earlier work (Ulinski et al., 1991a,b) suggests that the firing properties observed by Mazurkaya (1974) in her studies of neurons within turtle visual cortex result principally from neural interactions occuring within visual cortex itself. Thus, the shape of the P(t) curves that describe the responses of cortical neurons to sequential activation of two spatial channels within the geniculocortical pathway generally

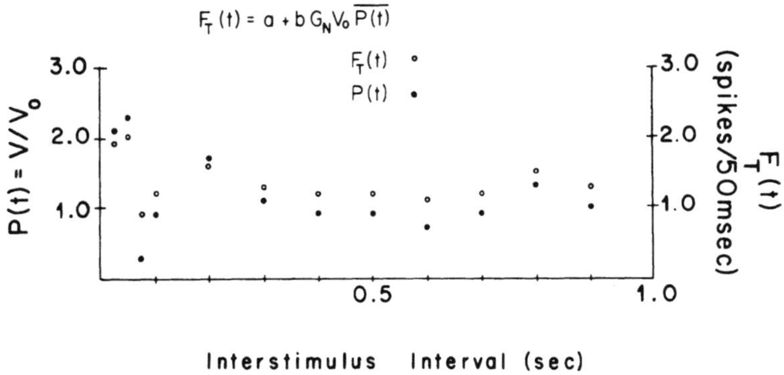

Figure 1. P(t) curve and predicted firing rate for a cortical neuron. The closed circles show the P(t) curve for a cortical neuron calculated from data obtained by synaptic activation of the cortical neurons as described in the text. The open circles show the average firing rate of the same cell calculated with equation (5) and the biophysical data for the cell. Note that the cell shows a fast facilitatory peak at 25 and 50 msec and a second facilitatory peak at 200 msec.

resemble the plots of average firing rate obtained by Mazurksaya *in vivo.* This implies that the firiring rate curves are determined by some interaction between the properties of the cortical networks and the biophysical propertires of the cortical neurons. The analysis outlined here provides an initial estimate of the relative contributions of network and biophysical factors to firing properties of cortical neurons. The similarity of shape between the P(t) curves (which are derived from measurements of synaptic events) and the firing rate curves (obtained from extracellular records) suggests that the basic shapes of the firing rate curves are determined by the properties of the cortical networks because, the parameters a, b, G and V_0 in equation (4) are constants for an individual cell. The average firing rate is then a linear function of the time average of the P(t) curve *for an individual cell* . The shape of the firing rate curve is, thus, the same as the shape of the P(t) curve. However, the interaction between b, G and V_0 is non-linear, so that small variations in these parameters *between* cells will be important in determining the value of the average firing rate at any given interstimulus interval. Thus, variations in the biophysical properties within populations of cortical cells are likely to be of considerable functional significance.

We plan now to extend Mazurskaya's data to include an analysis of the instaneous firing rates of cells in turtle visual cortex using the presentation of natural stimuli in an intact eye-brain preparation *in vitro,* and to use the analytic methods outlined here to relate the detailed, time varying properties of cells to their firing rates. However, it should be pointed out that this approach holds some promise as a general method of investigating the functional significance of interactions between the biophysical properties of individual cells and the properties of the networks in which they are embedded.

This work was supported by PHS grants EY-68352 (PSU) and NS-17489 (NTS). Luis Smith was immensely helpful in developing computer programs. We are grateful for useful comments on the manuscript provided by Dr. Lina Massone.

References

Cosans, C. E. and Ulinski, P. S. 1990 Spatial organization of axons in turtle visual cortex: Intralamellar and interlamellar projections. J. Comp. Neurol. 296: 548 - 558.

Hall, W. C. and Ebner, F. F. 1970 Thalamotelencephalic projections in the turtle (*Pseudemys scripta*). J. Comp. Neurol. 140: 101 - 122.

Heller, S. B. and Ulinski, P. S. 1987 Morphology of geniculocortical axons in turtles of the genera *Pseudemys* and *Chrysemys*. Anat. Embryol. 175: 505 - 515.

Hounsgaard, J. and Nicholson, C. 1990 The isolated turtle brain and the physiology of neuronal circuits. In: *Preparations of Vertebrate Central Nervous system in*

Vitro, edited by H. Jahnsen. New York: Wiley, pp. 155 - 182.

Kriegstein, A. R. 1987 Synaptic responses of cortical pyramidal neurons to light stimulation in the isolated turtle visual system. J. Neurosci. 7: 2488 - 2492.

Larson-Prior, L. J., Ulinski, P. S. and Slater, N. T. 1991 Excitatory amino acid receptor-mediated transmission in geniculocortical and intracortical pathways within visual cortex. J. Neurophysiol. 66: 293 - 306.

Mazurskaya, P. Z. 1974 Organization of receptive fields in the forebrain of *Emys orbicularis*. Neurosci. Behav. Physiol. 7: 311 - 318.

Mulligan, K. A. and Ulinski, P. S. 1990 Organization of geniculocortical projections in turtles: Isoazimuth lamellae in the visual cortex. J. Comp. Neurol. 296: 531 - 547.

Ulinski, P. S. 1990 Cerebral cortex of reptiles. In: *Cerebral Cortex*, edited by E. G. Jones and A. Peters. New York, Plenum Press, vol. 8A, pp 139 - 221.

Ulinski, P. S., Larson-Prior, L.J. and Slater, N. T. 1991a Cortical circuitry underlying visual motion analysis in turtles. In: *Visual Structures and Integrated Functions*, edited by M. A. Arbib and J-P. Ewert. Springer-Verlag, New York, pp 307 -324.

Ulinski, P. S., Larson-Prior, L. J. and Slater, N. T. 1991b Synaptic properties of neurons in turtle visual cortex in an apparent motion paradigm studied *in vitro*. Soc. Neurosci. Abstr., 17: 114.

Ulinski, P. S. and Nautiyal, J. 1987 Organization of retinogeniculate projections in turtles of the genera *Pseudemys* and *Chrysemys*. J. Comp. Neurol. 276: 92 - 112.

SECTION 4

AUDITORY SYSTEM

James Simmons and Prestor Saillant describe a deterministic neural-network model using peripheral auditory signal processing stages that can encode the phase of sonar emissions and echoes to produce acoustic images very similar to those perceived by the echolocating bat.

Fred Rieke et al. suggest a novel approach to characterize information transfer in non-phase locked auditory afferents in the bullfrog sacculus using a choice between possible nonlinear functionals.

John Pearson et al. present a model of binaural intensity difference (BID) calculation in the inferior colliculus (IC) of the barn owl. The model explains the functional roles of the two marrowly frequency-tuned cell types in the lateral shell of the central nucleus of the IC and predicts the pattern of their connectivity. Simulations are consistent with the altered responses of neurons due to the injection of meuromodulators into the input nuclei of the IC.

Clay Spence and John Pearson present a neuronal resonance model to explain high frequency binaural phase sensitivity in the barn owl. See also **William Lytton** for another system that uses coincidence detector neurons (Other Sensory Systems).

Bryan Travis describes a computational model of the ventral division of the cat medial geniculate body. The model is calibrated by comparisons to physiological recordings.

Donal Sinex uses an auditory model to examine neural responses that underlie speech discrimination.

Bruce Parnas and Edwin Lewis model auditory brainstem neurons using an integrate-and-fire spike model with an intrinsic noise source. The paper is in Section 1 (Analysis).

Reconstruction of Target Images in the Sonar of Bats

James A. Simmons
Prestor A. Saillant
Department of Psychology and Section of Neurobiology
Brown University
Providence, RI 02912

Abstract

The echolocating bat, Eptesicus fuscus, *emits frequency-modulated (FM) sonar signals and perceives target range from echo delay. Reconstruction of the bat's image of a point-target from behavioral data yields approximately the ambiguity (crosscorrelation) function of echoes, with additional sharpening of the function's central peak according to the signal-to-noise ratio of echoes. Bats perceive a shift in the phase of this function even though phase conventionally is thought to be discarded during auditory encoding of sounds in the ultrasonic frequency band used by the bat (20 to 100 kHz). The perceived range of target images undergoes amplitude-latency trading, indicating that the timing of neural discharges encodes the delay of echoes, including echo phase information. The bat's auditory code resembles a spectrogram of the transmitted and the received signals in which the instantaneous frequency of FM sweeps is represented by the timing of neural discharges to each frequency in succession. A computational model of echo processing using physiologically identified peripheral auditory mechanisms indeed can produce the crosscorrelation function of echoes from a spectrogram representation. The model correlates transmission and echo spectrograms to determine target range, and, as a parallel process, it also can transform the echo spectrum back into the time domain to estimate target shape. The combined algorithms carry out the operations of Spectrogram Correlation and Transformation (SCAT). The critical feature of auditory coding for retention of phase information in the spectrograms is* half-wave-*rectification of excitation at the receptor-nerve interface.*

INTRODUCTION

Bats have evolved flight for locomotion and a kind of biological sonar, called *echolocation*, for orientation in the dark.[6] Their ability to both move about freely and to navigate at night confers upon them great biological success. Many species of bats emit frequency-modulated (FM) sonar sounds and perceive objects as multi-dimensional images derived from echoes of these sounds.[3,6,14] The bat's auditory system acts as the sonar receiver, coding the waveforms of transmitted and received sounds[16,24,25] to produce spatial images.[19] The relations among the bat's sonar signals, prominent aspects of auditory coding of these sounds, and the images that these sounds yield reveal general principles of perception in a strikingly well-defined manner. In particular, the integration of information represented in a distributed form across a population of neurons having diverse response characteristics is starkly highlighted in echolocation because both the stimuli (echoes of the bat's well-described emitted sounds) and the perceptions these stimuli produce (the bat's sonar images) are directly measurable, thus exposing the intervening operations to scrutiny. This chapter describes the experimental analysis and modeling of mechanisms for echo signal-processing by bats in the familiar context of target ranging. The results illustrate how the brain carries out computations whose end-product (sonar images having crosscorrelation-like features) is very well understood but whose implementation is highly instructive about alternatives to conventional signal-processing techniques.

EXPERIMENTS ON SONAR IMAGES IN BATS

Insectivorous bats use their sonar to detect, identify, and track flying prey in interception maneuvers.[6] The bat's actions during pursuit of targets are guided by a stream of images derived from echoes. Recent research describing the sonar images perceived by the big brown bat, *Eptesicus fuscus*, provides a new view of how the information in echoes comes to be portrayed in these images, with important implications for perceptual processes in general.[19,20] Fig. 1 shows the spectrogram of a sonar sound emitted by *Eptesicus*. All of the information used to produce target images is conveyed to the bat by echoes of such sounds, and there is an intimate connection between the structure of the transmitted waveform and the shape of the image (see below). *Eptesicus* transmits short-duration (0.5 to 10 ms) FM signals containing two prominent harmonic sweeps to guide its flight and to pursue flying insects. The bat emits these sounds in a stream at repetition-rates of 5 to over 100 Hz, and each sound yields an acoustic image depicting the location and

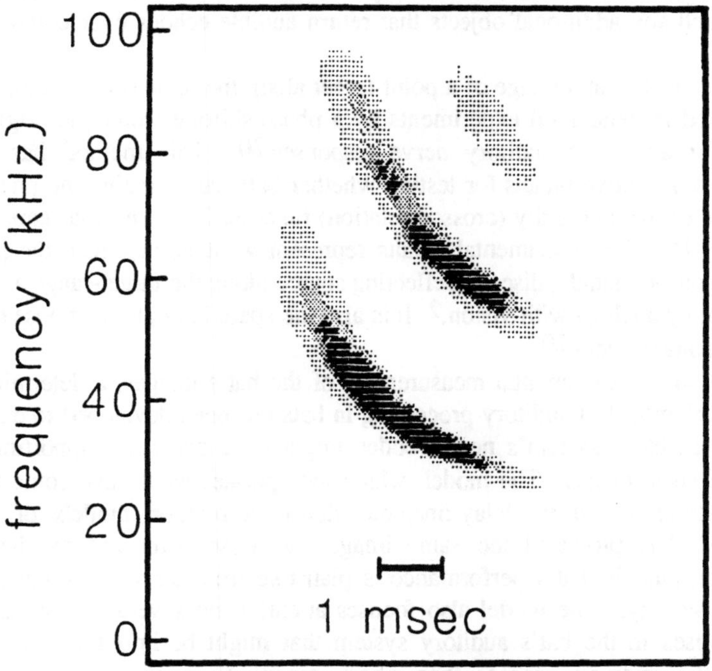

Fig. 1. **Spectrogram of a sonar sound emitted by the big brown bat, *Eptesicus fuscus*. The signal covers frequencies from about 23 to 95 kHz in two principal, harmonic FM sweeps.**

identifying features of objects located in the immediate environment.[6,19] Small, insect-sized objects are detected as far away as about 5 m, and the images probably include as well any additional objects that return audible echoes over comparable distances.

The shape of the bat's image of a point-target along the dimension of range has been measured in behavioral experiments by a phase-shifting technique originally developed for analyzing auditory nerve responses.[20] This method has been proposed as a definitive means for testing whether bats can perceive the phase or fine structure of the ambiguity (crosscorrelation) function between sonar emissions and echoes.[1,18] The experimental results represent what is, in effect, the point-spread function of a single, discrete reflecting source along the target-range axis for echolocation, by analogy with vision.[2] It is also the spatial impulse response of the bat's whole sonar system.[20]

In addition to experimental measurement of the bat's images, a deterministic neural-network model of auditory processing in bats has been developed to explore whether the auditory system's neural codes might be capable of supporting the formation of such images. The model, which incorporates what have come to be standard features found in delay-line/coincidence correlation models of pitch processing,[8,9] has produced the same images as those perceived by the bat, demonstrating that the bat's performance is plausible from current knowledge of auditory physiology. The model also focuses attention on several parameters of neural responses in the bat's auditory system that might be important for other aspects of the imaging process than phase sensitivity, such as carrying out frequency-to-time transforms for producing images of complex targets containing multiple reflecting sources.[22]

Method for Measuring Images

The experimental procedure for measuring the images perceived by the bat (Fig. 2A) presents bats with a choice between sonar echoes that *alternate* or *jitter* in their arrival-time by a controlled amount (Δt) from one emission to the next and echoes that *do not jitter*--are stationary--in arrival-time.[18,20] An electronic target simulator returns the bat's own sonar transmissions to its ears alternately at delays of a_1 or a_2 for the jittering condition and delay b for the stationary condition. The value of delay b, and the mean delay around which a_1 and a_2 are varied, nominally is 3.275 ms, which corresponds to a simulated target range of 56.47 cm. Bats (*Eptesicus fuscus*) were trained to discriminate the jittering from the nonjittering echoes for food reward (mealworm), which was offered when the bat crawled forward (*arrow* in Fig. 2A) towards the loudspeaker producing jittered echoes. The amount of jitter (Δt) was varied to assess the bats' performance (percentage correct or percentage errors) as a function of the amount of jitter, and the data appear as graphs of performance at various jitter values from zero to 50 μs.

The crucial technical feature of the simulator is that just one channel (left or right) is allowed to be active for any particular emitted sound (Fig. 2A). To compare echo delays for determining whether jitter is present, the bat must compare *remembered images* of the point-targets simulated by the echoes a_1, a_2 or b because only one stimulus echo is delivered for each sonar emission. During the experiment, the bat scans its head from left to right, aiming its sounds first at one simulated target and then at the other. A voltage comparator determines which microphone (*m* in Fig. 2A) receives the stronger signal from the bat's emission and activates only the loudspeaker (*s* in Fig. 2A) on the corresponding left or right side of the apparatus; the bat receives just one of the three stimulus echoes for that emission (a_1, a_2, or b) from the side (left or right) that it aims its sound towards. On the next emission, the bat receives a different echo depending on whether jitter is scheduled to be presented from that side of the simulator on that particular experimental trial. When the bat aims its head and sonar sounds toward the other channel, the echoes that are appropriate for that particular trial are delivered from the other loudspeaker. Due to this contingency, the bat's performance reflects its ability to retain an image from one sonar echo for comparison with the image from the next echo.[18,20] In perceiving jitter, the bat most likely detects changes in the delay of the stimulus echo in relation not only to the transmission but also to other echoes incidentally reflected by the apparatus, especially the loudspeakers and microphones. These other echoes do not actually overlap with the stimulus echoes, however, so spectral artifacts generated by overlap can be discarded as potential cues in the experiments (also, see below).

During experimental trials, the amplitudes of the sonar transmissions (see Fig. 1) of each bat were about 100 dB SPL p.-to-p. at the microphones, and the echoes were delivered at an amplitude about 30 dB below the level of the transmissions. In addition, small changes in echo amplitude were introduced at certain stages of the experiments to determine whether the bat perceived a different delay related to the amplitude change. The jitter discrimination data yield graphs that trace the probability of the bat perceiving a point-target at various ranges centered on the target's objective range, for different manipulations of the acoustic properties of echo a_1 or a_2. These graphs describe the apparent spread of the target's image along the range axis.[2,20] This type of psychophysical experiment, with jitter in the stimulus parameter of interest, offers advantages for relating performance to underlying signal-processing mechanisms because it yields a graph of the shape of the image itself.[2]

These experiments have measured not only the bat's ability to detect the presence of jitter in the delay of echoes, but they also have assessed the affects of echo phase-shifts on performance in the jitter-discrimination task. Earlier results indicated that the bat might perceive the phase of sonar echoes relative to its sonar transmissions,[12,18] and the 180° echo phase-shift introduced into the jitter

Fig. 2. (A) Diagram of two-choice discrimination procedure for studying delay perception by bats. Electronic target simulator picks up bat's sonar transmissions with microphones (*m*), delays the resulting electrical signals, and returns the delayed signals through loudspeakers (*s*) to the bat's ears.[8] Echo delay corresponds to desired target range. The bat is trained to choose between echoes (a_1, a_2) that jitter in arrival-time (Δt) and echoes that return with a fixed delay (*b*) by moving forward (arrow) for its reward towards the jittering echoes. The simulator determines whether the bat is aiming sounds at the left or the right microphone and activates only that particular channel to return echoes. The bat must remember the delay of one echo for comparison with that of the next echo to detect the jitter.

(B) Performance (percent correct) of three bats discriminating jitter intervals in nanoseconds.8 Bats perceive delay changes as small as 10 ns at 75% correct responses. Electronic delay increments were generated by delay-lines (solid data-points) or by varying cable lengths (open data-points). 40-60 trials per data-point.

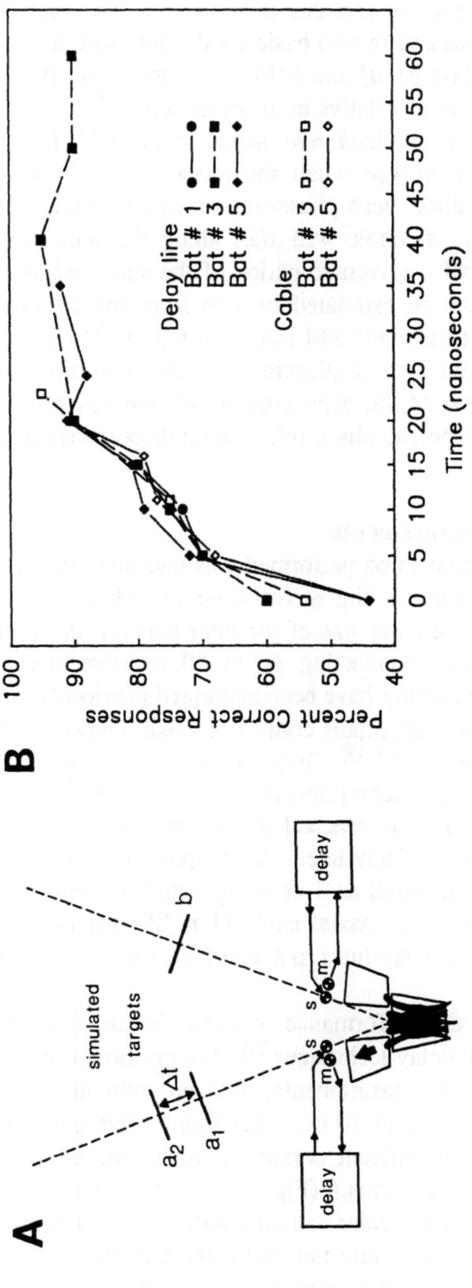
A
simulated
targets
a2
a1
Δt
b
m
s
s
m
delay
delay
B
Percent Correct Responses
Time (nanoseconds)
Delay line
Bat # 1
Bat # 3
Bat # 5
Cable
Bat # 3
Bat # 5
100
90
80
70
60
50
40
0
5
10
15
20
25
30
35
40
45
50
55
60

experiment was designed to test this possibility.[1] The bats' jitter-discrimination performance was measured in two basic conditions--with a fixed (0°) phase for all echoes arriving at delays a_1, a_2 and b (Fig. 2A), and with a phase-shift (180°) of the echoes arriving at delay a_1 relative to the other echoes.[20] If the bat is insensitive to phase--that is, uses a *semicoherent* sonar receiver,[11] its jitter discrimination performance will *not change* when the phase of a_1 is shifted relative to a_2. However, if the bat does sense phase--uses a *coherent* receiver,[11] the bat's jitter performance curve as a whole will *shift* along the time scale of jitter (Δt) by approximately one-half the average period of the sonar signals, which is 30 to 35 µs. (This period can be estimated by eye from the reciprocal of the average frequency in the first-harmonic FM sweep in Fig. 1.)[23] That is, the echo phase-shift will be perceived as a displacement in the time-of-occurrence of the echo according to the shape of the echo crosscorrelation function, which displays the "pulse-compressed" time and phase information theoretically available to determine echo delay.

Results of Image Measurements

The first experiment to be performed was measurement of jitter acuity, or the smallest detectable jitter in the arrival-time of echoes a_1 and a_2. This was accomplished by reducing the size of the jitter interval (Δt) in small steps from 50 µs down to zero while conducting 40 to 60 experimental trials at each step. Measurements of jitter acuity have been attempted previously with *Eptesicus*, but in each case the electronic apparatus could not make delay changes in steps smaller than about 0.4 to 0.5 µs.[12,13,18] These previous experiments found that bats were in fact able to discriminate echo jitters at least as small as 0.4 to 0.5 µs, but the bat's true perceptual limit was not reached due to the relatively coarse minimum jitter step-size provided by the delay-lines. In the present experiment, calibrated delay steps could be made as small as 3 ns using either lumped-constant analog delay-lines or variable lengths of coaxial cable (1 ns/30 cm) used in conjunction with longer digital or analog delay-lines to achieve the total desired echo delay of about 3 ms.[20]

Fig. 2B shows the performance of three *Eptesicus* detecting jitter intervals produced by different delay techniques.[20] The performance of these bats is stable over multiple series of measurements, with an estimated delay threshold (75% correct performance) of about 10 ns. This figure represents the bat's *hyperacuity* for echo delay in quiet ambient conditions under the extremely well-controlled circumstances of the experiment (Fig. 2A). The bat's "ordinary" delay acuity manifested in typical tasks under natural conditions presumably is not so extreme because conditions will be more uncertain, the bat will be moving, and noise is likely to be present. However, even this practical delay accuracy surely will be much better than the commonly-accepted figure of roughly 10 to 100 µs derived from earlier target-range discrimination experiments,[16,24,25] which, unfortunately, are unsatisfactory for measuring the bat's fine echo-delay accuracy owing to the bat's head movements during trials.[21] The bat's 10-ns echo-delay hyperacuity reflects the fundamental significance of specifying echo delay for its acoustic imaging system as a whole.[19,22] Control procedures and separate experiments (for example, see Fig. 3C, below) confirm that an auditory temporal representation of the delay of echoes itself cues the bat's responses rather than an amplitude or spectral effect of the delay-lines, or overlap of stimulus echoes with either the bat's transmissions or with any stray echoes from other objects located near the bat.[20] Two of the bats that were tested further (#3 and 5) achieved 40 ns jitter discrimination in controlled broadband noise where the echo signal-to-noise ratio $(2E/N_0)^{1/2}$ = 36 dB, thus approximating the performance of an ideal, coherent (phase-sensitive) sonar receiver.[11,20,]

The limiting accuracy of 10 ns for detecting changes in echo delay could not be achieved without explicit incorporation of the phase of echoes relative to transmissions into the images. A semicoherent (phase-insensitive) sonar receiver would perform about a factor of 10 worse than the bat in the presence of controlled noise,[20] so the use of phase or phase-like information is necessary to account for the results. Can the incorporation of phase information into the images be demonstrated by a more direct means, without recourse to so seemingly incredible an observation as the bat's perceiving a 10-ns shift in echo delay? The purpose of the phase-shift experiment was to show whether *Eptesicus* possesses phase sensitivity using data collected along a more convincing time scale of microseconds rather than nanoseconds.

The effects of changing the phase of echoes on the performance of *Eptesicus* are manifested in data collected for a much larger scale of jitter intervals than appears in Fig. 2B for fine acuity. For the 0° phase-shift condition (a_1 and a_2 are "in phase" with each other), Fig. 3A shows that *Eptesicus* readily perceives jitters as small as 5 μs, only making chance performance (50 percent errors) at a jitter of zero. (Fig. 2B shows how much the jitter interval has to be reduced below 5 μs before the bats fail to detect that jitter is present.) In addition to the expected poor performance at zero, the bats make a significant number of errors at jitters in the region of 30 to 35 μs (0° curve in Fig. 3A). *Eptesicus* evidently finds it *easier* to detect jitters of only 5, 10, or 15 μs than slightly larger jitters of 30 or 35 μs. The region of poorer performance at 30 to 35 μs was observed in the original echo jitter experiments with *Eptesicus* and served as the first evidence that bats might be able to perceive the phase of the ambiguity function of echoes.[18] This effect was missed in two recent experiments, however,[12,13] apparently owing to an unintended masking effect related to stimulus presentation in these particular replications of the original jitter procedure.[20]

Fig. 3. (**A**) Performance (percent errors) of four bats discriminating jittering echoes in the presence of a 0° or 180° phase shift of echo a1 relative to a2. Horizontal axis is jitter (Δt), and vertical axis combines 0° data (upward from zero) with 180° data (downward from zero) on same plot. 40-60 trials per data-point.[8]

(**B**) Compound performance curves derived from **A** by connecting extreme points above and below zero on vertical axis.

(**C**) Example of echo amplitude change (-1 dB on echo a_1) inducing a latency shift of whole compound performance curve. Bat's estimate of delay (location of image) moves to right due to neural response-latencies for echo a_1 relative to a_2 being retarded by amplitude-latency trading when echo amplitude decreases by 1 dB. 60 trials per data-point.[8]

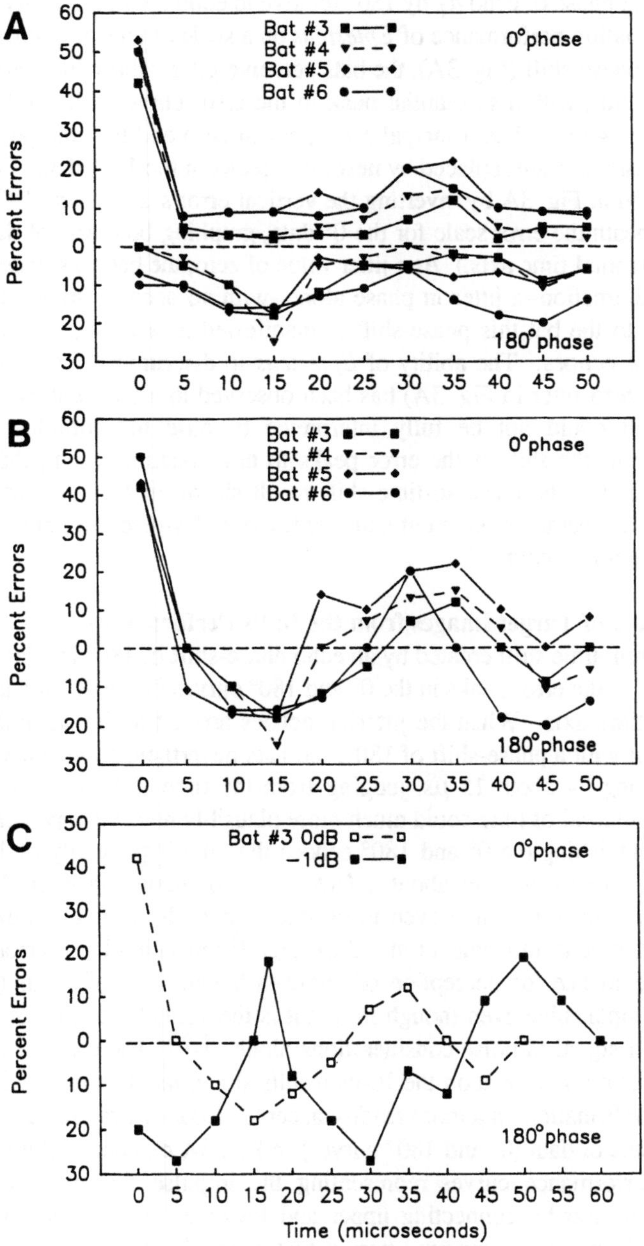
A
Bat #3
Bat #4
Bat #5
Bat #6
0°phase
180°phase
Percent Errors
B
Bat #3
Bat #4
Bat #5
Bat #6
0°phase
180°phase
Percent Errors
C
Bat #3 0dB
−1dB
0°phase
180°phase
Percent Errors
Time (microseconds)

Shifting the phase of echo a_1 by 180° relative to echo a_2 dramatically affects the jitter discrimination performance of *Eptesicus* on a scale of tens of microseconds.[20] For the 180° phase-shift (Fig. 3A), the bats perceive jitter values differently than for the 0° phase-shift, with a substantial peak in the error curve at 15 to 20 μs and a smaller peak at 45 μs. The principal error peak at zero and the side-peak at 30-35 μs in the 0° data thus are replaced by new error peaks in the 180° data. (These 180° data are plotted in Fig. 3A by inverting the vertical error scale for the 180° data and placing it beneath the error scale for the 0° data, graphing both sets of data along a common horizontal time axis.) At a jitter value of zero, the bats easily detect a pure 180° phase alternation--a jitter in phase alone--with no accompanying delay jitter. Furthermore, to the bat this phase-shift is manifested as a change in the apparent arrival-time of echoes. The ability of *Eptesicus* to discriminate echo phase-shifts alone (as for zero jitter in Fig. 3A) has been observed in a different experiment,[12] but this result could not be fully interpreted because the stimulus conditions required to show the shift of the error peaks to new locations along the time axis were not used.[20] The phase-to-time-shift result shown in Fig. 3A has now been confirmed in a separate experiment using ordinary delay-discrimination techniques without the jitter procedure.[10]

Reconstruction of Target Images from the Bat's Performance

The apparent time shift created by an echo phase-shift of 180° is relatively large. The locations of the error peaks in the 0° and 180° curves in Fig. 3A are about 15 μs apart on the time axis. When the jittering echoes arrived at the same delay (zero time jitter) but with a phase-shift of 180°, the bats nevertheless evidently perceived them as jittering by about 15 μs, judging from the time shift in the error peaks. Such a large interval of time could much more plausibly have been perceived by the bats for distinguishing the 0° and 180° echoes than the hyperacuity of 10 ns (Fig. 2B). That is, 15 μs is only about a factor of two smaller than the best delay-discrimination values obtained even in ordinary range discrimination experiments where the bat's head movements introduce significant echo-delay errors.[21] The other direct evidence for perception of phase--a hyperacuity of 10 ns (Fig. 2B)--might seem implausible even though in fact it is theoretically attainable from echo waveform and signal-to-noise considerations alone.[11,20] The case for echo phase perception thus does not rest on the 10-ns results alone; the data in Fig. 3A provide the desired confirmation on a more readily-accepted, coarser time scale.

The two sets of data (0° and 180° curves) in Fig. 3A can be combined to create compound performance curves representing the impulse response of the sonar system of *Eptesicus* by connecting upper and lower extreme points in a method analogous to forming a compound period histogram.[20] The resulting compound curves in Fig. 3B show the response of the bat's sonar system taken as a whole (transmitter, signals, and receiver) for determining the range to a point-target or single acoustic "glint" whose echoes arrive at a total delay (amounting to about 3 ms) corresponding to the zero origin of the time axis. The most probable location for the bat to perceive the target is at its objective range of about 56 cm for an echo delay of about 3 ms (corresponding to the main peak in the error curves at zero on the time axis in Fig. 3). In addition, there is a region of reduced but still significant probable perceived locations about 0.5-0.6 cm further away in range, at delays of 30 to 35 μs (corresponding to the side peak in the error curves in Fig. 3).

Reconstruction of the bat's image of the point-target can be carried further by folding the compound curves in Fig. 3B around the zero origin of the time axis to form a symmetrical graph.[20] Fig. 4B shows the average of the four compound curves from Fig. 3B arranged in this fashion. This final curve is a depiction of the image from all of the data obtained in Fig. 3. It still more dramatically shows the location of the central estimate of echo delay surrounded by two side estimates located earlier and later by 30-35 μs. The whole jitter performance curve assembled from 0° and 180° phase-shift data (Fig. 4B) now resembles the crosscorrelation function between the bat's sonar transmissions and the echoes (Fig. 4A). The pattern of a central peak and two principal side peaks in the crosscorrelation function is reproduced in the bat's behavior, indicating that the stored images of the alternating echoes in the jitter task contain the information represented by the fine phase structure of the crosscorrelation function.[20] Reconstitution of the echo crosscorrelation function entirely from the bat's performance in the jitter-discrimination task is a convincing demonstration that the bat uses a coherent sonar receiver to produce its images of targets.

TIME-FREQUENCY CODING OF ECHOES

The mammalian auditory system does not preserve the phase of sounds at frequencies above several kiloHertz, at least in terms of the repetitive phase-locking of neural discharges to cycles of a sinusoidal, pure-tone stimulus. It is therefore generally thought that bats cannot perceive the phase of echoes relative to transmissions because they use frequencies above 20 kHz. However, the performance of *Eptesicus* in jitter-discrimination and echo phase-shift experiments seems to require that the bat retain an image that incorporates the phase of echoes relative to transmissions (see Fig. 4). That is, reconstruction of the full crosscorrelation function of echoes requires knowledge of the true phase-shift of echoes relative to transmissions to specify the function's fine structure.[1]

The problem posed by the data in Figs. 3A and 3B is whether the complete crosscorrelation function of echoes indeed can be recovered from the type of neural representation of FM sounds thought to be used by the bat's auditory system. *Eptesicus* and other FM bats register the instantaneous frequency in the FM sweeps

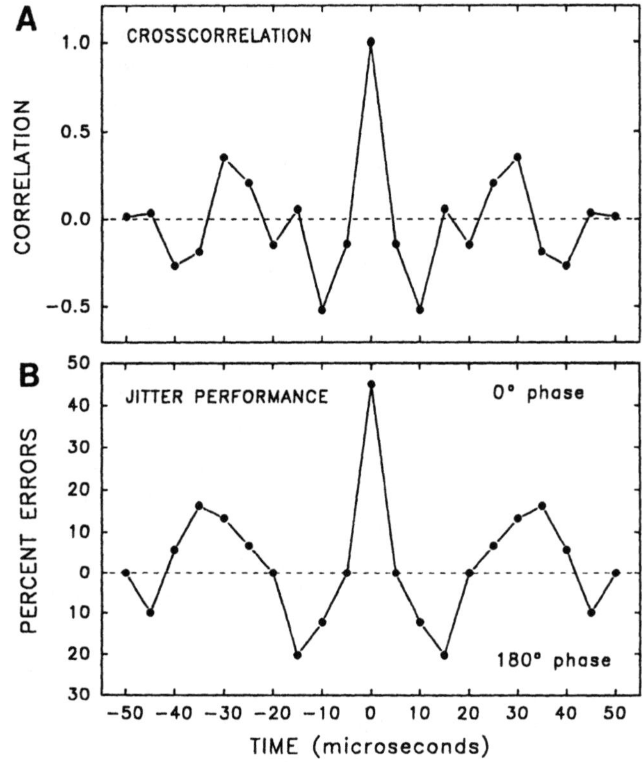

Fig. 4. (**A**) Typical crosscorrelation function between the bat's sonar transmissions (Fig. 1) and echoes from an ideal point-target (sampling interval 5 μs). (**B**) Mean compound performance curve (from curves in Fig. 3**B**) arranged symmetrically around zero time. Curve represents location of bat's estimate of echo delay at true delay (zero), with shape of whole image extending along delay or range axis. Small differences between curves in A and B are caused by simulator's loudspeaker frequency response. Echo delay hyperacuity (Fig. 2B) is manifested as a narrowing of the width of the central peak in the jitter data to ± 10 ns.

of sonar signals and echoes (see Fig. 1) by the timing of neural discharges to successive frequencies in the FM sweep.[16,17,24,25] Fig. 5 schematically illustrates the pattern of discharges that might occur in parallel neural paths within the bat's auditory system when stimulated by sounds reaching the bat's ears during echolocation. The diagram is a cartoon derived from physiological data obtained in *Eptesicus* and other species of FM bats.[4,7,16] Here, the peripheral auditory nervous system is shown schematically as thirty neural "channels" tuned to different ultrasonic frequencies from 20 to 100 kHz (vertical axis). The origin of band-pass filtering in each of these channels is the frequency tuning associated with the site of the receptor cells for each neural channel along the bat's cochlea, from its base (high frequency) to its apex (low frequency).[16,24,25] This is the familiar "place" dimension of auditory coding, in which stimulus frequency is represented by the location of activity across the tonotopic axis of the auditory pathway.

The stimuli for echolocation consist of *pairs* of sounds--the sonar transmission coupled with each of its echoes. In Fig. 5, the bat's FM sonar transmission (S_0) produces a volley of neural on-discharges (dots) in different frequency channels after being picked up by the bat's ears at the moment it is broadcast. When an echo (S_1) is received, it evokes a similar volley of on-discharges offset in time (to the right in Fig. 5) by the echo's delay (Δt_1). The volley of discharges evoked by the echo carries all the information available about the target, but it must be processed with reference to the volley evoked by the transmission to recover this information in a usable form. For example, both the delay and the spectrum of the echo signify specific features of the target, but they are defined respectively in terms of the time at which the transmission occurs and the spectrum of the transmission. The time separation of the discharge volleys in Fig. 5 constitutes the familiar time or "periodicity" dimension of auditory coding, in which the temporal organization of the stimulus is represented by the time-intervals between successive discharges.

The arrangement of neural discharges across the dimensions of time and frequency in each volley depicts the shape of the FM sweeps for the first and second harmonics in each sound; it thus functions as a kind of point-by-point spectrogram. The curvilinear sweeps that are present in the actual sound (Fig. 1) are "straightened" (sloping dashed lines in Fig. 5) by the correspondingly curvilinear scale of frequency along the bat's cochlea (dual vertical axes in Fig. 5). Observations of tonotopic representations in the auditory system of *Eptesicus* suggest that the frequency scale is approximately *hyperbolic*, or *linear with period*.[22] In addition to the curvilinear frequency scale in Fig. 5, note that the discharges (dots) for the two harmonics in the transmitted sound (S_0) do not exactly align with the sloping dashed lines tracing the FM sweeps. The discharges actually fall progressively later than the sweeps in the stimulus itself as the sound's frequency sweeps from high to low (going upward in Fig. 5). The slight retardation of neural responses at lower frequencies reflects the cochlea's action as a

Fig. 5. Schematic diagram showing arrangement of neural discharges (*dots*) evoked by a sonar transmission (S_0) and two sonar echoes (S_1, S_2) at different delays. Vertical frequency scale is expressed in terms of the location of excitation along the basilar membrane of the Organ of Corti, from the base of the cochlea to the apex. These neural spectrograms illustrate the pattern of on-discharges representing different excitatory frequencies in the FM sweeps of the sounds.[16,17] Sweeps are shown by sloping dashed lines, and the neural discharges lag behind the sweeps (to the right) at progressively lower frequencies (upward) as a consequence of the delay characteristics of traveling waves in the cochlea, a kind of phase distortion that is shared by all the signals. The delay of each echo is measured by the time that elapses between discharges to the transmission (S_0) and to the echo (S_1 or S_2). The shape of the distribution of delay estimates across frequencies defines the image of the target in terms of echo delay (Δt_1 or Δt_2). The objective delay of each echo is shown by a solid vertical line. The amplitude of echo S_2 is reduced by a small amount relative to echo S_1, retarding the latency of the discharges to S_2 relative to the discharges to S_1. The echo-delay estimate for S_2 (Δt_2) thus is artificially longer than the estimate for S_1 by an error (δt_2) equivalent to amplitude-latency trading.

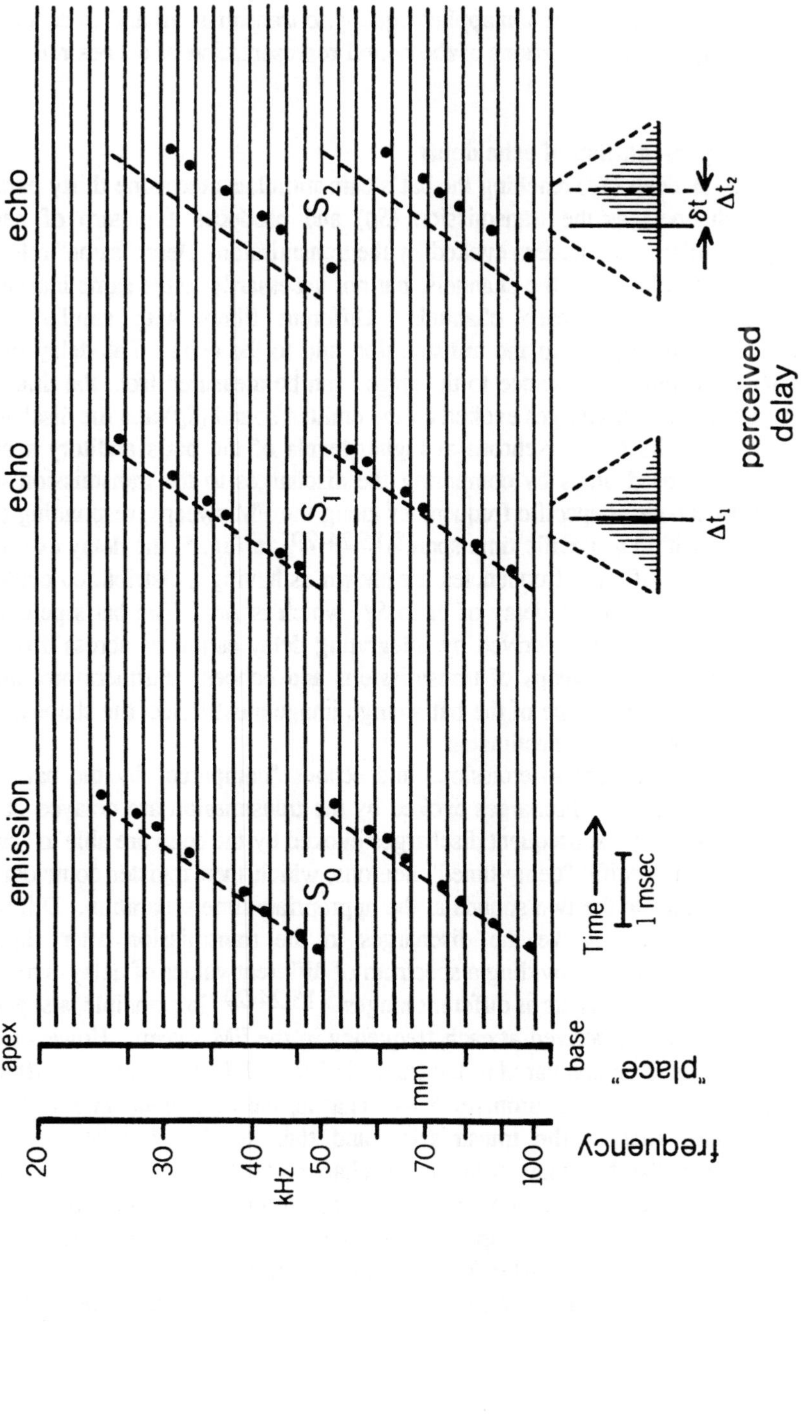

emission
echo
echo
S0
S1
S2
perceived delay
Δt1
δt
Δt2
Time
1 msec
apex
base
"place"
mm
frequency
kHz
20
30
40
50
70
100

frequency-dispersive delay-line, with high frequencies producing "outputs" before low frequencies. This dispersion applies equally to transmissions and to echoes, so there are no significant distortions of the neural representation of echoes relative to transmissions.

Temporal representation of echo delay

In Fig. 5, an echo (S_1) reaching the bat's ears stimulates the same thirty parallel frequency channels as the transmission (S_0) and produces a pattern of neural discharges similar to the pattern evoked by the transmission. Here, "same" refers to the frequencies defined for the channels, but not necessarily to the same individual neurons as constituting each channel. Different fibers with similar tuned frequencies could respond to the transmission and to the echo. The delay of the echo (Δt_1), and thus the distance to the target, can be measured from the time that elapses between the discharges evoked by the emitted sound (S_0) and the discharges evoked by the echo (S_1). Neurons at higher levels of the bat's auditory system extract these elapsed times by comparing the discharges to the transmission with discharges to echoes at specific frequencies along the FM sweeps, responding only if the delay is within a specific time span.[5,15,24-29] In Fig. 5, the delay estimates developed across different frequencies are shown as having a distribution centered on the best estimate of the delay of echo S_1, which is Δt_1. The bat's perceived estimate of target range is derived by integrating delay estimates across different frequencies in the FM sweeps of transmissions and echoes. Furthermore, under limiting conditions, the shape of the bat's range image must reflect the shape of the underlying distribution of Δt estimates.

How are discharges to emissions and echoes "compared" by the bat? To measure echo delay, the discharges evoked by the transmission are delayed in the nervous system so that subsequent discharges evoked by the echo are able to "catch up" with them at specific "delay-tuned" neurons, which then register coincidences between discharges to the two sounds at the appropriate time separation. Different delay-tuned neurons receive the discharges to the transmission after slightly different delay intervals, providing a spectrum of different values of delay-tuning to encode echo delay for targets at different ranges.[5,15,24-29] In principle, a separate echo-delay estimate is derived at each frequency in the FM sweep. However, the exact frequencies to be compared in the transmission and the echo may be different for any given delay-tuned neuron.[15,24-29] That is, the *horizontal lines* in Fig. 5 show responses first to the transmission and then to the echo at the *same* frequencies, but the bat may compare discharges to the transmission at one frequency with discharges to the echo at another frequency. Consequently, the neural delays that collectively specify the arrival-time of a given echo have individual delay values that differ from that of the whole echo they specify. Each neural delay must take into account the potentially disjoint relation between the location of the starting frequency in the transmission and the ending frequency in the echo at different points along the FM sweeps. Many of the transmission-echo coincidences thus would be better described by sloping rather than horizontal lines in Fig. 5. Due to registration of a coincidence only for a critical range of echo-delay values, many neurons at higher levels of the bat's auditory system respond selectively to specific delays. Such a delay-and-coincidence mechanism has previously been proposed to account for neural processing of periodicity pitch,[9] and it has been implemented in models of pitch processing.[8] The model of echo processing described below makes delay estimates at each frequency in the FM sweep of the transmission and each echo, grouping the discharges to each echo into the responses of hypothetical "neurons" that are selectively "tuned" to specific echo delays and specific FM sweep-rates, both for computational convenience and to utilize the capacity to recognize sweep characteristics as well as echo delays as shown in real neural responses.

Amplitude-Latency Trading as a Test for Temporal Coding of Echo Delay

The behavioral procedures already used to reconstruct the bat's image of a target (Figs. 3A and 3B) can be used as well to demonstrate that the timing of neural discharges carries the critical information required to perceive target range. The latency of neural responses to sounds depends on stimulus amplitude, and changes in latency with changes in amplitude are easily measured. In *Eptesicus*, auditory responses to echo-like FM sounds are retarded by about 15 to 18 μs for each decibel of decrease in stimulus amplitude.[22] In another species of FM bat, single-neuron discharges to FM sounds are also retarded by about 17 to 20 μs/dB at the stimulus levels used in the jitter experiments with *Eptesicus*.[17] In Fig. 5, the second echo (S_2) is shown as slightly weaker than the first echo (S_1), so the neural discharges it evokes are delayed by the affects of stimulus amplitude on latency as well as being delayed by the sound's propagation through the air. The bat's perception of the delay of S_2, which is based on these neural discharges and the corresponding distribution of delay estimates centered at Δt_2, accordingly will be made longer by a certain error interval (δt) past the objective delay of the echoes.

A critical test of whether the jitter results in Figs. 3A and 3B truly represent *temporally-coded* echo-delay perception by the bats is to determine whether the bat perceives echo delay to lengthen by 15 to 18 μs for each decibel of decrease in echo amplitude. That is, does amplitude-latency trading occur for the bat's *perceptions* as well as for its *neural discharges*? Fig. 3C shows the results obtained when the jitter experiments are carried out with a decrease of 1 dB in the amplitude of echo a_1 relative to echo a_2. The curves reveal a lengthening (rightward) shift of about 17 μs in the perceived delay of the image of a point-target, confirming predictions that amplitude-latency trading indeed should affect the bat's images.[17,20]

In Fig. 3C, not only does perceived delay undergo amplitude-latency trading, but the whole pattern of primary and secondary error peaks--the fine structure of the crosscorrelation function reconstructed from phase-shift experiments--slides to a later time by 17 μs. This result reveals that the bat encodes all the information it uses to determine echo delay, including the phase of echoes relative to transmissions, in the timing of the discharges evoked by different frequencies in the FM sweeps. The critical computations for producing the images (assembling the distribution for Δt_1 or Δt_2 in Fig. 5) must therefore take place in the *time domain* on events registering the different frequencies in the FM sweeps of emissions and echoes. The role of the *frequency-domain* representation is to define the various auditory channels that contain this time-domain information, thus converting the computationally-daunting problem of classical delay-and-multiply crosscorrelation in the time domain into the simpler task of time-coincidence detection across frequencies (Fig. 5).

Relation between Time and Frequency Codes

The highly purified perceptual situation supplied by the jitter-discrimination task thus offers an important lesson about the functional significance of the two primary dimensions of auditory coding--the *place* code for the tonotopic representation of frequency and the *temporal* code for the fine structure of complex sound waveforms over time. The initial separation of different frequencies into parallel auditory channels can be thought of as signal conditioning or *preprocessing* for subsequent analysis of the time structure of sounds. This conclusion is generally consistent with the success of parallel time-domain processes for reproducing crucial aspects of pitch perception.[8,9] However, it does not mean that the tonotopic representation developed at the periphery of the auditory system and preserved throughout the auditory pathways is only of peripheral significance as a substrate for time-domain processing. The content of the frequency representation appears as a specific component in the bat's images of targets.

In *Eptesicus*, information separately represented in terms of frequency about the spectrum of sonar echoes is explicitly used to encode the shape of targets.[19] The small size of flying insects (with linear dimensions of at most a few centimeters) insures that the time-scale of shape-like aspects of echoes is in the range of 0 to an absolute maximum of only about 100 to 200 μs, while the integration-time for echo reception by *Eptesicus* is 300 to 400 μs.[19,22] Furthermore, the bat's sonar signals are considerably longer even than that (Fig. 1). Consequently, the delay separation of echoes from different parts of complex targets can only be recovered from the interference spectrum of echoes, which is represented along the tonotopic axis of the auditory system.[16] However, *Eptesicus* perceives this delay separation in psychological units of *range* or, implicitly, *time*, which is only possible if the spectrum is transformed back into an estimate of the original distance or delay separation that gave rise to the observed interference spectrum.[22] Thus, the role of the frequency domain goes beyond preprocessing of stimuli. The frequency domain is central to the sophisticated nature of auditory representations for echolocation, even though its contents may be expressed along the range axis (that is, in the time domain) for perception.

COMPUTATIONAL MODEL OF ECHO-DELAY PROCESSING

The schematic diagram showing the neural spectrograms in Fig. 5 is a qualitative cartoon that can be expanded into a quantitative model of reception and coding of sonar transmissions and echoes by *Eptesicus*. The resulting computational model serves to demonstrate that the type of image perceived by the bat indeed can plausibly be created using neural representations known to occur in the mammalian auditory system. The model as a whole describes two parallel processes that the bat carries out in its auditory system: First, the delay of echoes is estimated from the time-intervals between volleys of neural discharges to the transmission and to echoes.[19,20,22] Because these volleys of discharges really are neural spectrograms of the sounds, this first process is called *spectrogram correlation*.[1] Second, the spectrum of echoes is converted into an estimate of the time separation of the echo components reflected by different parts of the same target. This process is *spectrogram transformation*. The complete model implements both processes and is called Spectrogram Correlation and Transformation (SCAT). This chapter focuses on the capacity of the spectrogram correlation stage to produce coherent sonar images of point-targets similar to those perceived by *Eptesicus* (Figs. 3 and 4).

In the present context, the SCAT model shows that the arrival-time and phase of broadband, high-frequency FM sounds can be encoded from the time-frequency structure of their waveforms entirely in spectrograms such as those illustrated in Fig. 5. Theoretically, the wideband sonar ambiguity function or echo crosscorrelation function can be recovered, phase and all, even at the ultrasonic frequencies used by *Eptesicus*, from the instantaneous frequency of their FM sweeps if some means are used to encode the real, constant phase-shift of echoes relative to transmissions.[1] Are the bat's neural spectrograms capable of capturing this echo phase-shift in the pattern of neural discharges?

Structure of the Model

Auditory processing for echo delay first separates the frequency span of the FM signals into numerous parallel frequency bands with an array of band-pass filters analogous to those found in the cochlea.[16,24,25] Except for details such as the actual ultrasonic frequencies of the filters, this aspect of the sonar SCAT model is comparable to the input stage of models for periodicity pitch.[8,9] The computational sonar model uses 81 parallel band-pass filters (10th-order Butterworth filters) with center frequencies from 20 to 100 kHz, which cover the band used by *Eptesicus* for echolocation (Fig. 1). The filters have fixed 3-dB bandwidths of 4 kHz across all frequencies, resulting in sharper tuning at higher frequencies and also overlapping of frequency-response curves for filters whose center frequencies are near each other. Sharpness of the model's band-pass tuning ranges from Q = 5 at 20 kHz to Q = 25 at 100 kHz. Each band-pass filter is followed by half-wave rectification and low-pass filtering (first-order filter at anywhere from 800 Hz to 3 kHz as desired) to supply the smoothed envelope of the filtered segment of the FM sonar signal in each auditory frequency channel. This configuration is intended to mimic the excitatory action of the auditory receptors in driving their associated auditory-nerve fibers.

The frequency scale defined by the filter array is *hyperbolic* with frequency, or linear with the *period* of the filter frequency. That is, the filters are spaced at intervals of 0.5 μs of period, from a period of 10 μs (at 100 kHz) to a period of 50 μs (at 20 kHz). Fig. 6 shows the smoothed envelope at the output of each filter channel for an FM sonar transmission and for two echoes arriving at delays of 2.3 and 4.6 ms. The transmitted sound is 2 ms in duration and sweeps downward hyperbolically with frequency from 100 kHz to 20 kHz. Such a signal sweeps linearly upward with period from 10 to 50 μs and is called a *linear period modulated* (LPM) signal.

In Fig. 6, vertical cursor lines delineate the beginning and end of the transmission at the input to the filter bank. The envelopes developed from the transmitted sound by the 81 band-pass filters appear as a ridge of amplitude that slopes downward to the right. This ridge is slightly delayed (offset to the right) relative to the vertical cursor lines as a consequence of the response-times of the band-pass and low-pass filters being excited in succession by the FM signal sweeping into their response areas. The two echoes at different delays are replicas of the FM transmission, and each echo is represented by envelopes forming a sloping ridge of amplitude at a corresponding delay after the ridge representing the transmission. The first echo (2.3 ms delay) is "in phase" with the transmission (0° phase-shift), while the second echo (4.6 ms delay) is "out of phase" with the transmission (180° phase-shift). In Fig. 6, the ridges formed by the envelopes derived from the FM transmission and both echoes have crests falling on straight lines because the hyperbolic shape of the FM sweeps and the hyperbolic frequency scale of the band-pass filters complement each other. The overall representation of the transmission-echo series is a spectrogram or time-frequency plot whose integration-time is given by the width of the envelopes along the horizontal dimension.

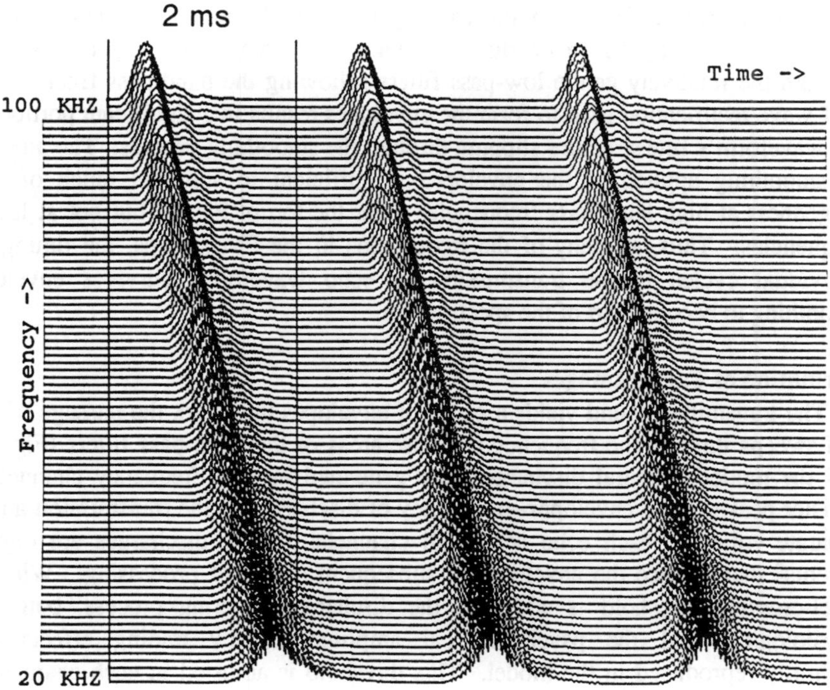

Fig. 6. Envelopes produced by 81 parallel band-pass filters in the SCAT model in response to a sonar transmission and two echoes arriving at different delays (2.3 ms and 4.6 ms), approximately as in Fig. 5. The FM transmission sweeps from 110 down to 15 kHz in 2 ms, and each echo is a delayed replica of this waveform, with a 0° phase-shift for the first echo and a 180° phase-shift for the second echo. The band-limited segments of the FM sweeps are half-wave rectified and smoothed (1 kHz first-order low-pass filters) by physiologically reasonable stages. Note that the degree of smoothing is gentle enough for high-frequency components to survive as ripple in the envelopes, especially at the lower frequencies. The SCAT model detects this ripple as part of the process of triggering neural discharges.

One additional feature of the envelopes in Fig. 6 deserves attention: Each envelope has a jagged shape due to "leakage" of high-frequency components through the relatively gentle low-pass filters following the band-pass filters. The half-wave rectification used to form the envelopes results in a significant portion of the waveform's fine structure slipping through the smoothing process. The amount of smoothing applied to the envelopes is sufficient to suppress much of this jaggedness at high band-pass frequencies (near the top of Fig. 6) but not at lower frequencies. Thus, at many of the frequencies in the FM sweeps, the timing of individual cycles in the transmission or echo waveforms survives envelope smoothing to become part of the spectrogram.

Formation of Target Images

The transmitted sound yields a series of envelopes that mark the passage of the FM sweeps through the frequency-response areas of the band-pass filters (Fig. 6), and the model's "neural discharges" are generated in each frequency channel to register peaks in the envelopes--one volley of discharges to the transmission and a separate volley to each echo across all channels. These discharges are digital "events" that replace the envelopes in subsequent stages of processing. When a level-crossing threshold is added to the discharge-triggering process, both the physiological and the psychophysical amplitude-latency trading effects are faithfully reproduced in the model. This threshold is adjusted in each channel to adapt the model to the level of ambient noise present in the environment. The exact amplitude-latency trading ratio in μs/dB is determined by interaction of the rate-of-sweep of the FM sound and the integration-time for smoothing the envelope from each band-pass filter. To measure echo delay, a type of neural-network autocorrelator is used,[9] modified to have adaptive sensitivity to different FM sweep-rates. Discharges to the transmission are delayed by various amounts and then applied to one of the two inputs of coincidence-detecting "neurons" which represent FM-FM delay-tuned cells found in the bat's midbrain and auditory cortex.[5,15,24-29] Subsequent discharges to echoes are detected by echo channels which are, in addition, sensitive to different FM sweep-rates. The outputs of the echo channels (each of which encodes a different FM sweep-rate) are then brought with a minimum of delay to each of the coincidence-detecting neurons distributed along the delay-lines. The simultaneous arrival of a discharge for the emitted sound and a discharge for the FM echo produces an "event" that registers an estimate of echo delay along the time scale provided by the delay-line taps. This feature of the SCAT model is functionally similar to extraction of periodicity pitch using delay/coincidence networks in models of pitch perception,[8,9] with added efficiency for processing sounds having a specific time-frequency or spectrogram pattern such as an FM sweep.

The SCAT model uses a computer-simulated delay-line in each frequency channel, with delay taps from zero to 6000 μs, at intervals of 1 μs, to store discharges to the transmission for comparison with discharges to echoes arriving later. The discharges traveling along the 81 delay-lines in the model constitute a dynamic memory for the transmitted sound, with a spectrogram-like format, that serves to transform the time that elapses after the transmission into an explicit scale of delay or range. Each delay tap feeds into an FM coincidence detector that integrates events at the same or closely-spaced delay taps across all frequency channels. The exact pattern of this integration from channel to channel is used to compensate for the staggered time-of-occurrence of successive frequencies in the FM sweep to "dechirp" the signals. A different pattern of integration from tap to tap across channels is used for each FM sweep-rate that is intended to be recognized. The actual implementation of dechirping can be either through connections from one tap on a particular channel's delay-line to a *different* tap on the adjacent channel's delay-line or else through additional delay-lines (vertically oriented in Figs. 5 or 6) used to deliver echo information to the *same* tap on adjacent delay-lines. A occurrence of coincidences between the delayed emission-evoked discharges and the more immediate discharges evoked by echoes registers the echo's delay by whichever set of taps in involved. Echo delays thus are mapped onto the series of delay taps in each channel, and the occurrence of coincidences at different taps is integrated across frequency channels to form a distribution of echo-delay estimates corresponding to Δt in Fig. 5. This delay-and-coincidence process amounts to *spectrogram correlation*, the first of two image-forming operations in the SCAT model. The delay axis is restricted only for convenience to 0 to 6 ms in the version of the model shown here.

For reading out the delay estimates, a threshold criterion determines whether enough coincidences have occurred at the same or closely-spaced delay taps across different frequency channels to indicate that an echo has in fact been received. The maximum number of coincidences at any particular delay step is 81, and the read-out threshold is set to some fraction of this total to reject spurious coincidences caused by noise. The threshold procedure locates each peak in a graph of integrated coincidences and decides whether enough coincidences have occurred in the same delay tap or in neighboring taps across different channels to assert that targets are present at various ranges. Only peaks in the value of integrated coincidences that rise above threshold are accepted as "targets," which appear as points along a composite map of target range. This map in fact is the range-axis image of targets. It awaits only modifications from the *spectrogram transformation* process to become a map of both range and shape.

Target Images for Phase-Shifted Echoes

At this point in the description of the SCAT model, the images it forms consist of distributions of coincidence-delay estimates across frequency channels, as illustrated schematically in Fig. 5. The shape of each image is represented by the shape of the corresponding distribution around its central estimate of delay (Δt), which amounts to an *interval histogram* between the model's discharges to the transmission and discharges to each echo. The procedure for using the SCAT model to evaluate the results of the jitter discrimination experiments with *Eptesicus* is to compute the images for echoes that are received in the same phase as the transmission (0° phase-shift) and for echoes that are received out of phase (180° phase-shift.) The distributions of delay coincidences for the first (0°) and the second (180°) echoes in Fig. 6 are taken to represent the desired images.

Fig. 7 shows the distributions of delay coincidences for the two echoes from Fig. 6, plotted as the number of coincidences registered at each delay value in 1-µs steps. (Recall that the maximum number of coincidences that can occur at each delay step is 81, with one coincidence in each frequency channel.) Each distribution as a whole reflects the occurrence of multiple discharges to the envelope in each channel, which is physiologically unrealistic if a single channel in the SCAT model is equated with a single auditory-nerve fiber. However, each auditory receptor cell in fact drives about 15 to 20 such fibers in parallel, and receptors located close to each other along the cochlea are very similar in tuning. Consequently, each model channel really represents a set of parallel auditory-nerve fibers with graded amplitude thresholds whose responses are correlated with each other by virtue of the identical or similar frequency responses of their receptor cells and the occurrence of "phase-locking" to the fine ripple passing through the excitation process (Fig. 6). (Note that synchronization of the model's neural discharges to this ripple is not required during steady-state stimulation with pure tones, only during the onset of the envelope emerging from each channel's low-pass filter.) The distribution of coincidences representing echo-delay estimates thus is derived across a population of physiologically more reasonable "neurons" that share very similar input signals than would be achieved if each model channel were supposed to represent only one auditory-nerve fiber.

The zero origin of the time axis in Fig. 7 stands for the objective arrival-time of each echo, where the actual delay is 2.3 ms for the first echo and 4.6 ms for the second echo in fig. 6. (This brings the two distributions to a common time origin for comparison by the same type of compound graph shown for the bat's performance in Fig. 3A.) The first echo in Fig. 6 arrives with 0° phase-shift, and its coincidences from the model are plotted in the top part of the graph in Fig. 7. The second echo arrives with 180° phase-shift, and its coincidences are plotted in the bottom part of the same graph. The model's distribution produces a sharp estimate

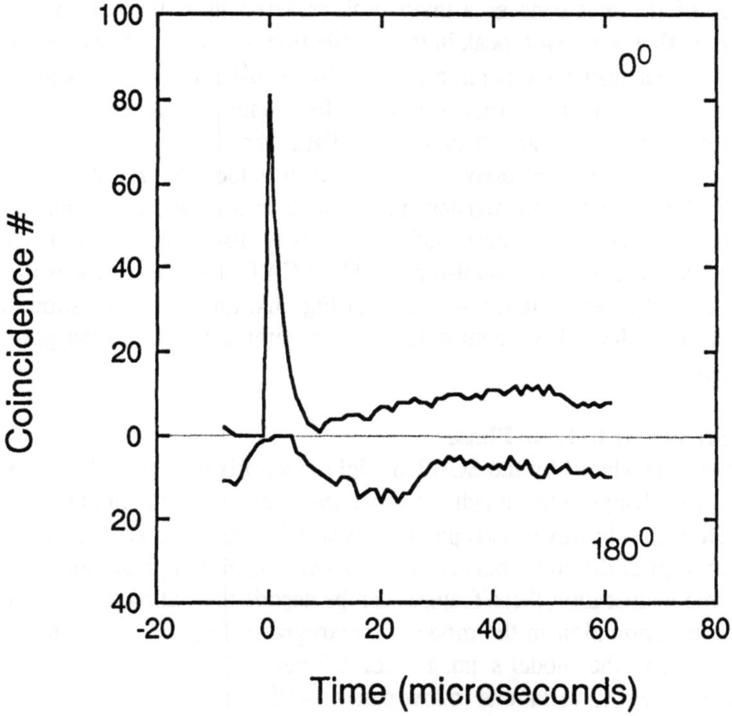

Fig. 7. Compound graph showing the distribution of transmission-echo coincidences in the SCAT model for the two echoes from Fig. 6, with *half-wave* rectification of band-limited sweep segments. The distribution for the 0° phase-shifted echo is shown on the top and the distribution for the 180° phase-shifted echo is shown inverted on the bottom (format analogous to Fig. 3B). The objective echo delays of 2.3 and 4.6 ms are aligned at zero on the time axis to better illustrate the shape of the distributions for the two phase-shifts. The model clearly preserves the phase difference between the echoes and represents it in the shape of the delay image compounded from the distribution of delay estimates.

of the delay of the first echo as a main peak centered upon the echo's objective arrival-time, with a broad side-peak in the distribution that covers delays from about 25 to 45 μs longer than the objective delay. The distribution for the second, 180° phase-shifted echo contains a broader, more diffuse main peak at about 15 to 25 μs, with lower coincidences from 30 to 45 μs. Thus, the SCAT model registers the phase-shift as a shift in echo arrival-time to a new value. At the objective delay (zero time), the second echo registers no coincidences at all, indicating that the model finds the 0° and 180° phase-shifts to be easily discriminable even when no difference exists in the echo arrival-times. The SCAT model reproduces the main features of the bat's performance as shown in Fig. 3A, raising the question of how the phase-shift in the echoes comes to be incorporated in the discharges of the model's neurons.

Origin of Sensitivity to Echo Phase

The images produced by the SCAT model are surprisingly robust--that is, they usually do not change very much as most parameters of the model itself are changed. There are, however, two parameters which directly affect the capacity of the model to register different images for echoes with different phase-shifts (Fig. 7), and with some justification these features can be described as allowing the retention of echo phase information in the model's spectrograms (Fig. 6). First, the crucial difference between the model's images for 0° and 180° phase-shifted echoes is surprisingly resistant to variations in low-pass cut-off frequency and low-pass filter order within a range that is physiologically realistic. First-order filter frequencies of 800 Hz to several kiloHertz are workable, for example. Only for unrealistically low low-pass cut-off frequencies does the ripple in the envelopes shown in Fig. 6 become too smoothed out to support registration of echo phase differences in the images. The degree of envelope smoothing thus is important for reproducing the images perceived by the bat (Figs. 3 and 4) in the model's images Fig. 7). Second, if the half-wave rectification that follows band-pass filtering is replaced by full-wave rectification, then the differential phase sensitivity of the model is abolished. Fig. 8 illustrates the images produced for 0° and 180° phase-shifted echoes when the envelopes used to trigger neural discharges are formed by full-wave rectification. In this condition, the two images are identical. The SCAT model still provides a very sharp registration of the delay of echoes by the location of the main peak in the curves, but the curves no longer change when the phase of the echoes is changed. In contrast, if the envelopes are formed by half-wave rectification, the image changes dramatically as the phase of the echoes changes (Fig. 7). Both full-wave rectification and extreme low-pass filtering remove the ripple on the envelopes that are to be coded in terms of the model's neural discharges, demonstrating that the synchronization of discharges to this ripple is the specific

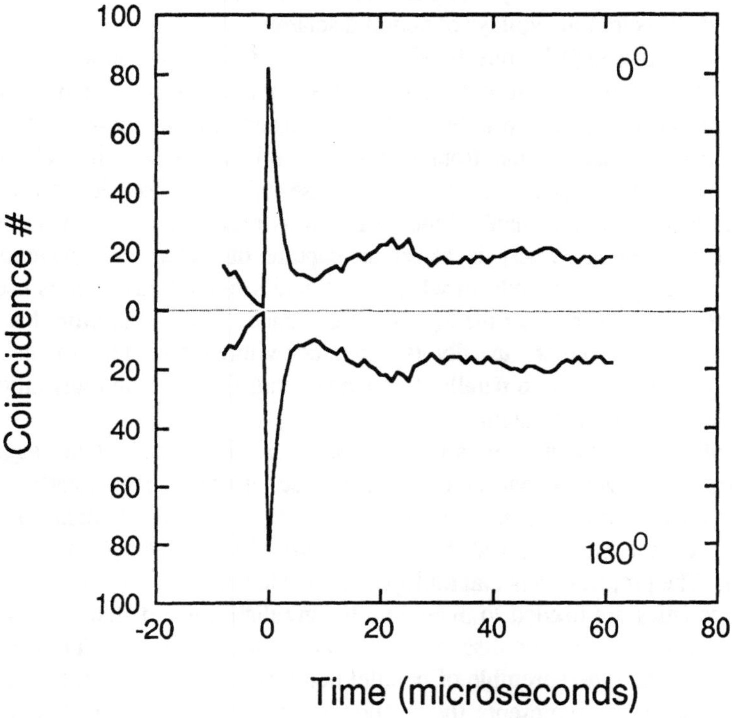

Fig. 8. Compound graph showing the distribution of transmission-echo coincidences in the SCAT model for the two echoes from Fig. 6, but with *full-wave* rectification of band-limited sweep segments. (Details as in Fig. 7.) Now, the model clearly represents the two echoes with the same shape of delay image because the echo phase difference is lost in the distributions.

feature of the model that conveys the phase of echoes relative to transmissions from the time waveforms into the volleys of neural discharges.

Full-wave rectification is considered "better" in engineering practice because it captures all of the energy present in the signal, both in positive-going and in negative-going peaks of the cycle-by-cycle time history of the signal. The area under the envelope that results from full-wave rectification contains all of the energy present in the original waveform. Because half-wave rectification throws away the energy present in half of the peaks, it is regarded as less than ideal. However, this viewpoint really is based on capture of signals by conventional single-channel sampling methods, in which the signal's filtered envelope is used to trigger a time-marker to register the signal's occurrence. The assumption that full-wave rectification is "better" usually is extended without comment to the case where the signal is broken into parallel frequency bands for separate registration of the occurrence of parts of the signal.

The auditory system uses parallel frequency channels, too, but the adjacent filters are so close together that their filter response curves overlap greatly. This overlap is typically described as oversampling in the frequency domain, and the underlying problem being solved by this oversampling is stated in terms of *redundancy*. The proposition is that multiple neural channels with highly-correlated filter characteristics are needed to overcome the unreliability and stochastic nature of single auditory neurons for discharging at the correct times in relation to the stimulus waveform. The ensemble of parallel neurons coming out of the receptor-cell filters collectively represents the occurrence and frequency composition of sounds even though individual neurons discharge to only a fraction of the potential triggering "events" in the sound. Any one neuron might not discharge to a given event, but it will discharge for *some* events, thus contributing to the ensemble code for the series of events in the sound as a whole. The enormous scale of auditory oversampling in this sense might also recover information or energy that would have been lost by half-wave rectification if no oversampling were present.

The use of half-wave rectification by the auditory system is widely understood to be useful for coding the cycle-by-cycle history of sounds at frequencies up to several kiloHertz, but no higher. The results presented here from experiments with bats and with the model suggest that the parallel filters, in conjunction with half-wave rectification of their outputs, might be doing more than supplying redundancy in the ordinary sense. They provide sufficiently accurate representation of the timing of individual frequencies in broadband sounds that the true instantaneous frequency of the signal is effectively represented in the pattern of neural discharges across frequency (Fig. 5). Instantaneous frequency formally is the time derivative of phase, and it thus is not surprising in principle that the phase structure of the echo can be recovered by integrating echo-delay estimates across frequencies. What *is* surprising, however, is that the bat actually may recover echo phase by this method.

The SCAT model identifies the use of half-wave rather than full-wave rectification and phase-locking of neural discharges to the fine structure of excitation envelopes as critical for this process. The model also draws attention to experiments that need to be carried out with bats to determine just how sharply instantaneous frequency is represented in individual auditory channels. Experiments of this type have not yet been adequately conducted in bats owing to the use of analog stimulus-generation techniques (triggered sweep-burst oscillators and function generators often produce too much phase jitter to stimulate the bat with rigorously phase-locked FM bursts), and owing to the use of simple level-crossing circuits to determine the time-of-occurrence of neural discharges (too much jitter is introduced into the recordings to specify spike timing very accurately). Experiments with the most rigorous application of these methods already indicate that the timing of discharges in single neurons in *Eptesicus* may register waveform events with an accuracy better than the periods of the lower frequencies in the bat's sounds,[4] which may be sufficient for the bat to reconstruct the entire phase history of echoes.[1] Modern digital techniques can solve both of these problems, and definitive experiments on the timing of the bat's neural discharges to FM transmissions and echoes should now be done.

ACKNOWLEDGMENTS

The research summarized in this chapter was supported by ONR Grant No. N00014-89-J-3055, by System Development Foundation Grant No. 57 and 57A, by NIMH Grant No. 7-K02-MH00521 (RSDA), and by a University Research Instrumentation Program Grant from the Department of Defense.

REFERENCES

1 Altes, R.A., Echo phase perception in bat sonar? *J. Acoust. Soc. Am.*, 69 (1981) 1232-1246.
2 Altes, R.A., Ubiquity of hyperacuity, *J. Acoust. Soc. Am.*, 85 (1989) 943-952.
3 Busnel, R.-G. and Fish, J.F., (Eds.), *Animal Sonar Systems*, Plenum Press, New York, 1980.
4 Covey, E. and Casseday, J.H., Parallel monaural pathways to the midbrain in an echolocating bat: The ventral lateral lemniscal complex (VLL), *Soc. Neurosci. Abstr.*, 15 (1989) 746.
5 Feng, A.S., Simmons, J.A. and Kick, S.A., Echo detection and target-ranging neurons in the auditory system of the bat *Eptesicus fuscus*, *Science*, 202 (1978) 645-648.

6 Griffin, D.R., *Listening in the Dark*, Yale Univ. Press, New Haven, CT, 1958; Cornell Univ. Press, Ithaca, NY, 1986.
7 Haresign, T., Ferragamo, M. and Simmons, J.A., Response properties of neurons in the big brown bat, *Eptesicus fuscus*, to closed field dichotic stimulation, *Soc Neurosci. Abstr.*, 17 (1991) 1404.
8 Lazzaro, J. and Mead, C., Silicon modeling of pitch perception, *Proc. Natl. Acad. Sci. USA*, 86 (1989) 9597-9601.
9 Licklider, J.C.R., A duplex theory of pitch perception, *Experientia*, 7 (1951) 128-134.
10 Masters, W. M., Tradeoff between echo crosscorrelation-peak delay and time delay in target ranging by big brown bats, *Soc. Neurosci. Abstr.*, 15 (1989) 512.
11 Menne, D. and Hackbarth, H., Accuracy of distance measurement in the bat *Eptesicus fuscus*: Theoretical aspects and computer simulations, *J. Acoust. Soc. Am.*, 79 (1986) 386-397.
12 Menne, D., Kaipf, I., Wagner, I., Ostwald, J. and Schnitzler, H.-U., Range estimation by echolocation in the bat *Eptesicus fuscus*: Trading of phase versus time cues, *J. Acoust. Soc. Am.*, 85 (1989) 2642-2650.
13 Moss, C.F. and Schnitzler, H.-U., Accuracy of target ranging in echolocating bats: Acoustic information processing, *J. Comp. Physiol. A*, 165 (1989) 383-393.
14 Nachtigall, P.E. and Moore, P.W.B., (Eds.), *Animal Sonar: Processes and Performance*, Plenum Press, New York, 1988.
15 O'Neill, W.E. and Suga, N., Encoding of target-range information and its representation in the auditory cortex of the mustached bat, *J. Neurosci.*, 47 (1982) 225-255.
16 Pollak, G.D. and Casseday, J. H., *The Neural Basis of Echolocation in Bats*, Springer, New York, 1989.
17 Pollak, G.D., Marsh, D.S, Bodenhamer, R. and Souther, A., Characteristics of phasic on neurons in inferior colliculus of unanesthetized bats with observations relating to mechanisms for echo ranging, *J. Neurophysiol.*, 40 (1977) 926-942.
18 Simmons, J.A., Perception of echo phase information in bat sonar, *Science*, 207 (1979) 1336-1338.
19 Simmons, J.A., A view of the world through the bat's ear: The formation of acoustic images in echolocation, *Cognition*, 33 (1989) 155-199.
20 Simmons, J.A., Ferragamo, M., Moss, C.F., Stevenson, S.B. and Altes, R.A., Discrimination of jittered sonar echoes by the echolocating bat, *Eptesicus fuscus*: The shape of target images in echolocation, *J. Comp. Physiol. A*, 167 (1990) 589-616.

21 Simmons, J.A. and Grinnell, A.D., The performance of echolocation: Acoustic images perceived by echolocating bats. In P.E. Nachtigall and P.W.B. Moore (Eds.), *Animal Sonar: Processes and Performance*, Plenum Press, New York, 1988, pp. 353-385.

22 Simmons, J.A., Moss, C.F. and Ferragamo, M., Convergence of temporal and spectral information into acoustic images of complex sonar targets perceived by the echolocating bat, *Eptesicus fuscus*, *J. Comp. Physiol. A*, 166 (1990) 449-470.

23 Simmons, J.A. and Stein, R.A., Acoustic imaging in bat sonar: Echolocation signals and the evolution of echolocation, *J. Comp. Physiol.*, 135 (1980) 61-81.

24 Suga, N., The extent to which biosonar information is represented in the bat auditory cortex. In G.M. Edelman, W.E. Gall, and W.M. Cowan (Eds.), *Dynamic Aspects of Neocortical Function*, Wiley, New York, 1984, pp. 315-373.

25 Suga, N., Auditory neuroethology and speech processing: Complex-sound processing by combination-sensitive neurons. In G.M. Edelman, W.E. Gall, and M.W. Cowan, (Eds.), *Auditory Function*, Wiley, New York, 1988, pp. 679-720.

26 Suga, N. and Horikawa, J., Multiple time axes for representation of echo delays in the auditory cortex of the mustached bat, *J. Neurophysiol.*, 55 (1986) 776-805.

27 Suga, N. and O'Neill, W.E., Neural axis representing target range in the auditory cortex of the mustached bat, *Science*, 206 (1979) 351-353.

28 Sullivan, W.E., Neural representation of target distance in auditory cortex of the echolocating bat *Myotis lucifugus*, *J. Neurophysiol.*, 48 (1982) 1011-1032.

29 Wong, D. and Shannon, S.L., Functional zones in the auditory cortex of the echolocating bat, *Myotis lucifugus*, *Brain Res.*, 453 (1988) 349-352.

Non-phase locked auditory cells and 'envelope' detection

Fred Rieke,[a,b] Walter Yamada,[c] E. R. Lewis,[c,d] and William Bialek[a,b,e]

[a] NEC Research Institute
4 Independence Way
Princeton, New Jersey 08540

[b] Department of Physics
[c] Neurobiology Graduate Group
[d] Department of Electrical Engineering and Computer Science
[e] Department of Molecular and Cell Biology
University of California
Berkeley, California 94720

Abstract

Many primary auditory neurons do not follow the temporal fine structure of waveforms within their passband; these are termed non-phase locked cells. Such cells do not encode the sound pressure waveform directly, but rather some nonlinear functional of the waveform such as the envelope. But which definition of the envelope — or, more generally, which nonlinear functional — does a particular cell implement? This difficulty is one example of a general problem: How do we characterize information transfer in a cell which codes the output of some unknown, nontrivial computation on the sensory stimulus? Here we suggest an approach in which one chooses from a set of possible nonlinear functionals by finding that functional of the stimulus which can be estimated most accurately from observation of the spike train. We apply these ideas to non-phase locked auditory afferents to measure the tuning of these cells for complex stimuli.

Introduction

The neural code defines the representation of sensory stimuli in neural spike trains. In simple systems we think we know which signals each cell is 'listening for' in the outside world, so the problem of neural coding is to understand *how* these particular signals are represented in spike trains. As we move away from the sensory periphery, especially in higher animals, we quickly reach a point where it is not clear *what* any particular cell is coding, let alone how it is encoded. We might find, for example, that a given cell hardly fires at all in response to simple stimuli, but gives robust responses to more complex signals. Alternatively the responses may be robust but seemingly ambiguous with respect to variations in different stimulus parameters. Ideally we would like a set of methods and analytic techniques which allow us to characterize what such cells are telling us about stimuli chosen from complex, naturalistic ensembles.

The essence of the problem is that cells in a sensory system need not directly encode the stimulus $s(t)$ which we control in an experiment. Instead the cell may encode some nonlinear transformation or functional $G[s(t)]$ of the stimulus. If the system is adaptive, as most sensory cells are, the choice of the functional G may even depend on the ensemble from which the signals $s(t)$ are chosen. Although these problems are probably most severe in the far reaches of mammalian cortex, a simple example can be found in the primary sensory neurons of lower vertebrate auditory systems. The firing pattern of many primary auditory afferents is not modulated by the phase of tones within the cell's passband; instead the cell responds to sinusoidal stimuli with only a DC change in firing rate. Thus the spike train at best informs the organism about some nonlinear functional of the signal, and this functional is insensitive to the signal phase. Which particular functionals are these cells encoding?

Reading the code — extended version

Traditional approaches to neural coding characterize the *encoding* of sensory stimuli in *average* properties of neural spike trains. Living creatures, however, do not have access to these average responses; instead they make behavioral decisions in real time based on single examples of the neural response. We approach neural coding from the point of view of the organism — we learn to *decode* neural spike trains in real time to estimate continuous sensory stimuli [1, 2].

This decoding approach has now been successfully applied in a number of systems — a movement sensitive neuron in the blowfly (*C. erythrocephola*) visual system [1], mechanoreceptors in the cricket (*A. domestica*) cercal system [3], and vibration sensors in the bullfrog (*R. catesbeiana*) sacculus [4]. In each of these systems we can estimate the stimulus $s(t)$ by linearly filtering the spikes: $s_{\text{est}}(t) = \sum_i F(t - t_i)$, where the spikes occur at times t_i. We choose the decoding filter $F(\tau)$ to minimize the mean square error, $\chi^2 = \int dt|s(t) - s_{\text{est}}(t)|^2$; this minimization results in an analytic expression for the filter in terms of the experimentally accessible stimulus $s(\tau)$ and spike occurrence times t_i.

What happens when a cell encodes not the sensory stimulus itself but instead a nonlinear functional* of the stimulus? The simplest such nonlinearity is of second order, so the time-varying signal coded by the spikes is of the form

$$s^{(2)}(t) = \int d\tau \int d\tau' K_2(\tau, \tau') s(t - \tau) s(t - \tau'). \tag{1}$$

K_2 describes the particular second order computation.

If we believe that the cell encodes the time-varying signal $s^{(2)}(t)$, then we should treat this as the 'stimulus' we try to estimate from the spike train. An arbitrary definition of $s^{(2)}(t)$ will result in very poor estimates, since the cell is tuned to a particular time-varying signal. In other words, the cell implements a particular K_2 (or set of K_2's — see below), and if we choose a very different K_2 we will find it impossible to reconstruct the signal. These observations suggest that we use the quality of the reconstructions as a measure of the tuning of the cell. Out of all the possible kernels K_2 we want to choose the one which defines the signal that is estimated with least error. Thus we propose that *what* a cell encodes can be defined as that transformation of the stimulus which can be estimated most accurately from observation of the spike train.

As before, we write our estimate as a linearly filtered version of the spike train $s^{(2)}_{\text{est}}(t) = \sum_i F^{(2)}(t - t_i)$. Now, rather than varying only the

*If the cell encodes some linearly filtered version of the stimulus there is no effect on the estimation strategy. The structure of the linear filter will be reflected in the fact that the reconstruction is more reliable in some frequency ranges and less reliable in others — the filtering which precedes encoding is manifest in the frequency dependence of the errors in the reconstructions.

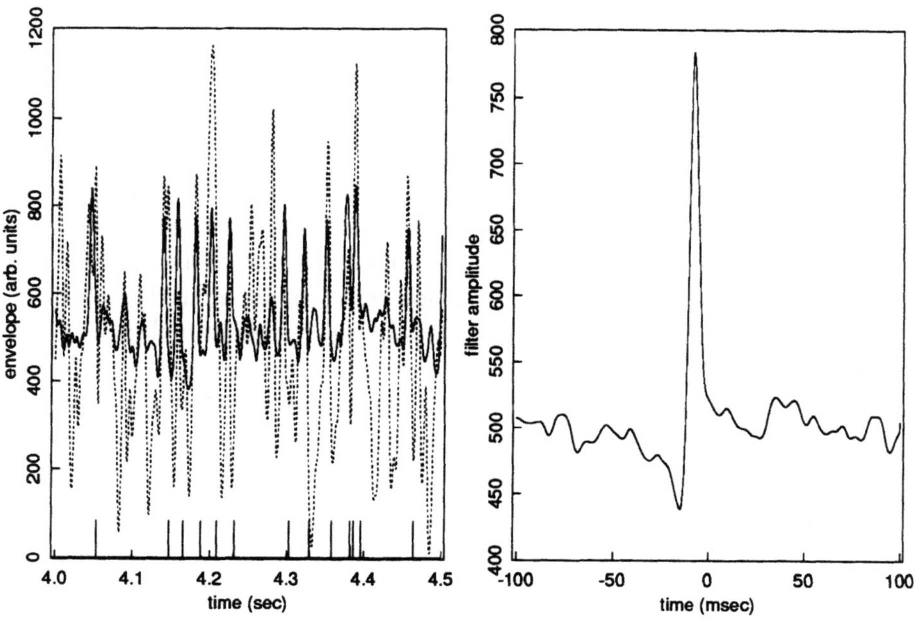

Figure 1: (left) Example of envelope, estimated envelope, and spike train for white noise stimulus. The estimate is asymetric because the cell exhibited very little spontaneous activity. (right) Decoding filter $F^{(2)}(\tau)$. The estimated envelope is calculated by convolving this filter with the spike train.

estimation filter $F^{(2)}$, we try to find the set of filters $\{K_2(\tau,\tau'), F^{(2)}(\tau)\}$ which minimize $\chi^2 = \int dt |s^{(2)}(t) - s^{(2)}_{\text{est}}(t)|^2$. Thus we are finding the best reconstruction of the signal which the cell best encodes. As stated this is an ill-posed problem, since one could simply set all the filters to zero — then the reconstruction is perfect, since the signal vanishes identically! To avoid this problem we constrain our search to the space of normalized kernels K_2; we implement this normalization by holding the signal power $P = \int dt [s^{(2)}]^2$ fixed. The solution to this constrained optimization problem is described in the Appendix.

An application — non-phase locked auditory afferents

The traditional description of a non-phase locked cell is that the cell spikes in response to power in a particular frequency range. More precisely, the cell responds to the envelope, in a particular frequency band, of the acoustic stimulus. Attempts to characterize the response properties of non-phase locked afferents by estimating the sound pressure waveform directly, or by using reverse-correlation methods [5], are doomed because of the phase ambiguity of the response — if the cell responds to $s(t)$ it responds essentially equally to $-s(t)$. An approach termed the 'dynamic spectrum analyzer' has been used to measure response characteristics in higher auditory centers in the grassfrog [6]. This approach provides interesting results, but is limited studying the coding of a particular functional of the stimulus, time-varying power in this case.

We record the spikes intracellularly from non-phase locked afferents in the auditory nerve of the bullfrog *Rana catesbeiana* [7]. The acoustic stimulus in these studies is either white noise (with a cutoff frequency of 2 kHz) or frog calls in white noise. How do we define envelopes for these stimuli? The idea is to take the original stimulus, band-pass filter it, rectify the result and smooth to remove high frequency terms introduced in the rectification:

$$e(t) = \int d\tau g(\tau) \left[\int d\tau' f(\tau') s(t - \tau - \tau') \right]^2 , \tag{2}$$

where f and g are, respectively, the band-pass and low-pass filters. The envelope is a specific example of a second order functional of the stimulus, where the kernel K_2 is related to the filters f and g as:

$$K_2(\tau, \tau') = \int dt\, g(t) f(\tau - t) f(\tau' - t). \tag{3}$$

We use the approach described in the preceeding section to calculate the set of filters $\{K_2(\tau, \tau'), F^{(2)}(\tau)\}$ for the non-phase locked cells. We then use (3) to recover the filters $f(\tau)$ and $g(\tau)$ from the kernel $K_2(\tau, \tau')$. Experimentally only some of the kernels K_2 which appear as solutions permit this decomposition — some of the solutions are envelopes of the stimulus, while others appear to be other nonlinear transformations. For solutions where we can recover the filters f and g, the power spectrum of the band pass filter $f(\tau)$ determines which frequency components of the original sound pressure waveform are included in the envelope. These methods allow us to measure the tuning of non-phase locked afferents using complex stimuli, as opposed to pure tones. Fig. 1 shows an example of the envelope, spike train and estimated envelope for the white noise stimulus. Also shown in Fig. 1 is the reconstruction filter $F^{(2)}$. Fig. 2 shows the power spectrum of f for the same experiment.

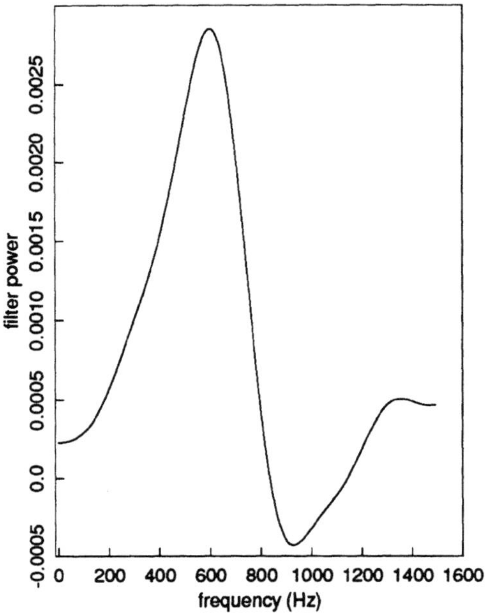

Figure 2: Power spectrum $|\tilde{f}(\omega)|^2$ of the band pass filter $f(\tau)$. This filter determines which frequency components of the stimulus contribute to the envelope, thus measuring the tuning for this cell.

What next?

We have developed an approach to neural coding which allows us to characterize the coding when we don't know what time varying signal a cell encodes. We feel that our initial application of this method is successful not just because we were able to find a well-reconstructed envelope, but because our methods expose some interesting questions. Probably the most interesting point is that when we try to find the 'best' kernel K_2 we always find that there are multiple solutions. This means that there are many different transformations of the stimulus waveform which can be reconstructed, and each of these different transformations is locally optimal, in that small changes in the transformation produce signals which cannot be reconstructed as accurately. This result implies that the central nervous system of the frog could extract different second order functionals of the stimulus waveform from the same afferent neuron by building different reconstruction filters $F^{(2)}$. Although much remains to be done in characterizing these multiple solutions (to what extent do they carry independent information?), it seems to us that the existence of multiple solutions will be a generic feature of the neural code in any cell which responds to some nonlinearly transformed version of the stimulus.

Appendix

Our problem is to find the combination of filters $\{K_2(\tau,\tau'), F^{(2)}(\tau)\}$ which minimize

$$E[K_2, F^{(2)}] = \left\langle \int dt |s^{(2)}(t) - s^{(2)}_{\text{est}}(t)|^2 - \lambda_{\text{P}} \int dt [s^{(2)}(t)]^2 \right\rangle , \quad (4)$$

where we use a Lagrange multiplier λ_{P} to impose the power constraint and the averages $\langle \cdots \rangle$ are over the stimulus ensemble.

We begin by expanding both filters in power series: $F^{(2)}(\tau) = \sum_\mu a_\mu l_\mu(\tau)$ and $K_2(\tau,\tau') = \sum_{i,j} b_{ij} h_i(\tau) h_j(\tau')$, where the $l_\mu(\tau)$ and $h_i(\tau)$ are power series functions (usually a Fourier series in a finite time window) and the a_μ and b_{ij} are variable coefficients. Our problem now is to find the coefficients $\{a_\mu, b_{ij}\}$ which minimize E — *i.e.* we set $\partial E/\partial a_\mu = 0$ and $\partial E/\partial b_{ij} = 0$. These conditions lead to a set of coupled linear equations:

$$\begin{aligned} \sum_k a_\mu N_{\mu\nu} &= \sum_{i,j} b_{ij} L_{ij\nu} \\ (1-\lambda_{\text{P}}) \sum_{i,j} b_{ij} G_{ijnm} &= \sum_\mu a_\mu L_{nm\mu} \end{aligned} \quad (5)$$

where the matrices L, N, G are defined as

$$\begin{aligned} N_{\mu\nu} &= \int dt \left\langle \sum_{i,j} l_\mu(t-t_i) l_\nu(t-t_j) \right\rangle \\ L_{ij\mu} &= \int dt d\tau d\tau' \left\langle \sum_n l_\mu(t-t_n) h_i(\tau) h_j(\tau') s(t-\tau) s(t-\tau') \right\rangle \\ G_{ijkl} &= \int dt d\tau d\tau' d\tau'' d\tau''' h_i(\tau) h_j(\tau') h_k(\tau'') h_l(\tau''') \text{ x} \\ &\quad \langle s(t-\tau) s(t-\tau') s(t-\tau'') s(t-\tau''') \rangle . \end{aligned} \quad (6)$$

These matrices can be computed numerically from the stimulus $s(\tau)$ (sound pressure waveform in this case) and the spike occurrence times t_i.

The coupled equations above can be solved in a number of ways. For example we can substitute for b_{ij} in the first set of equations to obtain an expression for a_μ:

$$(1-\lambda_{\mathrm{P}})a_\mu = \sum_{i,j,p,n,\nu,\beta} a_\nu L_{np\nu} G^{-1}_{npij} L_{ij\beta} N^{-1}_{\beta\mu} \tag{7}$$

The coefficients a_k are the eigenvector with eigenvalue $(1-\lambda_{\mathrm{P}})$ of the matrix

$$Q^{A}_{\nu\mu} \equiv \sum_{i,j,p,n,\beta} L_{np\nu} G^{-1}_{npij} L_{ij\beta} N^{-1}_{\beta\mu} \tag{8}$$

The b_{ij} are determined by substituting this solution for a_μ into (5). These coefficients then determine the best set of filters $\{K_2(\tau,\tau'), F^{(2)}(\tau)\}$ given the cost function E.

Acknowledgements

We thank Joseph Atick for helpful discussions and encouragement, Goeorge Zweig for insisting that some vague remarks be made rigorous, and David Warland for being an all around good guy. F. R. was supported in part by a Graduate Fellowship from the National Science Foundation. Additional support was provided by an INCOR grant through the Los Alamos National Laboratory.

References

1. W. Bialek, F. Rieke, R. R. de Ruyter van Steveninck, and D. Warland. *Science*, 252:1854–1857, 1991.

2. F. Rieke, D. Warland, and W. Bialek. Efficient coding in sensory systems, this volume.

3. D. Warland, M. A. Landolfa, J. P. Miller, and W. Bialek. In F. Eeckman, editor, *Analysis and Modeling of Neural Systems 1*. Kluwer, Norwell, Mass., 1991.

4. F. Rieke, W. Yamada, K. Moortgat, E. R. Lewis, and W. Bialek. In Y. Cazals, L. Demany, and K. Horner, editors, *Auditory Physiology and Perception: Proceedings of the 1991 European Hearing Workshop.* in press, 1991.

5. E. de Boer and H. R. de Jongh. *J. Acoust. Soc. Am.*, 63:115–135, 1978.

6. A. M. H. J. Aertsen, P. I. M. Johannesma, and D. J. Hermes. *Biol. Cybernetics*, 38:235–248, 1980.

7. X. Yu, E. R. Lewis, and D. Feld. *J. Comp. Physiol. A*, 169:241–248, 1991.

Model of the Origin of Neuronal Selectivity for Binaural Intensity Difference in the Barn Owl

J. C. Pearson[1], **C. D. Spence**[1] and **R. Adolphs**[2]
[1]David Sarnoff Research Center
Princeton, NJ 08543-5300
[2]Div. of Biology 216-76
CalTech, Pasadena, CA 91125

Abstract

A model is presented of the neural computation of binaural intensity difference (BID) in the lateral shell of the central nucleus of the inferior colliculus (ICL) of the barn owl. The ICL is the second stage of binaural intensity processing, following the nucleus ventralis lemnisci lateralis pars posterior (VLVp), and it projects to the external nucleus of the inferior colliculus (ICX). The model shows how monaural excitation from the cochlear nucleus angularis can be combined with inhibition from the VLVp to form a peak-like pattern of activation in the ICL from the wedge-like activation pattern in the VLVp. The model explains the functional roles of the two narrowly frequency-tuned cell types in the ICL, and it predicts their patterns of interconnection. An experiment was performed in which changes in BID selectivity in the ICX were measured following the injection of neural activity modulators into the VLVp. Model simulations of these experiments produce qualitatively similiar results.

Overview of Binaural Intensity Processing

Binaural intensity difference (BID) is the strongest cue of elevation for the barn owl (tyto alba) [Moiseff, 1989]. BID varies monotonically with elevation for frequencies above 4 kHz. This is caused by the asymmetrical pattern of dense ruff feathers surrounding the ears [Payne, 1970; Haresign and Moiseff, 1989].

A neural representation of BID is created in several stages, as depicted in Fig. 1. The blocks denote the four main nuclei involved, on both sides of the brain. The

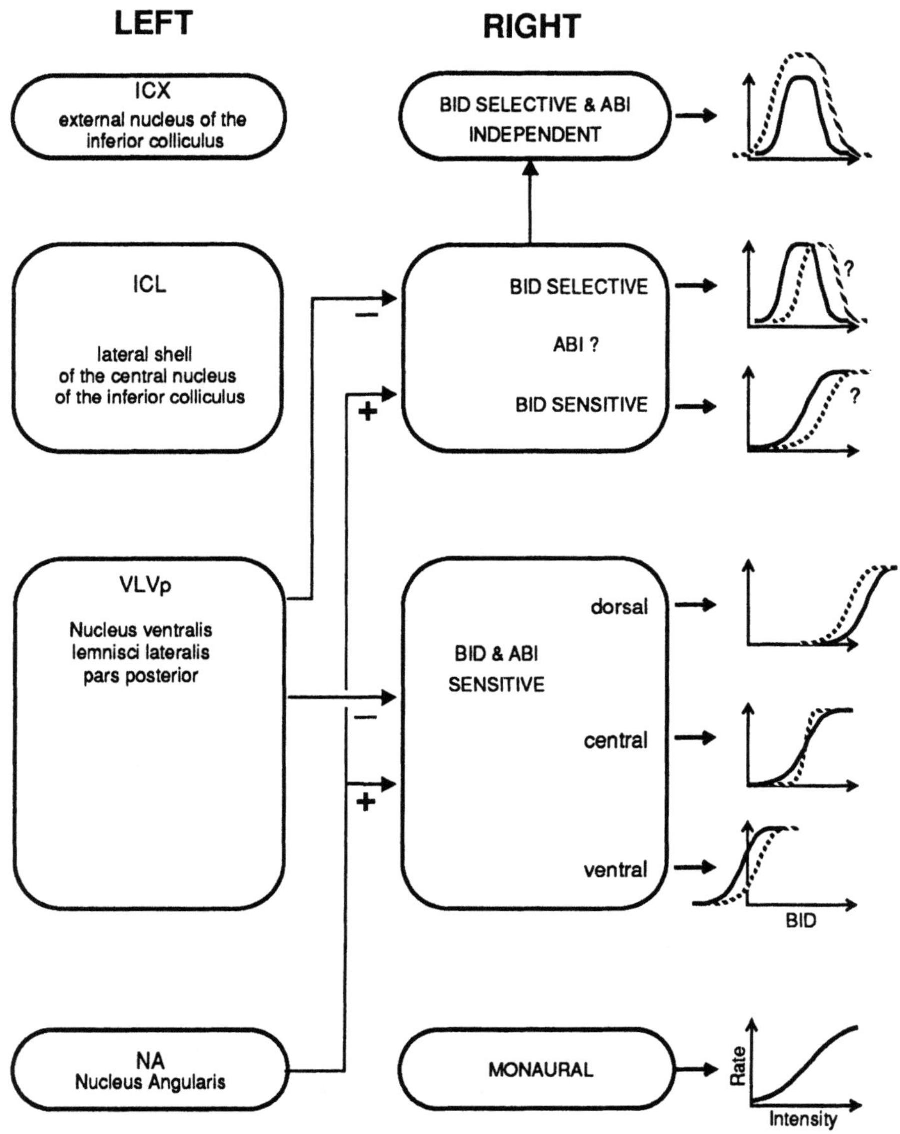

Fig. 1. The intensity processing system for sound elevation of the owl.

input to nucleus angularis (NA) (not shown) is from the sensory neurons of the 8th nerve. The names and abbreviations of these nuclei are given in the blocks on the left side, while their response characteristics are summarized on the right side.

NA neurons are monaural, responding only to ipsilateral stimuli within a narrow frequency range. Their firing rate increases montonically with intensity over a 30-50 dB range [Sullivan and Konishi, 1984].

At the other end of the processing stream, ICX neurons are exclusively binaural, firing only to broad-band stimuli with the appropriate BID. ICX neurons are relatively independent of the average binaural intensity (ABI), as their response curves simply broaden with increasing ABI [Knudsen, 1984]. ABI-independence is of course essential for the encoding of sound elevation, because it is principally a function of sound loudness.

The VLVp receives excitatory and inhibitory projections from the contralateral NA [Takahashi and Konishi, 1988] and VLVp [Takahashi, 1988], respectively . VLVp neurons are narrowly tuned in frequency, and are tonotopically arranged along the anterior-posterior axis. VLVp neurons are BID-sensitive, that is, their BID response curve is sigmoidal. This is in contrast to the BID-selective or bell-shaped response curves of ICX neurons. There is a systematic, map-like representation of BID along the dorso-ventral axis. The BID threshold, or BID at which the cell's response is half of its maximum value, decreases linearly from 20dB dorsally to -20dB ventrally. There is no systematic variation of any response characteristic along the medio-lateral axis. A tone of a given frequency and BID will thus excite a ventro-dorsal slab of neurons, the length of which is related to the BID, i.e., a thermometer-like representation of BID. However, VLVp neurons do not accurately represent BID, or sound elevation, as they are also sensitive to ABI. As the ABI is increased, the BID curves of dorsal cells shift to higher BID, those of ventral cells shift in the opposite direction, and those of central cells do not shift but the slope of the curve around threshold decreases [Manley, *et al.*, 1988]. There are three distinct anatomically defined cell classes [Carr, *et al.*, 1989] in the VLVp, however they do not appear to have different stimulus response characteristics.

The next stage in binaural processing is the ICL, which receives projections from the contralateral NA [Takahashi and Konishi, 1988] and VLVp [Takahashi and Konishi, unpublished]. The NA and VLVp inputs are presumably excitatory and inhibitory, respectively, since contralateral stimulation excites ICL neurons, while ipsilateral stimulation inhibits them [R. Adolphs and I. Fujita, unpublished observations]. ICL neurons have properties intermediate between those of the VLVp and the ICX. There are both BID-sensitive cells, as in the VLVp, and BID-selective cells, as in the ICX [Fujita and Konishi, 1989]. All of the BID-sensitive cells and some of the BID-selective cells are narrowly-tuned in frequency, as are VLVp neurons, however some of the BID-selective cells are broadly-tuned, like ICX neurons. The ABI response characteristics of these neurons have not yet been thoroughly investigated, nor has a systematic map of BID-thresholds been discovered. The BID-sensitive and BID-selective neurons are also tuned to binaural time delay which encodes stimulus azimuth. Frequency and binaural time delay are known to be mapped along the dorso-ventral and anterior-posterior axes, respectively [Wagner, *et al.*, 1987].

Modeling Objective

The objective of this work was to develop a network model of the BID processing of the ICL. The model was designed to produce both the BID-sensitive cell type and the narrowly-tuned (in frequency) BID-selective cell type. The broadly-tuned, BID-selective neurons are not addressed with this model. The model should make predictions concerning the ABI dependence of these two cell types. Of special interest is the ABI dependence of the BID-selective neurons, because they are presumably a major input to the BID-selective, ABI-independent, space specific neurons of the ICX. The model should clarify the functional role of the BID-sensitive cells in the ICL, which is puzzling because this response characteristic already exists in the VLVp. The model should predict the projection patterns of the incoming axons from the VLVp and NA. The model should also be realistic in the time domain, producing stable responses with short stimulus latencies.

Modeling Approach

The quantitative, mathematical network model of the ICL presented here is based upon an underlying qualitative model of the generation of BID-selectivity, called the "spatial-derivative model". This model, illustrated in Fig. 2, was developed independently by a number of investigators, including S. Volman, T. Takahashi, and the authors of this paper. The spatial-derivative model holds that a tone of a given frequency produces a wedge-like pattern of activation in the VLVp, and that for a fixed ABI, the position of the edge varies roughly linearly with BID. Such a pattern of activation is suggested by the observed dorso-ventral variation of BID-threshold depicted in the graphs of Fig. 1. The model predicts that a peak-like pattern of activation is created along the medio-lateral axis of the ICL, from the wedge-like pattern in the VLVp. The location of the peak of activation would vary roughly linearly with BID, which is clearly consistent with BID-selective responses.

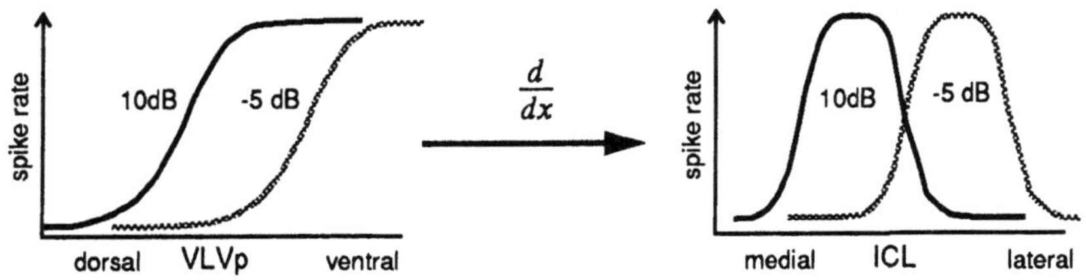

Fig. 2. The spatial-derivative model of BID-selectivity in the ICL. The curves illustrate the hypothetical activation patterns within the VLVp and ICL for stimuli with BID's of 10dB (solid) and -5dB (grey).

The degree of biological realism chosen for the model is based upon the objectives stated above, and the level of detail of existing neurophysiological data. The objective is to make predictions at the levels of anatomical connection patterns and qualitative, single cell response characteristics. The level of realism chosen is the same as that used in a previous model of the VLVp [Spence and Pearson, 1991].

The VLVp and ICL are modeled as large networks of discrete, time-dependent, interconnected, spike-producing units. The membrane potential, v, is modeled by

$$C\frac{dv}{dt} = -g_L(v - v_L) - g_E(v - v_E) - g_I(v - v_I) \qquad (1)$$

where v_L, v_E and v_I are fixed reversal potentials for some set of channels, g_L is a fixed passive leakage conductance, and g_E and g_I are excitatory and inhibitory conductances that are smoothly and linearly driven by incoming spikes in such a way as to approximate the time course of synaptic potentials. Spikes are randomly generated with an average rate given by a sigmoidal function of the membrane potential. (see [Spence and Pearson, 1991] for more detail on the methods of simulation used).

Neural response characteristics are computed as they are in corresponding biological experiments, i.e., through counting the spikes produced to a set of stimuli. Only tone stimuli are considered, as the cells being modeled at this stage are narrowly tuned in frequency.

The VLVp and NA are modeled and simulated in the same way as presented in [Spence and Pearson, 1991]. These models, as depicted in Fig. 3, are briefly summarized here for completeness.

The output from NA is very simply modeled as a constant spike rate that is linearly proportional to the stimulus intensity (5 spikes/ms at 50 dB).

An iso-frequency, dorso-ventral slab of the VLVp is modeled as a network of 1200 neurons (which is close to the actual number in such a slab), divided into 40 clusters of 30 neurons each, defined by their position along the dorso-ventral axis. Three cell types are modeled, as defined by their variation in density along the dorso-ventral axis (as suggested by [Carr, *et al.*, 1989]), and by their inter-connection patterns (as predicted by the model). One class projects in a criss-cross pattern to the contralateral VLVp, and its number density increases from the ventral (18 cells) to the dorsal (2 cells) surface. Another class provides local inhibition, and follows the opposite density gradient. The third class (not shown) is uniformly distributed along the dorso-ventral axis (10 cells in each cluster), and projects to the contralateral ICL. The excitatory input from the contralateral NA is distributed uniformly throughout the slab. For all connections, the strength of individual synapses is independent of dorso-ventral location. However, the cell number gradients effectively create dorso-ventral gradients of inhibition.

This VLVp model reproduces the experimentally measured dorso-ventral distribution of single-unit BID response curves (see Fig. 1, for details see [Spence and Pearson, 1991]), as well as the wedge-like pattern of activation across the VLVp, in

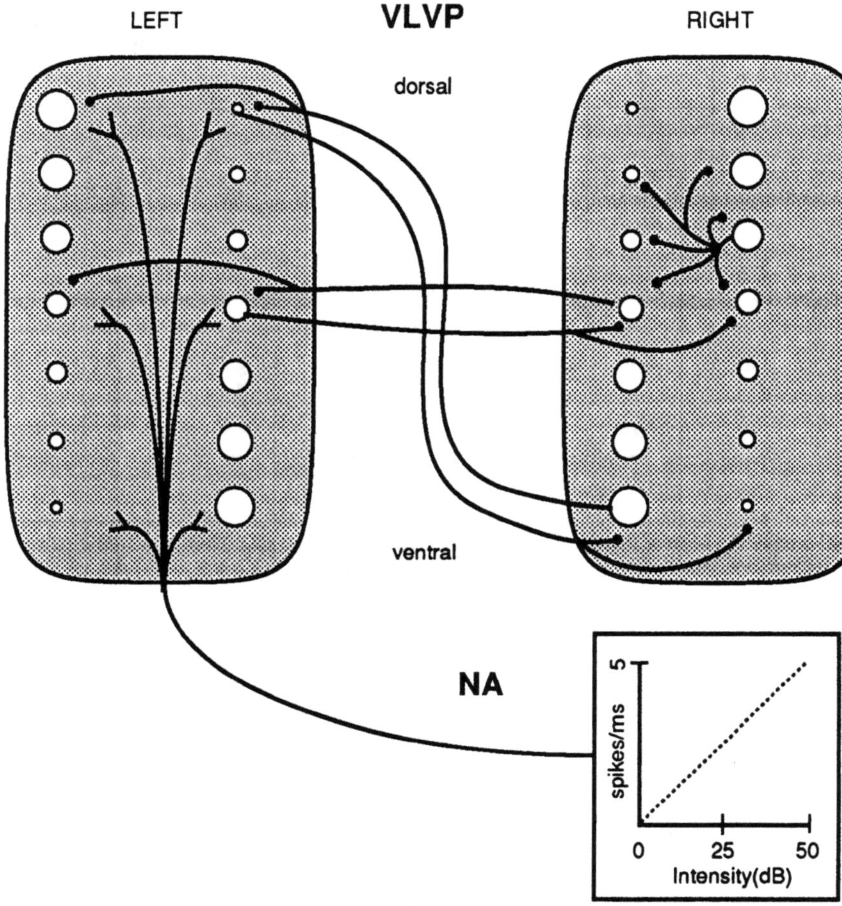

Fig. 3. Competitive edge model of the VLVp. The size of the circles denote the number density of the cells of a given type at that position

keeping with the underlying spatial-derivative model (see Fig. 2.). The pattern of dorso-ventral activation for a range of ABI and BID is shown in Fig. 4. Note that increases in ABI compress the range of edge movement, which remains linear (the physiologically relevant range of BID is from -10dB to +20dB).

The wedge-like pattern is the result of a competitive feedback interaction between the criss-cross cell classes of the two VLVp. The cluster with the greatest effective inhibition (roughly the product of NA activation level and cell number) squashes the response of its partner in the other VLVp. The local inhibitory class was found to be essential for insuring the stability and speed of convergence of this feedback process. Interestingly, the three cell classes in the VLVp model have identical response characteristics, due to their having identical input patterns, but unique functional roles, due to their differing output patterns.

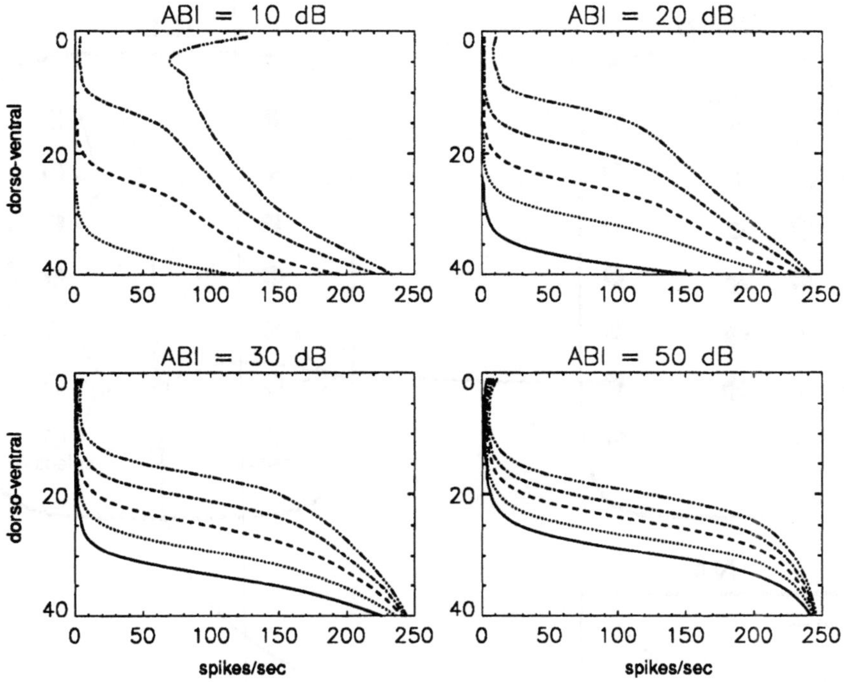

Fig. 4. Dorso-ventral patterns of activation in the VLVp. Each graph shows the response to 5 BID's for a fixed ABI. The y-axis is dorso-ventral depth (cluster number), with 0 corresponding to the dorsal surface. The BID's were -20dB (solid), -10dB (dotted), 0dB (dashed), 10dB (dot-dash) and 20dB (dot-dot-dot-dash). Note: for ABI=10dB, BID=+/- 20dB corresponds to monaural stimulation.

ICL Model

The architecture of the ICL model is illustrated in Fig. 5. There are two ICL cell types, defined by their connectivity within the ICL, with the sensitive cells inhibiting the neighboring selective cells to one side along the medio-lateral axis. There is a topographic, point-to-point, inhibitory projection from the dorso-ventral axis of the VLVp onto the medio-lateral axis of the ICL (the orientation of which is arbitrary), which treats both cell types the same. NA projects onto the medio-lateral axis of the ICL in the same way it projects onto the dorso-ventral axis of the VLVp — uniformly to all cell types. The cells are grouped into 40 clusters along the medio-lateral axis, with 10 cells of each type in each cluster.

The patterns of stable stimulus-driven activity formed in the ICL are indicated by the stipling in Fig. 5. As it does in the VLVp, the NA input serves to uniformly activate all of the cells, independently of their position along the medio-lateral axis. The inhibitory projection from the VLVp squashes the activity in the ICL that cor-

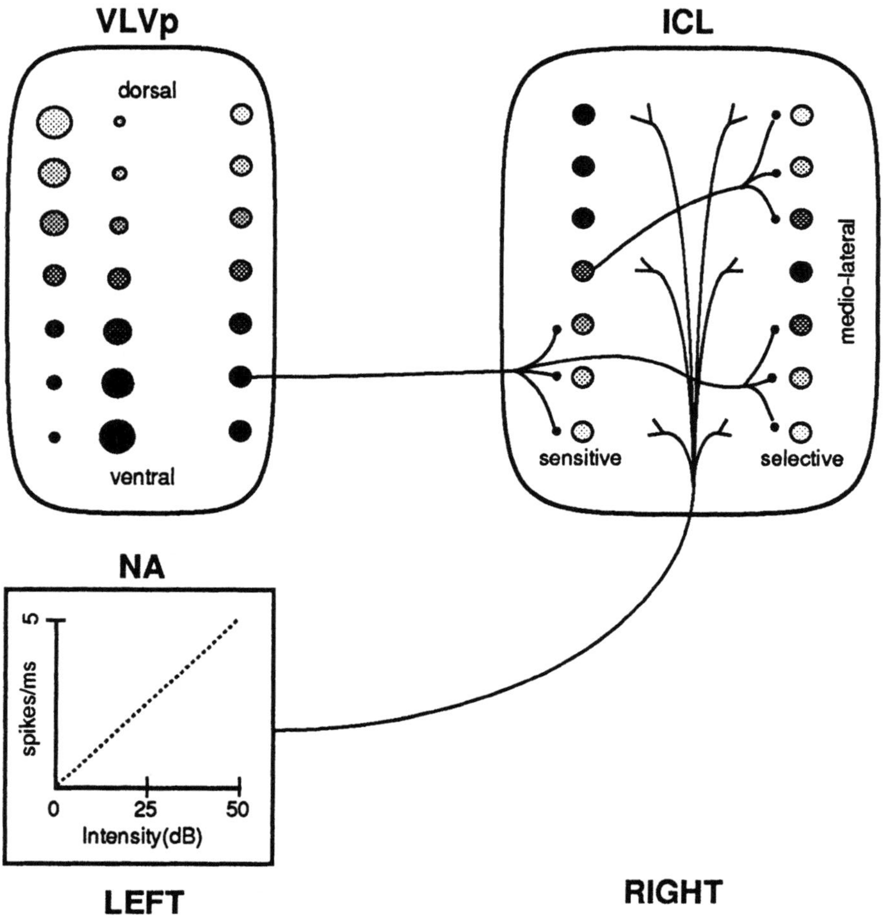

Fig. 5. ICL model architecture. The stipling within the cells indicates the degree of stimulus-driven activation (the darker the more active).

responds to the region of activity in the VLVp (ventral of the edge). This creates a wedge-like pattern of activation among the sensitive cells that is the complement of the pattern in the VLVp. The active sensitive cells then squash the activity in the corresponding selective cells. Only the selective cells in the position corresponding to the edge in the VLVp are free of heavy inhibition and remain active.

The dynamics of this pattern formation process are illustrated in Fig. 6. Note first that stable patterns are achieved in the ICL within the physiologically reasonable period of 6-10 ms. from stimulus onset (this does not include the response latency of NA itself). The rate limiting step in pattern formation is the feedback competition process in the VLVp. Such processes can be very slow to converge. An early version of the VLVp model [Spence and Pearson, 1990] that did not include the local inhibitory cell class exhibited very slow convergence of roughly 60 ms. The local inhibitory class in the VLVp establishes a gradient of spontaneous activity, as seen in the 0 ms. panel in Fig. 6. This provides an intitial bias to the competition that prevents the dorsal neurons from getting excited at stimulus onset, excitation that would otherwise be slowly removed during the competition. These neurons also add negative feedback which increases the stability of the patterns. A gradient of spontaneous activity in the VLVp was reported in [Manley, *et al.*, 1988].

The model produces the desired response curves for the sensitive and selective cells, as shown in Fig. 7 (compare with graphs in Fig. 1). Their ABI dependence can be understood entirely in terms of the ABI dependence of the VLVp (The following discussion concerns the BID-selective cells. Similar arguments can be made for the sensitive cells). The position of the peak of activity along the medio-lateral axis in the ICL is directly determined by the position of the edge of activity along the dorso-ventral axis in the VLVp. Since the edge in the VLVp shifts with ABI (see Fig. 4), so will the peak in the ICL. This shift in the peak of the activation pattern corresponds to the shift in the response curves of those neurons with non-zero optimal BID (i.e., the neurons not positioned at the middle of the medio-lateral

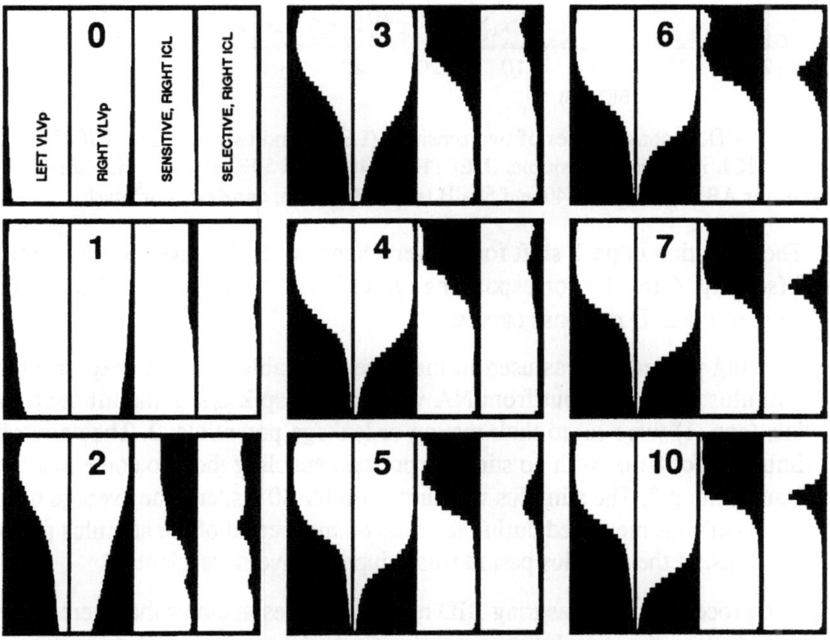

Fig. 6. Ten ms. sequence of firing patterns following stimulus onset in the left and right VLVp, and the BID-sensitive and BID-selective cells in the right ICL. The firing rates of the 40 cell clusters along the BID axis are depicted as a vertically oriented bar graph, with the bars pointing towards the left for all but the left VLVp, which point to the right. Frame "0" shows the spontaneous activity pattern prior to stimulus onset.

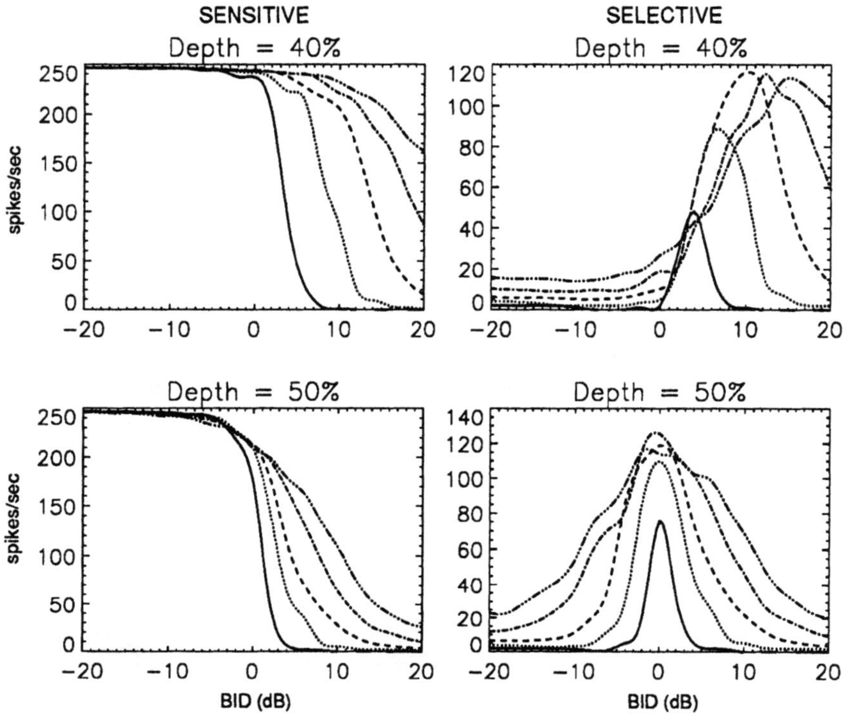

Fig. 7. BID response curves of two sensitive (LEFT) and two selective (RIGHT) model ICL neurons, at positions (DEPTH) of 40% and 50% of the medio-lateral axis, for ABI's 10, 20, 30, 40 and 50 dB (solid, dot, dash, dot-dash, dot^3-dash).

axis). The reduction in peak shift for a given change in BID caused by an increase in ABI (see Fig. 4 for the corresponding effect in the VLVp) corresponds to the broadening of the BID response curves.

The following procedure was used to measure the stable ongoing response to a given stimulus: 1. The output from NA was set to 0 spikes/sec and all neuronal potentials (eqn. 1) were set to their resting or leakage potentials; 2. The networks were simulated for 8 ms. with no stimulus present, enabling them to come to a quiescent equilibrium; 3. The stimulus was turned on for 80 ms. and the average firing rate of each cell was measured during the final 68 ms. period of the stimulus period (the first 12 ms. of the stimulus period was skipped to avoid transients).

The above procedure for measuring BID response curves assumes that there are no systematic shifts in the stimulus response during the 68 ms. period of spike averaging. Fig. 8 demonstrates that this assumption is valid. Such shifts do occur without the local inhibitory cell class within the VLVp, because of the slower convergence of the wedge-like patterns in the VLVp.

Fig. 8 also shows that there is a transient onset response in the VLVp and ICL model neurons, which is independent of BID, and which is more pronounced in the

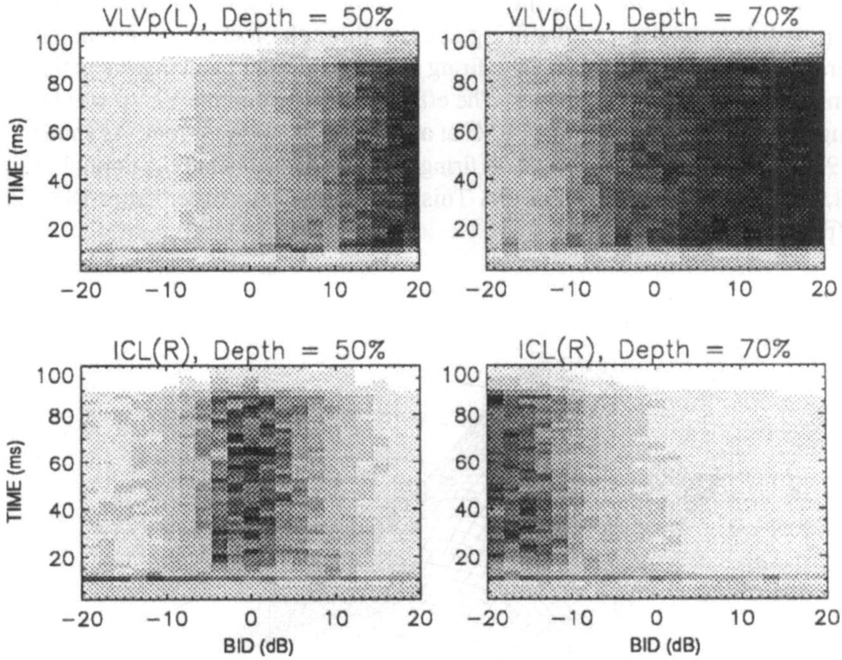

Fig. 8. Firing rate as a function of BID and time, for neurons at positions 50% and 70% along the BID axis of the left VLVp and the right ICL. The darker the point the higher the firing rate. Stimulus onset and offset occured at 8 and 88 ms., respectively.

ICL. These effects are due to the fact that the excitation from the NA is at first unchecked by inhibition. This effect is smaller in the VLVp due to the spontaneous activity which maintains a gradient of inhibition.

An earlier ICL model [Spence and Pearson, 1990] was based on the assumption that the output neurons of the VLVp were excitatory. However, recent unpublished work by R. Adolphs and I. Fujita has shown that ICL neurons are excited and inhibited by contralateral and ipsilateral stimulation, respectively. This implies that the VLVp input to the ICL must be inhibitory, and that the excitatory input must come from NA.

VLVp Activity Modulation Experiment and Simulation

Recent experiments by R. Adolphs provided an indirect way of testing the VLVp and ICL models. In these experiments, the neural activity modulators lidocaine and bicuculline (BMI) were globally injected into one VLVp, while BID response curves were measured in the contralateral ICX. Given that the ICL is the predominant source of BID information to the ICX (see Fig. 1), it is likely that any observed changes in ICX BID-tuning would be the result of similar changes in in the ICL.

Preliminary tests showed that these drugs produced the expected effects in the VLVp — lidocaine reduced the firing activity through blocking action potential generation, while BMI increased the firing activity, through blocking the action of the inhibitory transmitter GABA-A. The effect of BMI within the VLVp was tested through the iontophoresis of 5mM BMI at a continuous 10 nA current. As shown in Fig. 9, there is a reversible increase in firing that is proportional to the normal firing level, and which is not BID specific. This behavior was recorded from 18 of 22 VLVp neurons.

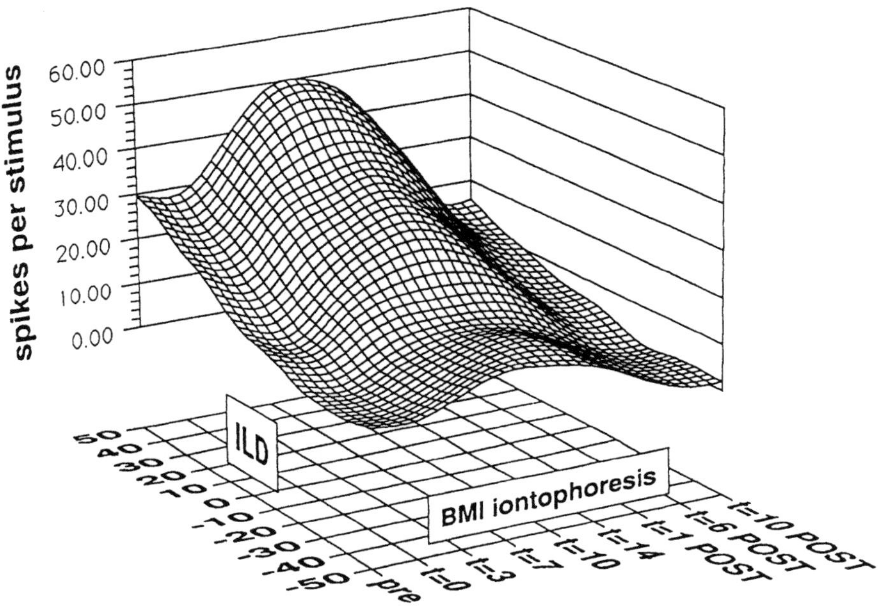

Fig. 9. Firing rate of a VLVp neuron as a function of BID (also known as ILD) before, during and after the iontophoresis of BMI.

The effects of global injections of lidocaine and BMI into the VLVp on the BID-curves of contralateral ICX neurons are shown in Fig. 10. The injections consisted of .2 microliters of a 5% xylocaine solution, or .2 microliters of a 5mM BMI solution. Lidocaine produces a reversible reduction in maximal response and a shift to lower optimal BID. BMI produces the opposite effect, in addition to a disproportionate increase in the response to greater than optimal BID's.

Similar changes are produced in the response of model ICL selective neurons, as shown in Fig. 11. Injections of lidocaine were simulated in the model by reducing the maximal firing rate of the "injected" VLVp to one-half that of the normal VLVp. The injection of BMI was simulated in the model solely by eliminating the inhibitory input from the local inhibitory cell class in the "injected" VLVp, without modifying the inhibition from the criss-cross cell class. This of course assumes that

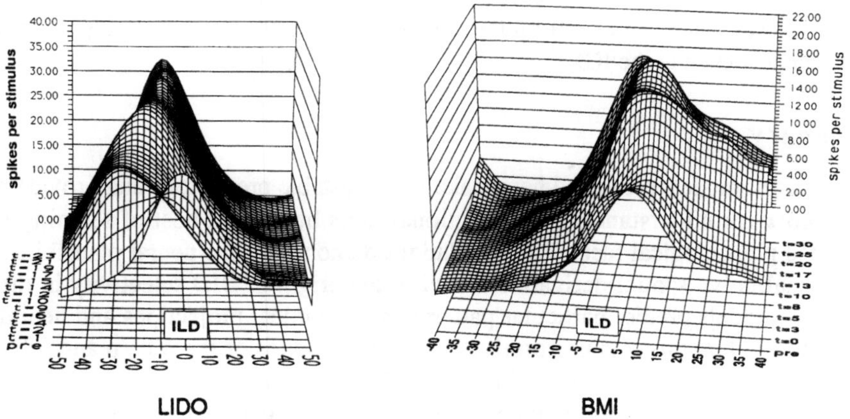

Fig. 10. BID (ILD) response curves from ICX neurons before, and following the global injection of lidocaine (LIDO) or bicuculline (BMI) into the contralateral VLVp.

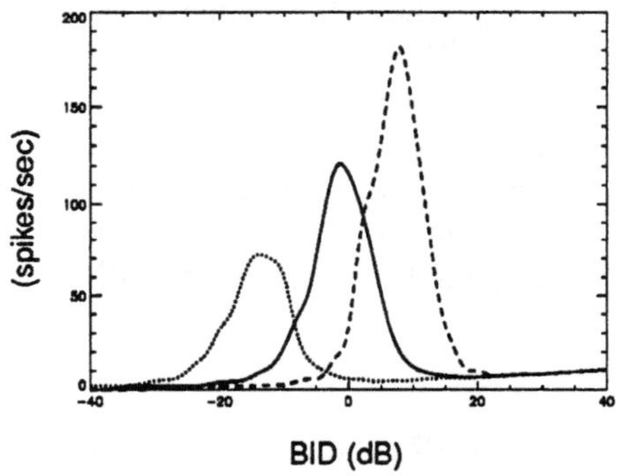

Fig. 11. BID-response curves of a model ICL selective neuron in its unperturbed condition (solid), and following simulated injections of lidocaine (dotted) and BMI (dashed) into the contralateral VLVp.

the local inhibitory class is GABAergic, while the criss-cross cells are not. This assumption is consistent with the results of [Carr, et al., 1989].

The effects of the injections can be simply understood in terms of the competitive balance between the two VLVp's. Injections of lidocaine into one VLVp will reduce the output of its criss-cross neurons, putting them at a competitive disadvantage. This will produce a ventral shift in the wedge-like pattern of the injected VLVp, and an equal dorsal shift in the other VLVp. These shifts in the VLVp will of course produce shifts in the peak-like patterns of the ICL, and corresponding shifts in the response curves of the BID-selective cells of the ICL. The model can not explain the disproportionate increase in the response to greater than optimal BID's under the application of BMI. This effect could be due to perturbations to non-ICL inputs and processes, such as those from the medial shell and the process that generates ABI independence in the ICX.

Summary

The network models presented here make many testable predictions and provide a framework for understanding the functional links between a host of seemingly unrelated physiological facts. The simulations demonstrate the neural plausibility of the qualitative spatial-derivative model of the origin of BID selectivity in the ICL. The models predict particular patterns of anatomical projections and neural circuitry. (1) NA projects uniformly along the BID axes of the VLVp and ICL. (2) The VLVp consists of three inhibitory cell classes, distinguished by their patterns of interconnection, their number density gradient, and their transmitter type. One class is non-GABAergic, projects in a criss-cross fashion to the contralateral VLVp, and its number density follows a dorso-ventral gradient, being high at the ventral end. A second class is GABAergic, connects locally, and its number density follows the opposite gradient from the criss-cross class. The third class is uniformly distributed along the dorso-ventral axis, and projects topographically onto the medio-lateral axis of the ICL. (3) The BID-sensitive cells of the ICL are inhibitory and connect to the neighboring BID-selective cells with greater BID thresholds. Temporally, the model predicts that the VLVp and ICL neurons have a short, BID-independent onset response, with the ICL transient being stronger. Future work will extend these models to include the broadly-tuned, BID-selective neurons in the ICL, the integration of the time and intensity pathways in the ICL, the projection from the ICL to the ICX, and the ABI independence of ICX neurons.

Acknowledgements

This work was supported by AFOSR contract F49620-89-C-0131 and DARPA contract F49620-90-C-0010. We thank W. E. Sullivan, T. Takahashi and S. Volman for valuable technical discussions, and K. Miller for helpful suggestions concerning the presentation.

References

Carr, C. E., I. Fujita, and M. Konishi (1989) Distribution of GABAergic neurons and terminals in the auditory system of the barn owl. *J.Comp. Neurol.* **286**:190-207.

Fujita, I. and M. Konishi (1989) Transition from single to multiple frequency channels in the processing of binaural disparity cues in the owl's midbrain. *Soc. Neurosci. Abstracts* 15:114.

Haresign, T. and A. Moiseff (1988) Early growth and development of the common barn-owl's facial ruff. *The Auk*, **105**:699-705.

Knudsen, E. I. (1984) Synthesis of a neural map of auditory space in the owl. In *Dynamic Aspects of Neocortical Function*, G. M. Edelman, W. E. Gall and W. M. Cowan, eds., Wiley, New York.

Manley, G. A., C. Köppl, and M. Konishi (1988) A neural map of interaural intensity differences in the brain stem of the barn owl. *J. Neurosci.* **8**:2665-2676.

Moiseff, A. (1989) Binaural disparity cues available to the barn owl for sound localization. *J. Comp. Physiol.* **164**:629-636.

Payne, R. S. (1971) Acoustic localization of prey by barn-owls (Tyto alba). *J. Exp. Biol.* **54:** 535-573.

Spence, C. D. and J. C. Pearson (1990) The computation of sound source elevation in the barn owl. In *Advances in Neural Information Processing Systems 2*, D. S. Touretzky, ed., Morgan Kaufmann, San Mateo, CA. .

Spence, C. D. and J. C. Pearson (1991) Models of the computation of sound elevation in the barn owl. In *Analysis and Modeling of Neural Systems*, F. Eeckman, ed., Kluwer, Boston.

Spence, C. D. and J. C. Pearson (1992) A resonance model of high frequency binaural phase sensitivity in the barn owl's auditory brainstem. In *Analysis and Modeling of Neural Systems II*, F. Eeckman, ed., Kluwer, Boston.

Sullivan, W. E. and M. Konishi (1984) Segregation of stimulus phase and intensity coding in the cochlear nucleus of the barn owl. *J. Neurosci.* **4**:1787-1799.

Takahashi, T. T. (1988) Commissural projections mediate inhibition in a lateral lemniscal nucleus of the barn owl. *Soc. Neurosci. Abstracts* **14**:323.

Takahashi, T. T. and M. Konishi (1988) Projections of the cochlear nuclei and nucleus laminaris to the inferior colliculus of the barn owl. *J. Comp. Neurol.*, **274**:190-211.

Takahashi, T. T. and M. Konishi (1988) Projections of nucleus angularis and nucleus laminaris to the lateral lemniscal nuclear complex of the barn owl. *J. Comp. Neurol.*, **274**:212-238.

Wagner, H., Takahashi, T. and M. Konishi (1987) Representation of interaural time difference in the central nucleus of the barn owl's inferior colliculus. *J. Neurosci.*, **7**:3105-3116.

A resonance model of high frequency binaural phase sensitivity in the barn owl's auditory brainstem

Clay Spence and John Pearson
David Sarnoff Research Center
CN5300
Princeton, NJ 08543-5300
e-mail: cds@sarnoff.sarnoff.com

ABSTRACT

The auditory system of the barn owl (tyto alba) contains neurons sensitive to the phase of sounds of remarkably high frequency, up to 9 kHz. Nucleus Laminaris represents phase differences as part of the computation of stimulus azimuth. The high frequency of the stimulus and the high level of noise in the input spike trains make the response properties of laminaris neurons hard to explain. We use simulations and semi-numerical analysis to show that the cellular and synaptic time constants must be unreasonably fast in order for ordinary biophysical mechanisms to reproduce the observed behavior. Several people have suggested that a resonance mechanism may exist in laminaris neurons to amplify the signal. We present a simple neuronal resonance model that can deal with realistic input.

INTRODUCTION

The barn owl (*tyto alba*) uses inter-aural phase difference as a cue for azimuth [Moiseff and Konishi, 1981; Moiseff, 1989]. At the neural level, action potentials in the eighth nerve, in the neurons of nucleus magnocellularis (NM), and in the neurons of nucleus laminaris (NL) are all phase locked to the stimulus [Sullivan and Konishi, 1984; Sullivan and Konishi, 1986]. They do not occur on every cycle of

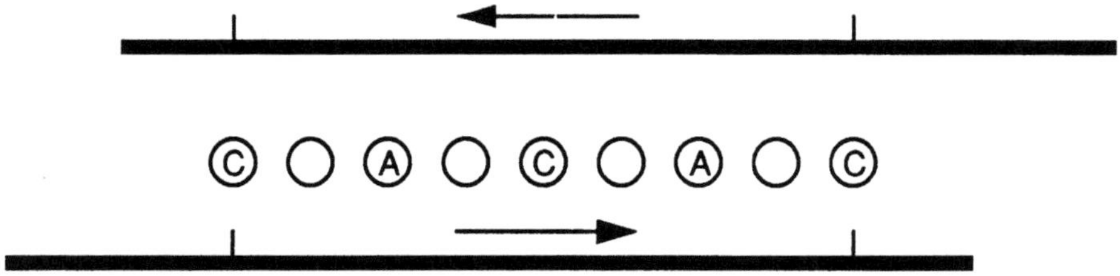

Figure 1. The Jeffress model. Circles labelled C and A represent neurons at coincident and anti-coincident positions for this stimulus condition. The heavy lines are axons with spike trains.

the stimulus, but a phase histogram shows a maximum at a particular phase and a minimum 180° from this phase. This phase sensitivity of the response decreases with increasing frequency, but is still present at nearly nine kHz. Nucleus magnocellularis (NM) receives synapses from the eighth nerve and is monaural. Axons from each NM innervate nucleus laminaris on both sides of the brain, penetrating the nucleus in opposite directions. NL neurons are sensitive to the binaural phase difference of the stimulus, showing a large response to a particular phase difference and very small response when the phase difference is 180° from the optimal, and in fact the response at this phase is less than the response to a monaural stimulus. There is evidence that the optimal phases of NL neurons vary systematically with their position, thus forming a map of inter-aural phase difference [Sullivan and Konishi, 1984; Carr and Konishi, 1988; Carr and Konishi, 1990].

Laminaris appears to be a perfect example of Jeffress' model for converting timing information at the ears into the azimuth of the source (Fig. 1) [Jeffress, 1948]. If a source of sound is to the right of the direction the animal is facing, its sound reaches the right ear before the left, and the difference of arrival times is a function of the azimuthal angle. In Jeffress' model, monaural neurons on each side of the head generate precisely timed action potentials in response to the sound. Thus the spike trains from each side of the head are very similar except for a relative time shift which reflects the azimuth of the source. These axons pass each other in opposite directions, making synapses onto binaural neurons. For each such neuron, there is an azimuth for which the difference in the time delays from a source to the ears is compensated by the difference in the neural time delays, so that the difference in the total time delays is zero. For this azimuth, spikes will arrive in coincidence at this neuron (a stimulus condition we will call in-phase). For the same stimulus, other neurons will have positions at which the spikes arrive first from one side and then the other, or in anti-coincidence (out-of-phase). If it can respond differently to coincident spike arrival than to anti-coincident arrival, the population of binaural neurons will form a map of inter-aural time difference and hence azimuth.

The essential features of the Jeffress model are the use of axons as delay lines and coincidence detection by binaural neurons. Although the evidence for the function of magnocellularis axons as delay lines is compelling, there is no real evidence that laminaris neurons can detect the coincidence of action potentials. There is strong reason to believe that they cannot do this, simply because laminaris functions at up to nine kHz, which should exceed the speed limitations of an ordinary neuron. Also the jitter or imprecise timing of magnocellularis action potentials represents a source of noise which can swamp the signal that laminaris is trying to detect. We will analyze a single compartment model of a laminaris neuron to illustrate this. Such a model works only if the time constants for the synaptic conductances and the membrane potential are extremely fast, and then it works poorly.

As an alternative to coincidence detection, several people have suggested that a resonance mechanism of some sort may be present in NL neurons. This would serve to amplify the signal a laminaris neuron is trying to detect while suppressing the noise. As a preliminary model of a resonator, we show that a damped harmonic oscillator has the same response behavior as NL neurons to several types of stimuli: a large response to binaural in-phase stimuli, a smaller response to monaural stimuli, and very little response to binaural out-of-phase stimuli. We will present results showing that such a system can perform quite well, but good performance with reasonable synaptic and membrane time constants may require the cell's latency to be longer than is experimentally observed.

This work is still preliminary. In the last section we discuss the limitations of these models, and what remains to be done.

LAMINARIS NEURONS ARE NOT COINCIDENCE DETECTORS: QUALITATIVE ARGUMENTS

In the simplest picture of a laminaris neuron as a coincidence detector, incoming spikes from nucleus magnocellularis excite the neuron. If the spikes occur in coincidence from each side of the brain, the excitation is relatively large, whereas if the spikes are anti-coincident, their effects do not add as effectively, and so the excitation is relatively small. One would only need to set a threshold to distinguish between the two stimulus conditions. However, there are laminaris neurons with best frequencies of around nine kHz. Synaptic and membrane time constants are usually thought to be on the order of one millisecond, so a typical neuron would badly smear signals with a frequency greater than about one kHz.

If the input to a laminaris neuron had no noise the membrane potential would have a DC component and a very small component following the input signal. A thresholding mechanism could still work if the threshold could be adjusted with sufficient accuracy. However, the input to laminaris is noisy. The magnocellularis neurons providing this input show little or no modulation of their firing rates with stimulus intensity, but phase lock to the stimulus. They do not fire spikes at exactly the same phase of the stimulus, rather the spikes are spread over some range of phases. Furthermore, they do not fire during every period of the stimulus, nor are their spikes spaced by a predictable number of stimulus periods. This randomness in the occurrence of spikes represents noise in the input to each laminaris neuron.

One might imagine that by making the synaptic and membrane time constants short enough the modulation of the spike rate will become detectable by the cell. This may require the time constants to be unreasonably fast. A more basic problem with this picture is that it does not really cure the noise problem. Because the stimulus period is so short, only a few spikes arrive during one such period on average, and the fluctuation in the number of spikes arriving is substantial, unless magnocellularis neurons communicate with each other so that the average number firing in a given period does not fluctuate much, a possibility which we think is unlikely.

THE SIGNAL-TO-NOISE RATIO IN LAMINARIS

To give a more quantitative analysis of the problem, we will estimate the mean squared amplitude of the noise fluctuations and of the signal in the membrane potential of a single compartment model of a laminaris neuron. We model the input to a laminaris neuron as a Poisson process $\xi(t)$ whose rate is modulated sinusoidally. (A periodic but non-sinusoidal modulation will not give better performance in a linearized model, since it differs from a sinusoidal modulation only by having non-zero harmonics of the fundamental frequency. These higher frequencies are suppressed even more than the fundamental.) In the in-phase condition the modulation is maximal, so ξ has average

$$\langle \xi(t) \rangle = R(1 + \cos(\Omega t + \varphi)) \tag{1}$$

where φ is a random phase, R is the long-time-averaged rate and Ω is the (angular) stimulus frequency. The auto-correlation function is

$$\langle \xi(t+\tau)\xi(t) \rangle = R^2(1 + \cos(\Omega(t+\tau) + \varphi))(1 + \cos(\Omega t + \varphi)) + R\delta(\tau)(1 + \cos(\Omega t + \varphi)) \tag{2}$$

Averaged over φ, this gives

$$\langle\langle \xi(t+\tau)\xi(t) \rangle\rangle_{\varphi} = R^2\left(1 + \frac{1}{2}\cos(\Omega\tau)\right) + R\delta(\tau) \tag{3}$$

The mean square fluctuation of a stochastic process is given by the total power of the process, so we would like the power spectrum of the model input process. The power spectrum of this process is (taking the Fourier transform)

$$P(\omega) = R + R^2\delta(\omega) + \frac{R^2}{4}(\delta(\omega - \Omega) + \delta(\omega + \Omega)) \tag{4}$$

The first term is noise and is present at all frequencies. The second term represents a DC offset. The third term is the signal. Notice that the total power in the noise is infinite.

The model neuron's membrane potential obeys the equation

$$C\frac{dv}{dt} = -g_L(v - v_L) - g_E(v - v_E) \tag{5}$$

where C is the membrane capacitance, g_L and v_L are the constant leakage conductance and reversal potential, g_E is the excitatory synaptic conductance, and v_E is the constant excitatory reversal potential. The response of g_E to a spike is just the commonly used function $Ate^{-\gamma_s t}$ [MacGregor, 1987]. The response of g_E to several spikes is just the sum of the responses to individual spikes. This gives g_E the value

$$\int_{-\infty}^{t} dsA \cdot (t-s)\, e^{-\gamma_s (t-s)} \xi(s) \qquad 6$$

The conductance response to the spike train is a just a linearly filtered version of it. The power spectrum of the conductance is the power spectrum of the input multiplied by the squared-amplitude of the fourier transform of the response of the conductances impulse response, which is

$$\frac{A^2}{(\omega^2 + \gamma_S^2)^2} \qquad 7$$

The equation for the membrane potential is non-linear, but for time constants which are not too fast the fluctuations in the membrane potential are small compared to the DC offset, so we can linearize about the constant parts of the potential and synaptic conductance. Denoting the fluctuating part of the potential by u, we get

$$C\frac{du}{dt} + g_L u + G_E u + g_E (U - v_E) = 0 \qquad 8$$

where the capital letters denote constant parts and small letters denote the fluctuating parts. The impulse response of u to g_E is

$$\frac{(U - v_E)^2 / C^2}{\omega^2 + \left(\frac{g_L + G_E}{C}\right)^2} \qquad 9$$

$(g_L + G_E)/C$ is just the inverse of the average time constant of the membrane potential, so we will call it γ_C in what follows. Thus the power spectrum of the noise in the membrane potential is

$$\frac{RA^2 (U - v_E)^2 / C^2}{(\omega^2 + \gamma_S^2)^2 (\omega^2 + \gamma_C^2)} \qquad 10$$

The total power is

$$\int \frac{d\omega}{2\pi} \frac{RA^2 (U - v_E)^2 / C^2}{(\omega^2 + \gamma_S^2)^2 (\omega^2 + \gamma_C^2)} = \frac{RA^2 (U - v_E)^2}{C^2} \times \frac{\gamma_C + 2\gamma_S}{4\gamma_C \gamma_S^3 (\gamma_C + \gamma_S)^2} \qquad 11$$

The total power of the signal in the membrane potential is trivial to evaluate

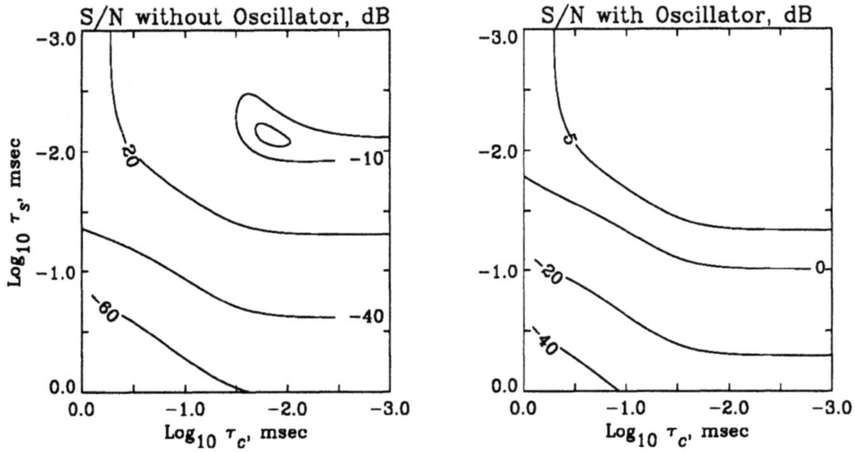

Figure 2. The signal-to-noise ratio of a passive compartment as a function of the synaptic and membrane time constants. Note the maximum without an oscillator.

because of the delta-functions. It is

$$\frac{A^2R^2(U-v_E)^2/C^2}{4\pi(\Omega^2+\gamma_S^2)^2(\Omega^2+\gamma_C^2)} \tag{12}$$

The signal-to-noise ratio is

$$\frac{R\gamma_C\gamma_S^3(\gamma_C+\gamma_S)^2}{\pi(\gamma_C+2\gamma_S)(\Omega^2+\gamma_S^2)^2(\Omega^2+\gamma_C^2)} \tag{13}$$

Figure 2 is a contour plot of the signal-to-noise ratio versus γ_C and γ_S, with $\Omega = 2\pi \times 5$ kHz, and $R = 30$ spikes per msec (100 synapses times 0.3 spikes per msec per synapse). Note that the function has a maximum. The signal-to-noise ratio there is -9.6 dB. Furthermore, this maximum occurs for $\gamma_C^{-1} = 0.015$ msec and $\gamma_S^{-1} = 0.0074$ msec, which we think are unreasonably small.

RESONATORS IN LAMINARIS NEURONS?

A resonator can help a laminaris neuron by suppressing the noise while amplifying the signal. A linear resonator has some of the response properties of a laminaris neuron:

> Real laminaris neurons have a binaural response that is roughly a sinusoidal function of the phase difference. In a resonator model, the average input spike rate is the sum of two periodic signals. The amplitude of the modulation at the stimulus frequency is a smooth function of the phase difference, and so the response of a linear filter at this frequency will also be a smooth

function of the phase difference.

Real laminaris neurons have a monaural response that is roughly half of the in-phase binaural response. In a resonator model, the input spike rate would have two pieces: a constant piece due to the unstimulated ear and a maximally modulated piece from the stimulated ear. The modulation of the total rate at the stimulus frequency has half of the amplitude of the in-phase binaural stimulus condition, so a linear filter will also have half of the response at the stimulus frequency.

A simple and familiar model of a resonator is a damped harmonic oscillator. The oscillator's response at the resonance frequency can be made arbitrarily large, but at the cost of increasing the time it takes for the oscillator to reach its steady-state response. This means a oscillator model neuron could have an unacceptably long latency.

We repeat the same calculations as the last section, except with the cell voltage driving a damped harmonic oscillator. In the simplest model, there is no feedback to the cell voltage. So we have simply added an extra equation,

$$\frac{d^2q}{dt^2} + \gamma_O \frac{dq}{dt} + \Omega^2 q = Bv \qquad 14$$

The power spectrum of q is just the power spectrum for v, but with the additional factor

$$\frac{B}{(\omega^2 - \Omega^2)^2 + \gamma_O^2 \omega^2} \qquad 15$$

In principle, the resulting power spectrum can also be integrated, but it is more tedious than the case without the oscillator, so we have computed the total power by integrating the power spectrum numerically. Since the damping constant γ_O is the inverse of the oscillator's latency, we choose the constant to be one kHz. The results are shown in figure 2. Obviously, the oscillator improves the performance quite a bit, but for this value of γ_O the synaptic and cellular time constants must still be rather short.

SIMULATIONS

In sections 3 and 4, we had to use very fast time constants in order to get acceptable behavior from the model neuron. At very fast time constants, the linear approximation used breaks down, so one might question the relevance of the results. Also, while the signal-to-noise ratio may give the relative magnitudes of the signal and noise fluctuations, this is still not the same as spike rates. In this section we present the results of simulations of the model laminaris neuron, with and without the harmonic oscillator.

The differential equations for the membrane potential (without linearizing it) and

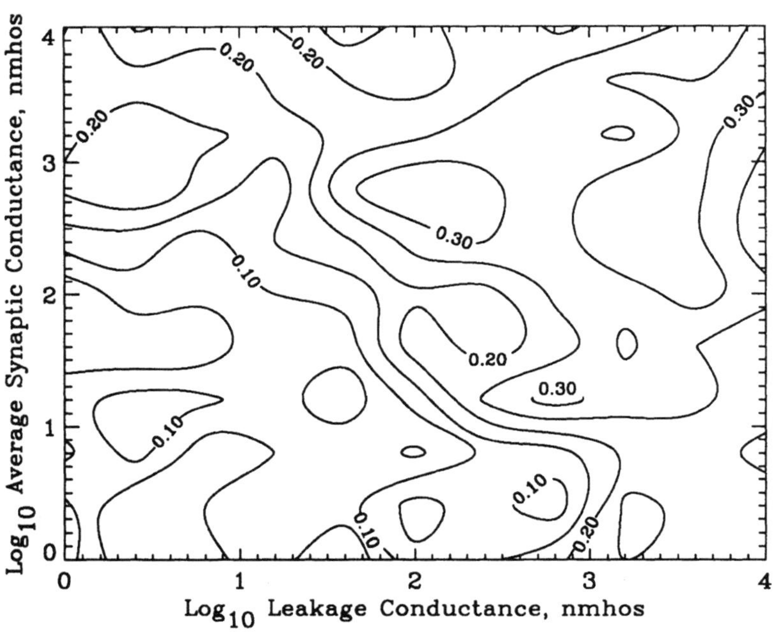

Figure 3. Rate difference between the in-phase and out-of-phase stimulus conditions from a simulation of a passive compartment model. The rate difference is expressed asa fraction of the maximum possible firing rate. The synaptic time constant was 0.01 msec.

the oscillator were integrated in time steps much shorter than the stimulus period or any of the time constants. The synaptic conductance obeys the equation for a critically damped harmonic oscillator driven by the input spikes:

$$\frac{d^2 g_E}{dt^2} + 2\gamma_S \frac{dg_E}{dt} + \gamma_S^2 g_E = A\xi(t) \qquad 16$$

A input spike just increments dg_E/dt by A. This is just a convenient way of realizing equation 3.6. Spikes only arrived at the beginning of time steps. Each independent variable's differential equation was integrated by using the exact expression for the solution of the same equation with the other variables held constant. An action potential was considered to be generated when the membrane potential exceeded a fixed threshold, or when the oscillator coordinate q exceeded a threshold in the model with an oscillator. The threshold which gave the largest difference in the firing rate in a 100 msec interval was used. The cell's capacitance was 30 pF (from a specific capacitance of 1 $\mu F/cm^2$ and a cell diameter of 30 μm). The upper limits were chosen to give similar time constants as in figure 2. For comparison, note that if the soma had a high density (500 / μm^2) of sodium channels as in the Hodgkin-Huxley model and all of them were open (conductance ~ 0.01 nmho), the conductance would be about 50000 nmho, giving a time constant of about

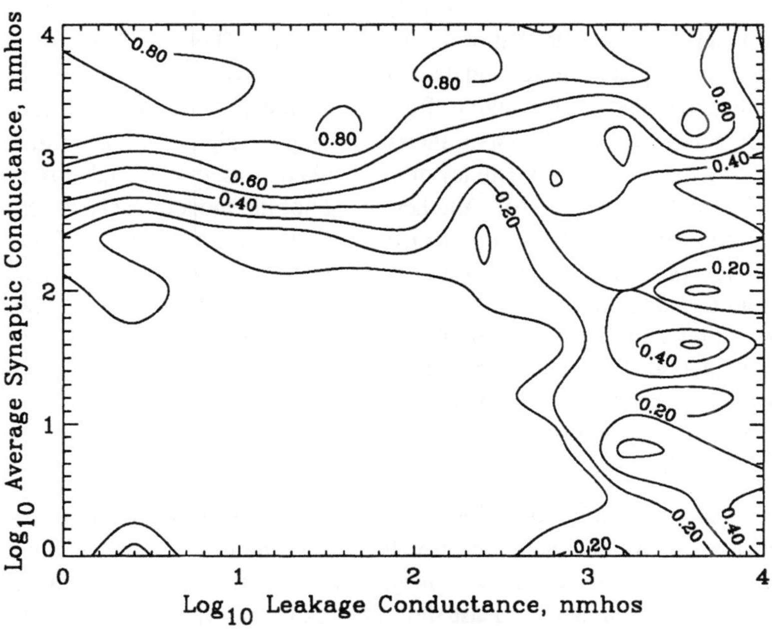

Figure 4. Rate difference between the in-phase and out-of-phase stimulus conditions from a simulation of a passive compartment model driving a damped harmonic oscillator. The rate difference is expressed as a fraction of the maximum possible firing rate. The synaptic time constant was 0.25 msec.

0.0006 msec. Figures 3 and 4 show the resulting difference between the in-phase and out-of-phase firing rates, shown as a fraction of the maximum firing rate. Without an oscillator, the rate difference is significant, but only at very high conductance values. With an oscillator there is substantial improvement, but it is still questionable whether a neuron can have such high conductances and short synaptic time constants.

CONCLUSIONS AND FUTURE WORK

We have presented the results of calculations of a passive single compartment model for a laminaris neuron. The results show that such a model cannot work at the high frequencies and noise levels which are typical of the input to a laminaris neuron. Faster time constants for the synaptic conductance and membrane potential cannot cure the problem, which arises from the noisy nature of the input. There must be some mechanism for filtering the neuron's input to suppress the noise relative to the signal. We have also presented calculations for a resonance model of such filtering. This model has substantially improved performance, and in fact can work well if the synaptic and membrane time constants are fast enough. It can work well even if those time constants are slow, but only by increasing the latency of the cell's response.

Part of the purpose of this work was to determine if ordinary neuron models can show behavior similar to that of a laminaris neuron. There are several fairly conventional changes to these models which could improve their performance. One such change would be an adaptive effective threshold. This would suppress low frequency components in the input, thereby raising the signal-to-noise ratio. Active channels as in the Hodgkin-Huxley model could do this, or one could use a simpler model like Hill's [Hill, 1936]. Another change of the model which could improve its performance would be a more realistic input spike train model. The Poisson process model of the input is the noisiest possible choice, and is almost certainly noisier than the actual input. For example, the refractory periods of magnocellular neurons prevent more than a certain number of spikes from arriving at laminaris in a given interval. There may also be a lower limit on the number of spikes which can arrive in a given interval, if magnocellularis neurons can't go longer than a certain interval without spiking. These properties are not true of a Poisson process, and may reduce the noise.

There is also the question of the biophysical mechanism of such high-frequency resonance. Active channels can show behavior similar to resonance, but they are normally thought to respond too slowly. In addition, they would form a non-linear system whereas laminaris at least shows hints of being linear in the magnitude of the modulation of the input firing rate. Hudspeth proposed a resonator model of cochlear hair cells [Hudspeth, 1985]. However, this model requires different external ionic concentrations on different parts of the resonating cell's membrane. The structure of laminaris gives no hint that it can maintain such differences. Also, Hudspeth's model operates at very low frequencies, on the order of one hundred Hertz.

We do not yet have a mechanism to suggest, we can only say that some enhancement of the signal relative to the noise is necessary. In order to make further progress, the resonator hypothesis needs to be tested experimentally. The voltage response to step current injections and/or sinusoidal current injections at various amplitudes and frequencies around the neuron's best frequency should reveal the presence of a resonator and provide insight into its mechanism, if in fact a resonator is found.

ACKNOWLEDGMENTS

This work was supported by AFOSR contract F49620-89-C-0131 and DARPA contract F49620-90-C-0010.

REFERENCES

Carr, C.E. and M. Konishi (1988). Axonal delay lines create maps of interaural phase difference in the owl's brainstem. Proc. Natl. Acad. Sci. USA 85:8311-8315.

Carr, C.E. and M. Konishi (1990). A circuit for detection of interaural time differences in the brainstem of the barn owl. J. Neurosci. 10:3227-3246.

Hill, A.V. (1936). Excitation and accommodation in nerve. Proc. R. Soc. London, Ser. B, 119:305-355.

Hudspeth, A.J. (1985). The cellular basis of hearing: the biophysics of hair cells. Science, 230:745-752.

Jeffress, L.A. (1948). A place theory of sound localization. J. Comp. Physiol. Psychol. 41:35-39.

MacGregor, R.J. (1987). Neural and Brain Modelling. Academic Press, San Diego.

Moiseff, A. and M. Konishi (1981). Neuronal and behavioral sensitivity to binaural time differences in the owl. J. Neurosci. 1:40-48.

Moiseff, A. (1989). Bi-coordinate sound localization by the barn owl. J. Comp. Physiol. 164:637-644.

Sullivan, W.E. and M. Konishi (1984). Segregation of stimulus phase and intensity coding in the cochlear nucleus of the barn owl. J. Neurosci. 4: 1787-1799.

Sullivan, W.E. and M. Konishi (1986). Neural map of interaural phase difference in the barn owl's brainstem. Proc. Natl. Acad. Sci. USA 83:8400-8404.

A Computational Model of the Cat Medial Geniculate Body Ventral Division

Bryan J. Travis
MS-F665
Los Alamos National Laboratory
Los Alamos, NM 87545

Abstract

The medial geniculate body of the thalamus receives input from a variety of sources, both auditory and non-auditory. The ventral division (MGv), however, appears to be a purely auditory center, that is, it receives from lower (and higher) centers which are primarily involved in encoding auditory stimuli. It has some of the characteristics of auditory cortex, e.g., layering of neurons and EE/EI bands, but has a much simpler architecture. The MGv plays a central role in analysing auditory stimuli. A wealth of data is available now on its anatomy, connectivity and responses. A detailed computational model of the ventral division of the cat medial geniculate body has been constructed. This model is being compared to physiological recordings for validation purposes, and should prove a useful tool for testing theories of auditory sensory representation and functioning.

INTRODUCTION

Value of Modeling Neural Structures

Recent advances in recording from neurons and in staining techniques are revealing the details of a key structural level in the brain -- that of circuitry or micro-architecture. This is the level that is believed to hold the secret of how the brain performs such complex functions as vision and audition and motor control, and it is also the level at which computational models may make their greatest contribution.

The need for realistic computational models of neural micro-architecture is growing increasingly apparent. While traditional neural networks have made inroads to understanding cognitive functions, more realism (in the form of structural, dynamic and connectivity constraints) is likely required to explain processes such as vision or audition. In the past few years, a number of highly detailed architectural models of neural systems have been developed. Rather than model a function or behavior itself, either mathematically or computationally, these models mimic the actual network architecture of a system to determine what contributions various aspects of structure -- including connectivity, synaptic weighting, and geometry -- make to function.

Models serve several purposes. They provide a mechanism for uniting in a single operation a great deal of diverse information about neural systems. In addition, they allow for testing the consistency of such data. Further, they allow us to determine the relative contributions of various pathways and processes, and to test theories about the relationship between function and structure. Finally, they can make predictions which can be tested experimentally and thereby guide as well as interpret experimental research.

Modeling The Auditory System

The sense of hearing serves two functions in higher animals. It is an essential part of communication (speech in humans, sound signals and utterances in lower mammals), and it is used to determine spatial location of sound sources. A great deal of experimental work has been reported in the last three decades on the structure and functioning of the mammalian (in particular the cat) auditory system. There has not been as much attention from a modeling perspective as there has been for vision and motor control, however. It is an area ripe for exploration through computational modeling.

In this short paper, only the anatomy and connectivity of the thalamic MGv is presented, and simple calculations to reveal some of the possible dynamics in that body are described. This study is part of a larger effort to model an entire auditory system. A detailed computational model of the cat auditory system, including the basilar membrane and hair cells of the cochlea, plus selected divisions of the cochlear nucleus and inferior colliculus, as well as the ventral division of the medial geniculate body, and a rudimentary cortical AI model, has been constructed, is being tested and will be described elsewhere (earlier and simpler versions have been described, e.g., in Travis, 1987). It is faithful to what is known about the anatomy and neural connectivity of these systems (see, e.g., Britt and Starr, 1976a,b; Pickles, 1982; Imig and Morel, 1983; Morest and Oliver, 1984; Oliver and Morest, 1984; Brugge and Reale, 1985; Oliver, 1984; Oliver, 1987; Mitani and Shimokouchi, 1985; Mitani et al., 1985; Matsubara and Phillips, 1988; Diependaal and Viergever,

1989; Westerman and Smith, 1988; McKenna, Weinberger and Diamond, 1989; Oliver et al., 1991), and includes models of physiological responses for the identified neuron types. It is restricted to nuclei that appear to be strictly auditory as opposed to those that are clearly multi-modal.

MEDIAL GENICULATE BODY VENTRAL DIVISION

The thalamic medial geniculate body (MGB) plays a central role in analysing auditory stimuli. A wealth of data is available now on its anatomy, connectivity and responses (see, e.g., Morest, 1964, 1965, 1975; Aitken and Webster, 1972; Majorossy and Kiss, 1976; Middlebrooks and Zook, 1983; Imig and Morel, 1985). The medial geniculate body of the thalamus receives input from a variety of sources, both auditory and non-auditory. The ventral division (MGv), however, appears to be a purely auditory center, that is, it receives only from lower (and higher) centers which are solely involved in encoding auditory stimuli. The MGv is the principal source of inputs to the AI auditory cortical area.

The structure of MGv is simple in comparison to AI -- it has only two neuron types: principal neurons (PN), whose output proceeds up to AI, and Golgi II cells (GII) which are inhibitory interneurons. In the cat, the principal neurons (PN's) are arranged in lamina (see Fig. 1; the laminae have been unrolled to make the underlying structure more apparent ; there are approximately 50 laminae). Each lamina is an isofrequency sheet, that is, all the neurons on a sheet have the same characteristic best frequency. The neurons in MGv as well as in the cortical region to which MGv projects (AI) are characterized by fairly sharp tuning curves. The dendritic field of a PN extends approximately 240 microns in both directions within its isofrequency sheet, or lamina, but only 45 microns in the direction perpendicular to the lamina. Laminae occur at about a 30 micron spacing. Spacing between PN somas within a sheet is about 180 microns (Morest, 1964). The overall dimensions of the MGv, then, are about 1,500 microns in x (perpendicular to laminae), and roughly 2,000 microns in y and z (sheet coordinates). Based on these dimensions, it should contain about 6,000 PN's, and approximately 24,000 GII's. The GII cells are dispersed among the PN's. Their dendritic fields are roughly 100 micron in diameter and extend over 2 or 3 laminae. Their dendrites can enter into glomerular structures, nests, with PN dendrites and ascending fibers from the ipsilateral inferior colliculus central nucleus.

There appear to be two classes of ascending fibers -- the thicker fibers which contact the PN-GII glomerular nests, and fine fibers, which synapse on distal branches of PN's and on distal or proximal regions of GII's. Ascending fibers enter the MGv and course along an isofrequency sheet, sending collaterals to the lamina on either side of the target sheet (see Fig. 2) as well as within the target lamina. GII cells send axon collaterals to PN-GII glomerular nests (Morest, 1975), to PN distal

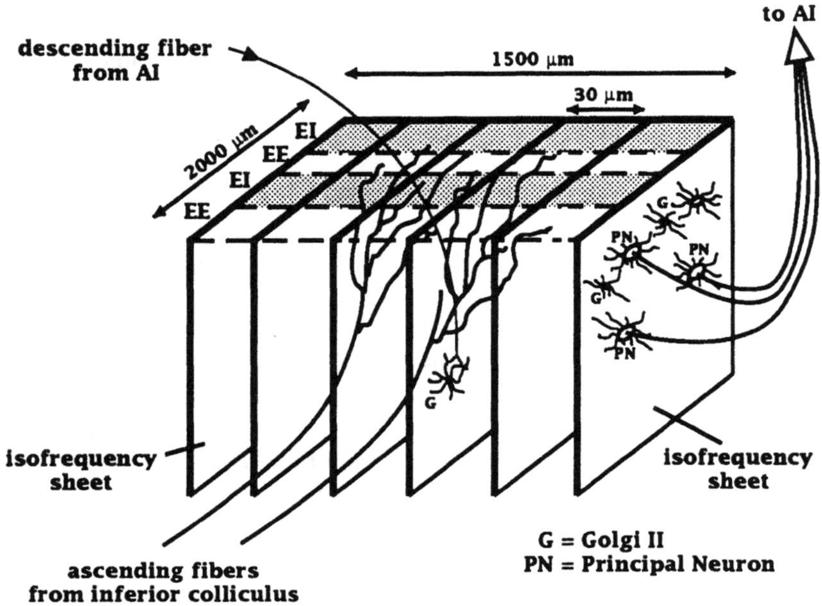

Figure 1. Basic features of cat Medial Geniculate Body ventral division structure. Isofrequency sheets have been flattened.

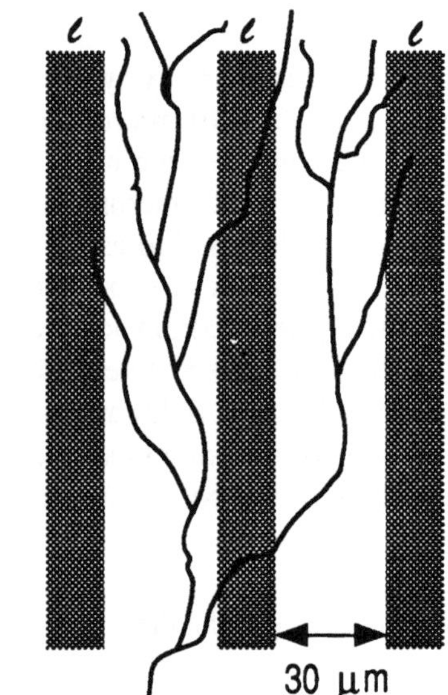

Figure 2. Pattern of branching of ascending fibers from inferior colliculus. Dark bands represent isofrequency sheets.

dendritic branches, and to proximal and somal regions of other GII's. Morphological evidence suggests that GII's are inhibitory neurons. There is an additional set of inputs to the MGv, i.e., descending fibers from AI. These project to distal parts of PN's and especially to proximal and somal regions of GII's. Fig. 3 summarizes the connections seen in the MGv. The descending fibers tend to course parallel to the laminae in a direction perpendicular to the ascending fiber paths. The descending fibers can exert a significant control over the MGv by activating the strongly inhibitory GII's. The fine ascending fibers however probably play more of a modulating role; their synapses on GII's tend to occur in the distal processes rather than near or on the soma.

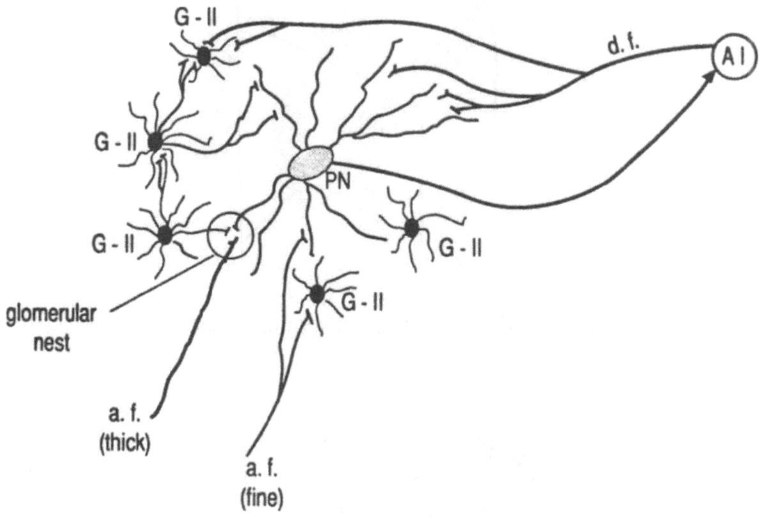

Figure 3. Principal connections between ascending and descending fibers and MGBv neurons.

In addition to the isofrequency laminae, an additional level of structure exists in the MGv: EE and EI bands. These bands run perpendicular to the laminae (see Fig. 4). About 5-6 bands have been identified; these correspond to similar bands in AI. Cells within an EI band are stimulated by contralateral input and are inhibited by simultaneous input to both ears. EI bands are generally very distinct only for frequencies above 3-5 kHz. Below 2.5 kHz, e.g., AI (and presumably other auditory centers) is strictly EE (i.e., neurons are excited by stimulation to one or both ears). This banded structure is seen in most cortical auditory maps, in the MGv and in the inferior colliculus central nucleus. Generally, neurons in EE bands project to EE neurons in the next higher station, and similarly for EI neurons. There is both convergence and divergence in projections between bands, as illustrated in Fig. 5. Based on retrograde labelling studies (Middlebrooks and Zook, 1983) cells in an AI EI band, e.g., receive projections from neurons in all of the MGv EI bands (at a given frequency) and most neurons in an MGv-EI band project across all the AI-EI bands. The EE bands do not exhibit this same degree of convergence/divergence in projections from MGv to AI. Each EE band in AI receives projections from a different mix of EE bands in MGv, suggesting a qualitative difference in properties (Middlebrooks and Zook, 1983; Kitzes et al, 1980; Phillips and Irvine, 1981).

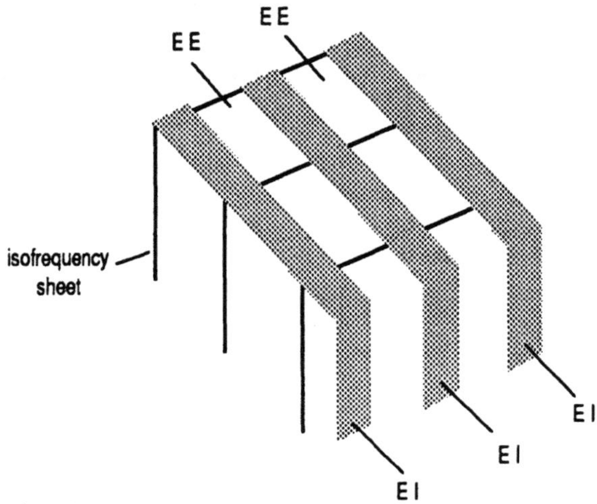

Figure 4. Relationship of EE and EI bands to isofrequency sheets.

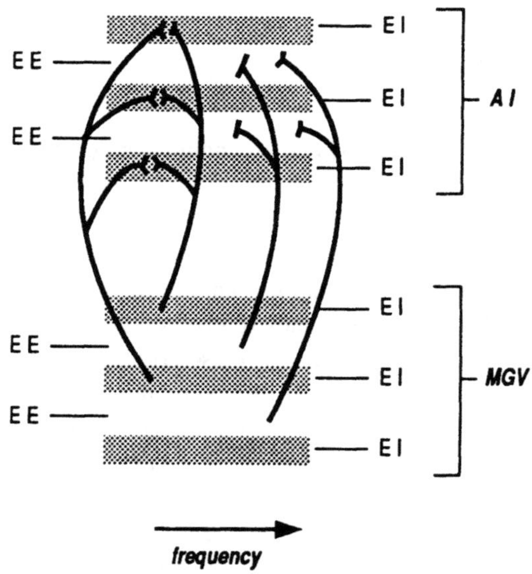

Figure 5. Convergence and divergence of projections from EI and EE bands in MGBv to cortical AI area.

There is evidence (e.g., Middlebrooks and Zook, 1983; Andersen et al, 1980) that the binaural bands in AI receive input from different brainstem nuclei through separate pathways. The EE bands in AI sample different parts of the EE regions in MGv, and presumably represent information from different lower brainstem nuclei. However, the EI bands in AI clearly receive input from all the EI bands in the MGv. The EI bands may be involved in spatial localization while sound spectral analysis could be handled by EE bands. The functional unit may be an EE/EI combination.

Additional dimensions in response properties exist within each band. For example, in EE regions, binaural response neurons at a given frequency can vary in differences in relative thresholds for the two ears, strength of facilitation, and aural dominance (see Phillips and Irvine, 1981). Neurons in EI bands at a given frequency can vary in threshold intensity, strength of suppression or inhibition, and width of tuning curve, for example.

SIMULATIONS WITH A NUMERICAL MODEL

There is almost always a problem with trying to model real neural structures. Invariably, there is missing information, either regarding structure, or connectivity, or physiological response, or nature of the input signals. In this particular case, the details of the MGv are actually relatively well defined in terms of structure and connectivity. There are also available data on recordings from individual neurons in this structure, so that the dynamics of the neuron types are not a complete unknown. The nature of the inputs to the system is, however, not clear. The MGv receives input from the inferior colliculus central (ICC) nucleus. The ascending thick and fine fibers may correspond to the axons of the disc-shaped cells and the stellate cells, respectively, in the ICC. The information coded in the incoming signals from those cell types is not known. Similarly, the descending fibers from AI modulate the output of the MGv, but why that feedback exists is not known. The descending fibers are a good example of another kind of difficulty often encountered. There is apparently conflicting information on their role; on one hand, these projections are known to make asymmetric synaptic contact on their target cells, a usual morphological signature of excitatory synapses; on the other hand, the descending fibers are known to suppress activity in the MGv. In this case, these conflicting reports could be reconciled in a model in which the descending fibers excite the Golgi II cells, which have a strong inhibitory effect on the Principal Neurons.

To really understand the role of the MGv, what its function is, we may very well need to model the entire auditory system, at least up through AI. However, we can consider the MGv in isolation and determine what effect its structure will have on whatever signals it receives. That is, we may determine what it does, if not why. In fact, this is a common strategy in science: take something apart, see how the components work, and derive from that an idea about how the total system functions.

The present study is concerned with (1) validating the MGv model by comparing model neuron responses to observed responses for qualitative agreement, and (2) conducting numerical experiments to see what happens to signals that enter the structure. Unfortunately, there is only space here for a sparse sampling of results.

MODEL DYNAMICS

The steps in a simulation consist of: (1) specifying the numbers, types and spatial distributions, spatial dimensions and conductive properties of neurons; (2) determining the connectivity matrix, based on sizes and locations of dendritic trees and axonal projections; (3) computation of the state of each neuron for each of a sequence of time steps (timestep size is typically 0.1 msec), in response to a given input pattern of stimulation; (4) graphical and statistical analysis of the simulation results.

In the present version of the model, neurons in MGv are represented as three-compartment neurons: a basal/somal compartment plus a proximal and a distal compartment. A special compartment is also designated for glomerular nests. (The simulator used in this study can treat neurons having a large number of branching, compartmentalized dendrites, but generally when many neurons are involved, the number of compartments per neuron must be reduced because of computer time limitations.) In each compartment, inputs from axons synapsing onto the neuron in that compartment are summed and convolved; each synapse is represented by a set of one or more conductances. Each time-dependent conductance is characterized by a peak conductance value, an equilibrium voltage and a time constant; the transient response of each conductance is computed and tracked. For excitatory synapses, a single conductance is used -- a peak conductance of 0.1 nanoSiemens, an equilibrium voltage of +55 mV, and a time constant of 1.5 msec. For inhibitory synapses, a single conductance is also used -- a peak conductance of 0.1 nanoSiemens, an equilibrium voltage of -65 mV, and a time constant of 10.0 msec. For all neurons, membrane capacitance Cm = 1 microFarad/cm^2, membrane resistance Rm = 100,000 ohm-cm^2, and axial resistance Ra = 200 ohm-cm. Current diffuses between compartments and into the soma, where Hodgkin-Huxley dynamics (voltage dependent conductances), linearized at low stimulus intensities (see, e.g., Theunissen, Eeckman and Miller, 1992), determine the total neuron response. This model, then, contains about 100,000 compartments. Simulations with the full MGv model contain about 3,000,000 synapses. Incoming axons will synapse in one of the neuron compartments, determined by the characteristics of the donor and acceptor neuron types. A probability factor can be applied, so that, for example, the ascending fibers will synapse on a certain fraction of principal neurons in their range of influence; the neurons that actually receive the input can be determined on a random basis.

The overall structure of the MGv has already been explained. The 30,000 neurons in the model are distributed according to the dimensions previously summarized. Each model neuron receives about a 100 synapses on average; this is still considerably less than reality. Connection strengths in the model must be multiplied by a factor which is approximately equal to the ratio of actual number of synapses between the MGv neuron types to the number of synapses between the model neurons. The actual average numbers of synapses can be estimated from microphotographs taken in cell-staining experiments. This provides a handle on the range of connection strengths that yield interesting behavior. Model neurons are also limited in that they have only three compartments (in large system simulations), so that much of the processing that probably occurs in real MGv neuron dendrites is averaged out in this model.

In one set of sample simulations, a fixed firing rate is established in ascending fibers innervating one isofrequency sheet. Figure 6a shows the axon hillock potential vs. time for one principal neuron in the target isofrequency lamina, for the case of 50 Hz firing rate in the ascending fibers, and 40 Hz firing rate in the descending fibers to the target isofrequency lamina. There is a quiet period of about 60 msec between the initial burst of spikes and subsequent spiking. This is seen in many of the model PN neuron responses. Fig. 6b shows the typical response of a model PN on an isofrequency lamina at some distance from the target sheet, corresponding to a lower characteristic best frequency than the target. Only one burst of two spikes is seen, at a latency of about 120 msec. Both figures are very similar to the kinds of measured responses found for neurons in MGv (e.g., compare to Fig. 10b-e of Aitken and Webster, 1972). The initial firing of PN's is cut off by the action of GII's. But the GII's inhibit each other as well. As the inhibition fades, the incoming signal can once again excite PN's to discharge. The interval between PN spikes depends both on the characteristic time constant of the GII membrane channels and the intensity (firing rate) of the stimulus, and also on the complex interaction of GII's. At higher stimulus intensity (100 Hz firing rate in the ascending fibers), the PN response typically becomes less regular (Fig. 6c). Spiking tends to occur in clusters separated by 30 to 60 msec silent periods. (This same response is also seen at lower stimulus intensities at distances from the target isofrequency sheet less than that represented by Fig. 6b.) When the descending fiber input is removed (as should be the case in a cortical ablation or heavy anesthesia experiment), the PN cell response is primarily a series of regularly spaced bursts of spikes (Fig. 6d). This might be the basis for the reverberations recorded in MGv units in heavily anesthetized cats as reported by Aitkin et al (1966). Finally, the reponse of GII cells can vary greatly. Fig. 6e shows the reponse of a GII neuron in the isofrequency target lamina at high stimulus intensity. It receives so much inhibitory input from surrounding GII's that it can only fire at roughly 100 msec intervals. However, another GII, on an adjacent lamina, fires very rapidly (Fig. 6f). This suggests that

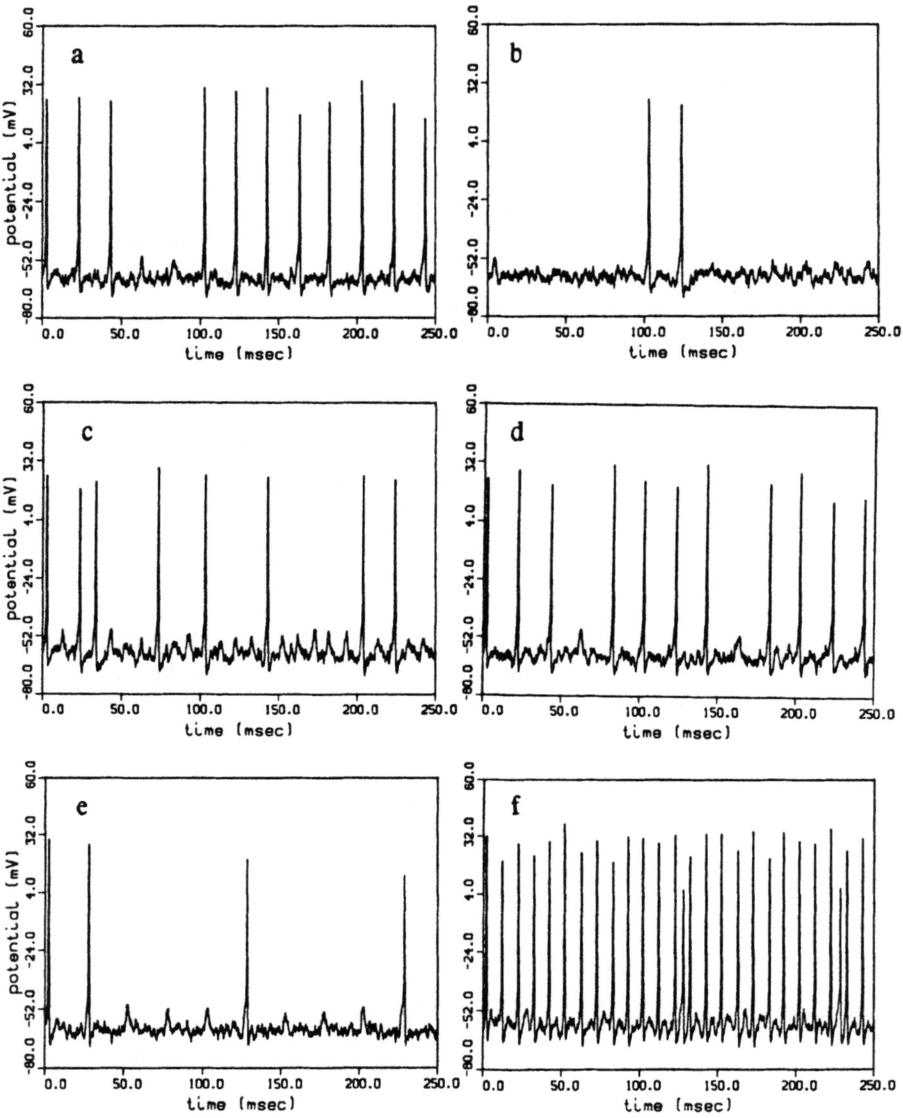

Figure. 6. Responses in model PN and GII neurons, as described in the text.

the target lamina is surrounded by strong inhibition in the adjacent lamina due to GII activity. GII reponses can vary between these two extremes. Many GII's fire with sporadic bursts of spikes. There is some evidence for plasticity in the synapses of GII's -- they may play a role in some form of auditory learning or adaptation in the MGv. Although these model responses appear realistic, their validity needs more support from additional comparisons with phsyiological data to insure that the calibrations are reasonable.

CONCLUSION

Neural modeling is feasible now, in the sense of capturing structure, connectivity and neuron number density of sensory subsystems, such as the MGv, and linking those with structure at smaller scales, via compartmentalization in individual model neurons. The level of detail in the current model allows comparison to a variety of physiological data. The effort at present is focused primarily on checking the self-consistency of the model and calibrating the major unknown parameter -- the average strength of connection between classes of neurons. As computer power increases, finer details can be resolved by increasing the number of compartments and capturing the dendritic geometries of neurons. In fact, one important question that modeling should be able to answer is: How much detail in individual neurons is needed for the higher level model to accurately reproduce the primary functions of a sybsystem such as the MGv, or, in other words, how does performance degrade as resolution of dendrites coarsens? There are other important questions that modeling may help answer -- e.g., what is the specific role of efferent projections from higher auditory nuclei to lower level ones, and how does the brain use the convergent/divergent nature of projections between nuclei, such as between the inferior colliculus central nucleus and the MGv, or between MGv and AI, when it would seem that such projections would smear out or degrade specific information from lower levels? The present MGv model can be used to investigate these questions, but only in the context of a more comprehensive lower auditory system model (cochlea through MGv). That will be the subject of a future paper.

REFERENCES

Aitken, L. M., C. W. Dunlop, and W. R. Webster (1966). "Click-evoked response patterns of single units in the medial geniculate body of the cat", J. Neurophysiol., 29, 109-123.

Aitkin, L. M., and W. R. Webster (1972). "Medial Geniculate Body of the Cat: Organization and Responses to Tonal Stimuli of Neurons in Ventral Division", J. Neurophysiol., 35, 365-380.

Andersen, R. A., R. L. Snyder and M. M. Merzenich (1980). "The Topographic Organization of Corticocollicular Projections From Physiologically Identified Loci in the AI, AII, and Anterior Auditory Cortical Fields of the Cat", J. Comp. Neurol., 191, 479-494.

Britt, R., and A. Starr (1976a). "Synaptic Events and Discharge Patterns of Cochlear Nucleus Cells. I. Steady-Frequency Tone Bursts", J. Neurophysiol., 39, 162-178.

Britt, R., and A. Starr (1976b). "Synaptic Events and Discharge Patterns of Cochlear Nucleus Cells. II. Frequency-Modulated Tones", J. Neurophysiol., 39, 179-195.

Brugge, J.F., and R. A. Reale (1985). "Auditory Cortex", chapter 6 in *Cerebral Cortex*, A. Peters and E. G. Jones, eds., Plenum Press, New York.

Diependaal, R. J., and M. A. Viergever (1989). "Nonlinear and active two-dimensional cochlear models: Time-domain solution", J. Acoust. Soc.Am., 85, 803-812.

Imig, T. J., and A. Morel (1983). "Organization of the Thalamocortical Auditory System in the Cat", Ann. Rev. Neurosci., 6, 95-120.

Imig, T. J., and A. Morel (1985). "Tonotopic Organization in Ventral Nucleus of Medial Geniculate Body in the Cat", J. Neurophysiology, 53, 309-340.

Kitzes, L. M., K. S. Wrege, and J. M. Cassady (1980). "Patterns of Responses of Cortical Cells to Binaural Stimulation", J. Comp. Neurol., 192, 455-472.

Majorossy, K., and A. Kiss (1976). "Specific Patterns of Neuron Arrangement and of Synaptic Articulation in the Medial Geniculate Body", Exp. Brain Res., 26, 1-17.

Matsubara, J. A., and D. P. Phillips (1988). "Intracortical Connections and Their Physiological Correlates in the Primary Auditory Cortex (AI) of the Cat", J. Comp. Neurol., 268, 38-48.

McKenna, T. M., N. M. Weinberger and D. M. Diamond (1989). "Responses of single auditory cortical neurons to tone sequences", Brain Res., 481, 142-153.

Middlebrooks, J. C., and J. M. Zook (1983). "Intrinsic Organization of the Cat's Medial Geniculate Body Identified by Projections to Binaural Response-Specific Bands in the Primary Auditory cortex", J. Neuroscience, 3(1), 203-224.

Mitani, A., and M. Shimokouchi (1985). "Neuronal Connections in the Primary Auditory Cortex: An Electrophysiological Study in the Cat", J. Comp. Neurol., 235, 417-429.

Mitani, A., M. Shimokouchi, K. Itoh, S. Nomura, M. Kudo and N. Mizuno (1985). "Morphology and Laminar Organization of Electrophysiologically Identified Neurons in the Primary Auditory Cortex in the Cat", J. Comp. Neurol., 235, 430-447.

Morest, D. K. (1964). "The neuronal architecture of the medial geniculate body of the cat", J. Anat., Lond., 98(4), 611-630.

Morest, D. K. (1965). "The laminar structure of the medial geniculate body of the cat", J. Anat., Lond., 99(1), 143-160.

Morest, D. K. (1975). "Synaptic Relationships of Golgi Type II Cells in the Medial Geniculate Body of the Cat", J. Comp. Neur.,162,157-194.

Morest, D.K., and D. L. Oliver (1984). "The Neuronal Architecture of the Inferior Colliculus in the Cat: Defining the Functional Anatomy of the Auditory Midbrain", J. Comp. Neurol., 222, 209-236.

Oliver, D. L. (1984) "Dorsal Cochlear Nucleus Projections to the Inferior Colliculus in the Cat: A Light and Electron Microscopic Study", J. Comp.Neurol., 224, 155-172.

Oliver, D. L. (1987) " Projections to the Inferior Colliculus From the Anteroventral Cochlear Nucleus in the Cat: Possible Substrates for Binaural Interaction", J. Comp. Neurol., 264, 24-46.

Oliver, D.L. and D.K. Morest (1984). "The Central Nucleus of the Inferior Colliculus in the Cat", J. Comp. Neurol., 222, 237-264.

Oliver, D. L., S. Kuwada, T. C. T. Yin, L. B. Haberly, and C. K. Henkel (1991). "Dendritic and Axonal Morphology of HRP-Injected Neurons in the Inferior Colliculus of the Cat", J. Comp. Neurol., 303, 75-100.

Phillips, D. P., and D. R. F. Irvine (1981). "Responses of Single Neurons in Physiologically Defined Primary Auditory Cortex (AI) of the Cat: Frequency Tuning and Responses to Intensity", J. Neurophysiology, 45, 48-58.

Pickles, J.O. (1982). *An Introduction to the Physiology of Hearing*, Academic Press, London.

Theunissen, F. E., F. H. Eeckman, and J. P. Miller (1992). "Linearization by Noise and/or Additional Shunting Current of a Modified Fitzhugh Nagumo Spiking Model", this volume.

Travis, B. J. (1987). “A Computational Model of the Auditory System”, Proc., IEEE International Conf. on Neural Networks, 4, 67-74.

Westerman, L. A., and R. L. Smith (1988). "A diffusion model of the transient response of the cochlear inner hair cell synapse", J. Acoust. Soc. Am., 83, 2266-2276.

SIMULATION OF NEURAL RESPONSES THAT UNDERLIE SPEECH DISCRIMINATION

Donal G. Sinex
Boys Town National Research Hospital
555 North 30th Street
Omaha, NE 68131

INTRODUCTION

Psychophysical acuity for speech sounds with differing voice onset time (VOT) has often been found to be nonmonotonic (Abramson and Lisker, 1970; Keating *et al.*, 1981; Kuhl, 1981; Rosner, 1984; MacMillan *et al.*, 1988; but see Kewley-Port *et al.*, 1988). Acuity for VOT is usually best for the syllables that bracket the boundary between voiced (/b, d, or g/) and voiceless consonants (/p, t, or k/) (Abramson and Lisker, 1970; Kuhl, 1981). This result was once thought to reflect the operation of a neural system that was specialized for processing speech. However, it is now known that chinchillas exhibit the same relation between identification and acuity for VOT (Kuhl and Miller, 1978; Kuhl, 1981). It has been suggested that some continua of speech sounds, such as those differing in VOT, may interact with general auditory mechanisms in a way that naturally produces a region of high sensitivity (Miller, 1977; Stevens, 1989). Until recently, however, specific hypotheses about the nature of this interaction or about neural responses that underlie enhanced psychophysical acuity or category formation had not been proposed.

The first detailed description of neural and spectral correlates of nonmonotonic acuity for a VOT continuum was provided by Sinex *et al.* (1991). They found that the variance of response latencies within small populations of primary auditory neurons was smallest for the same alveolar VOT syllables for which psychophysical acuity was best. For syllables with longer or shorter VOTs, latency variance was larger. Sinex *et al.* hypothesized that the location of the voiced-voiceless boundary for VOT might be determined by this pattern of latency variance. However, their conclusions were based on a study of one stimulus continuum and one sound pressure level (SPL). A complete description of VOT coding should be able to account for the known variability of psychophysical VOT boundaries, although the time required to examine the coding of other VOT continua in neurophysiological experiments would be prohibitive. In the present study, a model of the auditory periphery was used to simulate responses to additional speech stimuli for which the voiced-voiceless boundary is known to occur at different VOTs. The hypothesis that psychophysical VOT boundaries lie in the region of low neural variance was tested for three VOT continua differing in place of articulation. The primary goal of this study was to apply a fairly standard auditory model to a new problem in neural coding. The application of existing models to important neurophysiological questions provides an essential complement to the development of the models themselves.

STIMULI

Three VOT continua based on those first studied by Abramson and Lisker (1970) were used. The formant trajectories were appropriate for labial (/ba/-/pa/), alveolar (/da/-/ta/), and velar (/ga/-ka/) places of articulation. Each continuum was generated by "cutting back" or delaying the onset of the first formant (F1). Prior to VOT, higher formants (F2 and above) were excited by aspiration noise. At the onset of voicing, all formants were excited by periodic glottal pulses. For these syllables, the trajectory of F1 varied such that the location of the voiced-voiceless boundary shifted systematically with place of articulation (Lisker and Abramson, 1970).

THE AUDITORY MODEL

The model used here is representative of a class of auditory models (i.e., Geisler, 1989). Two stages, approximating basilar membrane vibration and transduction at the inner hair cell (IHC), were included. Each stage in the present model was developed by others and has been described in detail by the original investigators; only an overview is provided here.

The first stage of the model was a linear bandpass filter that approximated the frequency selectivity of a primary auditory neuron. Filter characteristics were computed from equations derived by Carney and Yin (1988). In a preliminary study, filters were computed to match individual chinchilla tuning curves (Sinex and Geisler, 1984). However, that procedure requires more computation, and the results presented below do not depend strongly on the exact shape of the filters. The present study used twelve filters (six steps per octave from 0.25 - 0.89 kHz) that covered the range over which the best neural cues to voicing onset were obtained (Sinex and McDonald, 1988, 1989).

The output of a filter was assumed to represent the input to an IHC. Neurotransmitter release at the IHC synapse was simulated with an implementation of the algorithm described by Meddis *et al.* (1990). This algorithm generates accurate approximations of responses to other stimuli and is very efficient (Hewitt and Meddis, 1991). All parameters for the synapse model were fixed except for parameter B, which governs the permeability of the IHC membrane. Two values of B were used to simulate neurons whose spontaneous discharge rates (SRs) were approximately 55 and 15 spikes/sec. These SRs are representative of the high and medium SR categories observed in recordings from the auditory nerve (Liberman, 1978). The thresholds of simulated neurons with medium SRs were approximately 9 dB higher than the thresholds of high SR simulated neurons. Comparable threshold differences are typical of real neurons (Liberman, 1978).

RESULTS

The model produced spike trains in response to VOT syllables that closely resembled the spike trains recorded from chinchilla auditory nerve fibers (Sinex and McDonald, 1988, 1989) and could be analyzed with the same methods used by Sinex *et al.* (1991). To produce quantitative measurements of response variability, the latency of a discharge rate increase was identified in each rate profile. The means and standard deviations of these latencies for the 24 simulated neurons are shown as a function of VOT in Fig. 1. The means and standard deviations of 39-54 real neurons are also shown for comparison. Except at the shortest VOTs, the mean latencies closely matched the stimulus VOTs (perfect agreement is indicated by the dotted line with slope = 1). The mean simulated latencies were nearly identical to the mean latencies of real neurons. The latency standard deviations of the simulated neural responses were always smaller but followed the same nonmonotonic pattern as the real standard deviations. The difference probably reflected the fact that the population of real neurons included a larger range of thresholds and SRs than did the simulation.

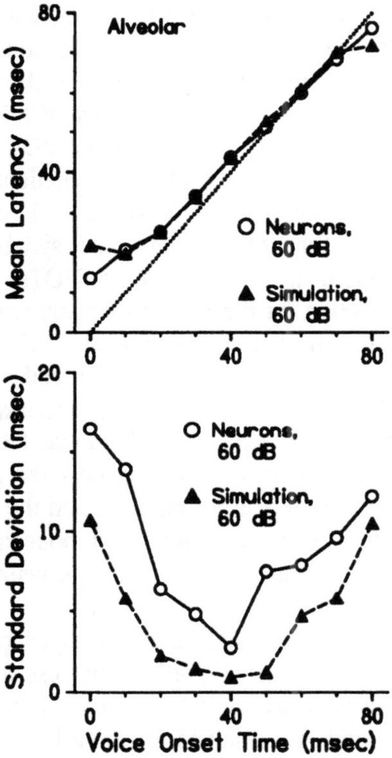

Figure 1

A normalized measure of the difference between the responses to pairs of syllables was computed as

$$D = (L_2 - L_1) / (V_1 + V_2)^{0.5},$$

where D is the index of response difference, L_n is the mean response latency, and V_n is the variance of the latencies. The index D, which we will refer to as representing "neural acuity", is analogous to the psychophysical measure d' from Signal Detection Theory (Green and Swets, 1974).

The shape of the neural acuity function closely resembled the shape of the psychophysical acuity function reported for similar syllables. Figure 2 compares neural to psychophysical acuity for constant VOT differences along an alveolar continuum. The neural acuity function was obtained from the response latencies of 24 simulated neurons driven by the alveolar VOT continuum presented at 70 dB above the threshold of the high-SR neurons. The psychophysical acuity data were obtained by Rosner (1984) for a similar alveolar continuum. Each curve has been normalized to its own peak. In each case, acuity was greatest for the syllables with VOTs of approximately 30-40 msec, and declined for shorter and longer VOTs. The overall shapes of the functions were similar, indicating that the neural acuity function provides an accurate prediction of the region of enhanced psychophysical acuity. Because the region of enhanced psychophysical acuity is related to the location of the identification boundary (Abramson and Lisker, 1970; Kuhl, 1981), the location of the neural acuity function can be compared to the location of the psychophysical voiced-voiceless boundary.

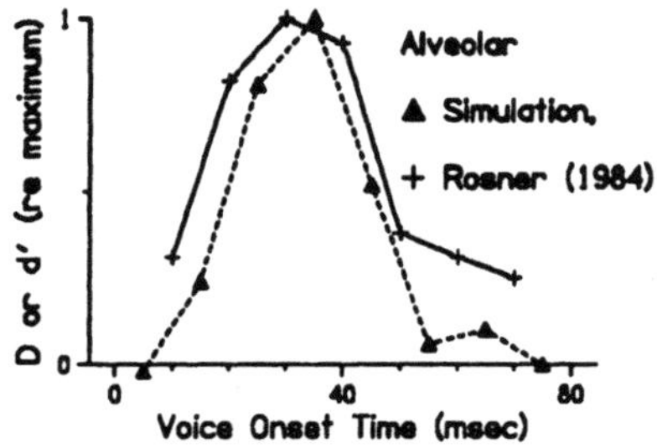

Figure 2

A region over which VOT syllables were encoded with low variance was observed for each of the three VOT continua described above. The data are shown as normalized neural acuity functions in Fig. 3. For the labial continuum, the acuity peak occurred for VOTs near 20-30 msec. The function for alveolar syllables was shifted to slightly longer VOTs, and the function for the velar continuum was shifted still further. Psychophysical labeling boundaries obtained by Lisker and Abramson (1970; English speakers) are shown as open symbols above the curves. The boundaries occurred near the peaks of the neural acuity functions, and the ordering of the psychophysical VOT boundaries was the same as the ordering of the neural acuity functions.

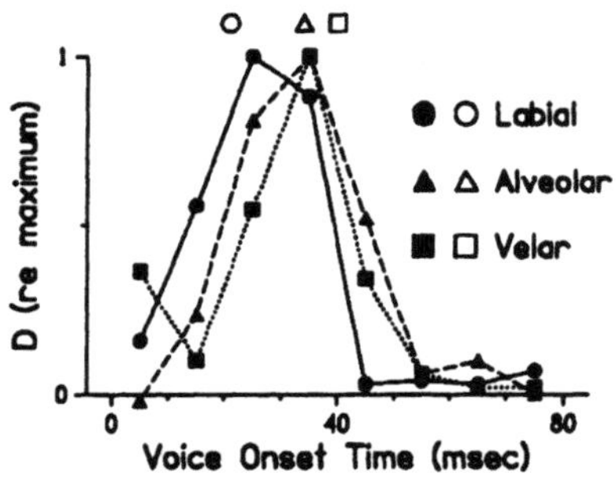

Figure 3

The location of the psychophysical voiced-voiceless boundary is not fixed, even across experiments that use identical sounds. Some of the variability is a consequence of the language spoken by the experimental subjects. However, boundaries obtained for English speakers (who have been studied most often) also vary across studies, even when nearly identical stimuli are used (Lisker and Abramson, 1970; Kuhl and Miller, 1978; Soli, 1983). Figure 4 presents neural acuity functions obtained at 70 dB above threshold for each of the three VOT continua. The + symbols plotted with each curve are 11-15 psychophysical labeling boundaries reported by Lisker and Abramson (1970), Caramazza *et al.* (1973), Kuhl and Miller (1978), Soli (1983), and Keating *et al.* (1981), and obtained informally by Sinex (unpublished). These experiments all used Lisker and Abramson's original VOT continua or syllables resynthesized from the same specifications. Data from speakers of five languages (English, Spanish, Thai, French, and Polish) and two species (humans and chinchillas) are represented in the figure. It should be noted that some Spanish and Polish boundaries that were strongly influenced by the presence of prevoiced syllables have been omitted, since the hypothesis under study here is relevant only for the short VOT lag region of the continuum. There was extensive variability in the VOT boundaries at each place of articulation, but it is clear that the boundaries obtained from psychophysical experiments tended to fall within the VOT region over which the neural acuity function was elevated. Thus, our view that the location of the VOT boundary is constrained by the variance of neural responses in the peripheral auditory system is supported. One interpretation of these results is that speakers of a particular language may learn to divide a voiced-voiceless continuum anywhere within the broad region over which neural acuity is elevated.

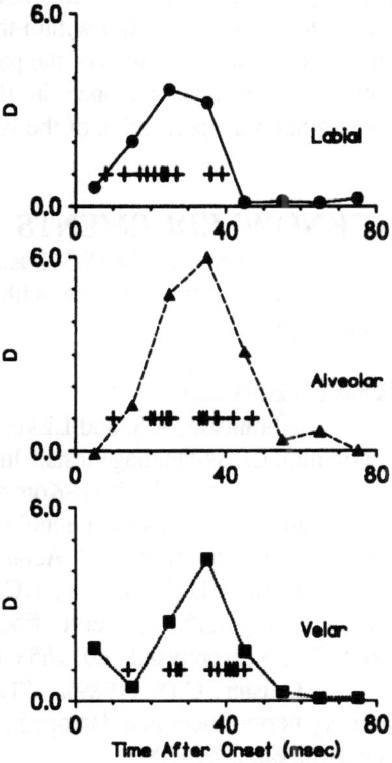

Figure 4

In a second experiment, simulated neural responses to the VOT continua were obtained at different levels. Although the effects of stimulus level on the perception of voicing have been studied infrequently, the data that are available show a small but consistent increase in the boundary VOT with decreases in level (Simon, 1978; Van Tasell and Crump, 1981; Kluender and Jenison, 1991). In simulations, decreasing the stimulus level increased the variability of latencies to short VOTs , and decreased the variability of latencies to long VOTs. As a result, the peak in the neural acuity function shifted to longer VOTs with decreases in level. According to our hypothesis, this pattern predicts that the voiced-voiceless boundary should increase at lower SPL, and the simulation correctly predicts the direction of the shift obtained in the three psychophysical studies. However, the quantitative agreement between the simulated neural and psychophysical data was not good, since the magnitude of the psychophysical effect was much smaller than the shifts predicted by the simulation. This lack of agreement may be a consequence of the use of linear filters in the model. A compressive nonlinearity after the filter might reduce the slope of the level effect.

SUMMARY and CONCLUSIONS

The auditory model generated responses to VOT syllables that closely resembled the responses of real neurons to the same sounds. The simulated responses were used to estimate neural acuity for VOT. The location of the peak in neural acuity for VOT shifted with changes in place of articulation and with changes in stimulus level. In each case, the neural acuity function shifted in the same direction as the psychophysical labeling boundary. Although the psychophysical voiced-voiceless boundary can vary across experiments and across languages, it tended to fall within the region of increased acuity as predicted from simulation of the responses of the peripheral auditory system. We conclude that the pattern of response variance in the auditory periphery provides an important constraint on the perception of the voicing contrast for stop consonants.

ACKNOWLEDGEMENTS

Supported by DC00341 and DC00215 from NIDCD. The author thanks Donna Beehler for assistance with the simulations and with preparation of the manuscript.

REFERENCES

Abramson, A.S. and Lisker, L. (1970). "Discriminability along the voicing continuum: Cross-language tests," In Proc. 6th Intl. Cong. Phon. Sci., 569-573.

Caramazza, A., Yeni-Komshian, G.H., Zurif, E.B., and Carbone, E. (1973). "The acquisition of a new phonological contrast: The case of stop consonants in French-English bilinguals," J. Acoust. Soc. Am., **54**, 421-428.

Carney, L.H. and Yin, T.C.T. (1988). "Temporal coding of resonances by low-frequency auditory nerve fibers: Single-fiber responses and a population model," J. Neurophysiol., **60**, 1653-1677.

Geisler, C.D. (1989). "The responses of models of high spontaneous auditory nerve fibers in a damaged cochlea to speech syllables in noise," J. Acoust. Soc. Am., **86**, 2192-2205.

Green, D.M. and Swets, J.A. (1974). *Signal Detection and Psychophysics*, (Krieger, Huntington, NY).

Hewitt, M.J., and Meddis, R. (1991). "An evaluation of eight computer models of mammalian inner hair-cell function," J. Acoust. Soc. Am., **90**, 904-917.

Keating, P.A., Mikos, M.J., and Ganong, W.F. III (1981). "A cross-language study of range of voice onset time in the perception of initial stop voicing," J. Acoust. Soc. Am., **70**, 1261-1271.

Kewley-Port, D., Watson, C.S., and Foyle, D.C. (1988). "Auditory temporal acuity in relation to category boundaries: Speech and nonspeech stimuli," J. Acoust. Soc. Am., **83**, 1133-1145.

Kluender, K.L. and Jenison, R.L. (1991). "Perception of voicing in syllable initial stops across intensity: Implications for synchrony encoding of voice onset time," Abstracts of the 14th Mtg. of the Assoc. for Res. in Otolaryngol.

Kuhl, P.K. (1981). "Discrimination of speech by nonhuman animals: Basic auditory sensitivities conducive to the perception of speech-sound categories," J. Acoust. Soc. Am., **70**, 340-349.

Kuhl, P.K. and Miller, J.D. (1978). "Speech perception by the chinchilla: Identification functions for synthetic VOT stimuli," J. Acoust. Soc. Am., **63**, 905-917.

Liberman, M.C. (1978). "Auditory nerve responses from cats raised in a low-noise chamber," J. Acoust. Soc. Am., **63**, 442-455.

Lisker, L. and Abramson, A.S. (1970). "The voicing dimension: Some experiments in comparative phonetics," Proc. 6th Intl. Cong. Phon. Sci., 563-567.

MacMillan, N.A., Goldberg, R.F., and Braida, L.D. (1988). "Resolution for speech sounds: Basic sensitivity and context memory on vowel and consonant continua," J. Acoust. Soc. Am., **84**, 1262-1280.

Meddis, R., Hewitt, M.J., and Shackleton, T.M. (1990). "Implementation details of a computational model of the inner hair-cell/auditory-nerve synapse," J. Acoust. Soc. Am., **87**, 1813-1816.

Miller, J.D. (1977). "Perception of speech sounds in animals: Evidence for speech processing by mammalian auditory mechanisms," in *Recognition of Complex Acoustic Signals*, T.H. Bullock, ed. (Abakon Verlags-Gesellschaft, Berlin), 49-58.

Rosner, B.S. (1984). "Perception of voice-onset-time continua: A signal detection analysis," J. Acoust. Soc. Am., **75**, 1231-1242.

Simon, C. (1978). "On the use of comfortable listening levels in speech experiments," J. Acoust. Soc. Am., **64**, 744-750.

Sinex, D.G. and Geisler, C.D. (1984). "Comparison of the responses of auditory nerve fibers to consonant-vowel syllables with predictions from linear models," J. Acoust. Soc. Am., **76**, 116-121.

Sinex, D.G. and McDonald, L.P. (1988). "Average discharge rate representation of voice onset time in the chinchilla auditory nerve," J. Acoust. Soc. Am., **83**, 1817-1827.

Sinex, D.G. and McDonald, L.P. (1989). "Synchronized discharge rate representation of voice onset time in the chinchilla auditory nerve," J. Acoust. Soc. Am., **85**, 1995-2004.

Sinex, D.G., McDonald, L.P., and Mott, J.B. (1991). "Neural correlates of nonmonotonic temporal acuity for voice onset time," J. Acoust. Soc. Am., **90**, 2441-2449.

Soli, S.D. (1983). "The role of spectral differences in the discrimination of voice onset time differences," J. Acoust. Soc. Am., **73**, 2150-2165.

Stevens, K.N. (1989). "On the quantal nature of speech," J. Phonetics, **17**, 3-45.

Van Tasell, D.J. and Crump, E.S.A. (1981). "Effects of stimulus level on perception of two acoustic cues in speech," J. Acoust. Soc. Am., **70**, 1527-1529.

SECTION 5

OTHER SENSORY SYSTEMS

This section includes models of the electric organ of "Eigenmannia" and the olfactory and vestibular systems of vertebrates. **Fred Rieke et al.** discuss the cricket cercal system (Analysis). A model of the olfactory cortex is presented by **Bill Baird and Frank Eeckman** (Analysis).

Walter Heiligenberg describes the computational rules and their neural implementation for the Jamming Avoidance Response in the electric fish "Eigenmannia".

William Lytton explores the origin of the small cell's temporal disparity sensitivity using a compartmental model with Hodgkin-Huxley kinetics. The small cell is a coincidence detector similar to the cells in the nucleus laminaris of the barn owl. See **Clay Spence and John Pearson** (Auditory Systems).

Thomas Chimento et al. discuss detailed compartmental models based on measurements obtained from 3D reconstructions of various afferent processes in the vestibular macula.

Koji Shimoide et al. demonstrate the generation of chaotic dynamics using a multilayer model of primary and secondary olfactory structures.

The Jamming Avoidance Response (JAR) of the electric fish, Eigenmannia: Computational Rules and Their Neuronal Implementation

Walter Heiligenberg
Neurobiology Unit, Scripps Institution of Oceanography
University of California at San Diego, La Jolla, CA 92093-0202

Tel (619) 534 2127
Fax (619) 534 7313
email: wheilige@ucsd.bitnet

This article is reprinted from *Seminars in the Neurosciences,* Vol 3, (1991), pp 3 1- 18, with the permission of Academic Press Inc. (London) Ltd.

Abstract

Studies of the neuronal organization of the jamming avoidance response (JAR) have revealed a distributed system of local computations of sensory information. A variety of electrosensory behaviors involved in social communication and object detection share networks with the JAR at the sensory and at the motor level within the hindbrain and the midbrain. Their respective flows of information, however, separate in the diencephalon, where nodes dedicated to a single type of behavior provide sensory-motor interfaces. A comparison of neurons across different levels of sensory information processing reveals a gradual increase in the specificity and sensitivity of their responses to stimulus patterns guiding the JAR. Neurons with response properties similar in specificity and sensitivity to those of the intact behavior are encountered ultimately in the diencephalon.

INTRODUCTION

Electric fish generate electric fields by periodically discharging an electric organ located in their tail. They sense their own currents as well as those of neighbors through arrays of electroreceptors distributed over their body surface. Specific modulations in the rate of their electric organ discharges (EODs) serve as social signals. Electric fish perceive the EODs of their neighbors either as discrete events or, more commonly, as perturbations in the feedback from their own EODs. In addition, an object in the fish's vicinity distorts the electric field and alters the current pattern perceived by arrays of electroreceptors closest to the object. This alteration in the current pattern on the fish's body surface represents the electric image of the object, and the analysis of such images enables electric fish to 'electrolocate' objects in their environment. Whereas the central processing of spatial electric images resembles a form of vision (1), the temporal analysis of periodic current signals (Fig. 1) shares aspects with audition (2).

Object-induced changes in current patterns may be masked by modulations caused by a neighbor's EODs interfering with the fish's own EODs, and the fish may avoid or at least minimize the detrimental effects of signal interference, or 'jamming', by altering its own firing pattern accordingly. These so-called jamming avoidance responses (JARs) are purely 'electrical' behaviors that persist in individuals paralysed by the injection of curare-like drugs and, therefore, can conveniently be explored at the intracellular level. Behavioral and cellular performance can be directly related, and the labelling of physiologically identified neurons leads to the functional and anatomical reconstruction of underlying neuronal circuits. In addition, the intactness of a behavioral response has proved to be an important indicator for the intactness of the neuronal substrate. Losses of a specific behavioral response occur with the failure of certain classes of neurons to respond in their usual way; such losses may occur without any obvious signs of deterioration in the overall metabolic state of the animal.

This review will focus on the JAR (3) of the gymnotiform genus, Eigenmannia. This response has been explored in great detail at the neuronal level so that all behavioral properties can now be traced to specific features at the cellular level. Most importantly, sensory information processing can be linked almost completely with the control of associated behavioral responses via a conversion of a representation of sensory space to a representation of motor commands within the diencephalon.

The computational rules of the JAR

Eigenmannia generates nearly sinusoidal EODs of individually different but highly stable frequencies in the range of 200 to 500 Hz. When exposed to a signal of a frequency within a few Hz of its own, however, the fish will shift its EOD frequency

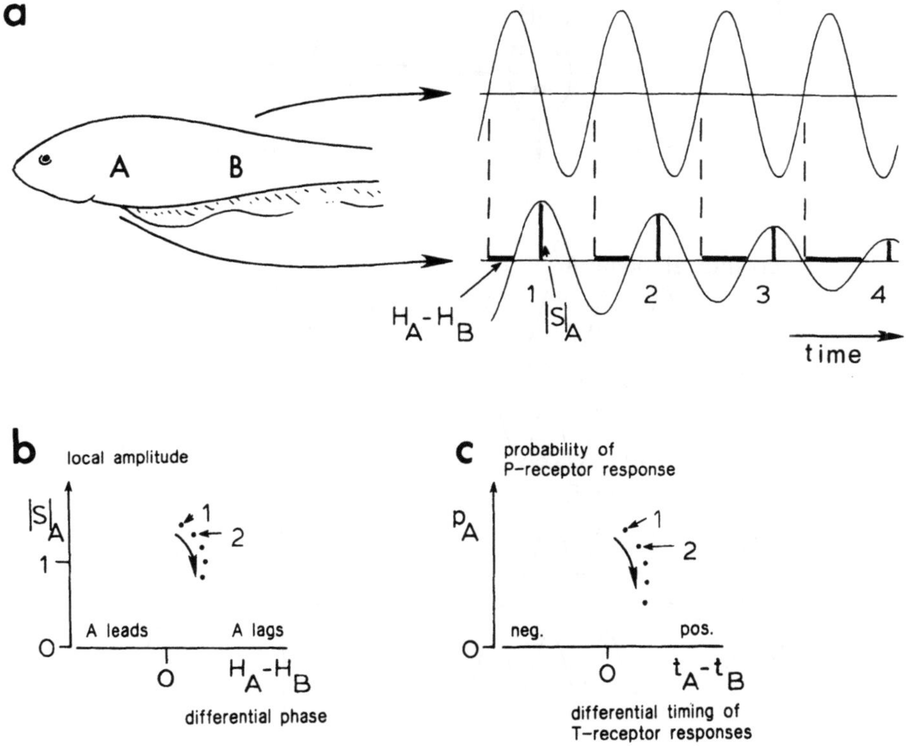

Fig.1: Objects in the vicinity of the fish's body surface may locally alter the electric signal generated by its electric organ discharge (EOD), causing a change in amplitude (| S |) as well as a shift in phase (H). (a) The EOD is idealized as a sinusoidal signal and is recorded as voltage in locations A and B, respectively. The signal in location A is assumed to be distorted by the passage of a small object, whereas the signal in location B is undisturbed. Depending upon the nature and direction and speed of motion of the object, the distortions induced in position A may have the following form as a function of time: As the object passes location A, the local amplitude of the signal, $|S|_A$ (height of thick bar), is gradually attenuated, while the 'differential phase', H_A-H_B (length of thick bar), measured as the difference in the timing of the positive zero-crossings of the signals in A and B, increases progressively. The sizes of the modulations in amplitude and phase have been exaggerated for demonstration purposes. In the absence of the perturbation in A, the differential phase, H_A-H_B, would be zero, i.e. the signals in A and B would cross the zero line simultaneously. Successive cycles of the signal in A have been numbered 1,2,3,4. As pairs of values of $|S|_A$ and H_A-H_B, sampled for successive cycles, are plotted in a two-dimensional plane (b), a graph is obtained which reflects the motion of the object. As will be explained in detail later in this article, the information plotted in the two-dimensional plane in (b) is coded by two classes of electroreceptors (c) in that each data point in the plane of (b) is mapped onto a corresponding point in the plane of (c). The probability, p_A, of P-type receptor firing in location A reflects the value of the local amplitude. T-type receptors fire one spike on each EOD cycle and at a fixed latency with reference to the timing of the local positive zero-crossing. If, for each cycle of the signal, the timing of T-type receptor spikes in A and B is measured in reference to the same zero point, such as the most recent zero-crossing in B, and written as t_A and t_B, respectively, then the difference in their timing, t_A-t_B, reflects the differential phase, H_A-H_B. Very similar modulations in amplitude and phase are generated by interference with EODs of a neighbor, which cause the JAR.

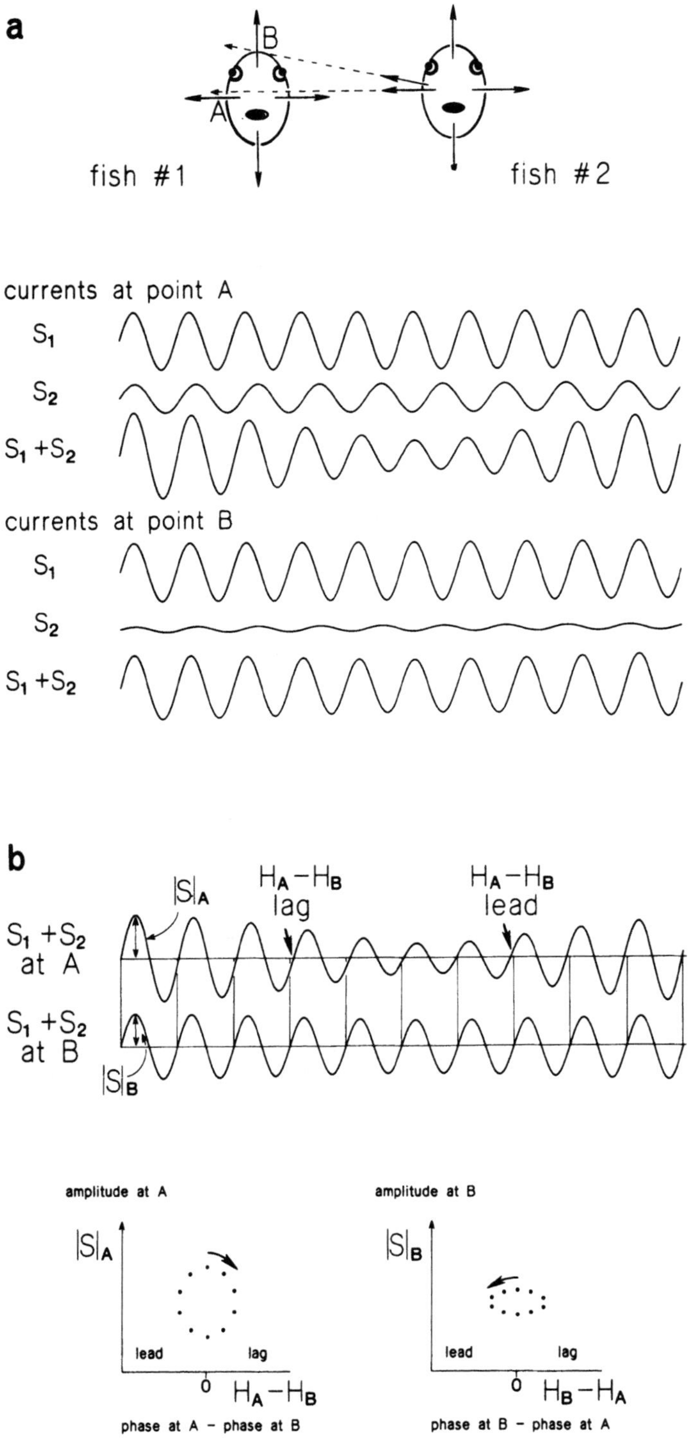
a
B
A
fish #1
fish #2
currents at point A
S1
S2
S1 +S2
currents at point B
S1
S2
S1 +S2
b
|S|A
HA−HB
lag
HA−HB
lead
S1 +S2
at A
S1 +S2
at B
|S|B
amplitude at A
|S|A
lead
lag
0
HA−HB
phase at A − phase at B
amplitude at B
|S|B
lead
lag
0
HB−HA
phase at B − phase at A

Fig.2

Fig.2: **(a) The addition of the EOD currents, S_1 and S_2, of two fish results in a nearly sinusoidal signal, S_1+S_2, whose amplitude, or envelope, is modulated at a rate equal to the difference in the frequencies of the two signals. For simplicity, the EODs are portrayed as sine wave signals. The signal frequency of fish #2 is assumed to be lower than that of #1, and as time progresses, the two signals alternately pass through periods of constructive and destructive interference, resulting in peaks and troughs in the envelope of the combined signal. The effect of EOD interference is shown for two points, A and B, on the body surface of fish #1. Electroreceptors sense only the current component oriented perpendicularly to the local body surface. Since the fish's own EOD current, indicated by solid arrows, has an internal source, it penetrates the body surface rather perpendicularly and at similar intensity at all points. The neighbor's current (indicated by broken lines for two directions), however, hits the fish's surface at various angles. As a consequence, point A is more strongly affected by the neighbor's current than point B, and the combined signal, S_1+S_2, is therefore more strongly modulated in A than in B. The modulation in the combined signal's envelope is commonly referred to as a 'beat pattern'. For demonstration purposes, fewer cycles of the signal have been plotted per beat cycle than would be observed in a typical JAR situation. With the two fish having EOD frequencies of 400 and 404 Hz, respectively, the beat rate would be 4 Hz, and approximately 100 EOD cycles would be contained in one beat cycle.**

The diagram in (b) compares the signals perceived in A and B and shows that differences in the timing of their zero-crossings, or 'phase', are modulated over the course of the beat cycle. In reference to the signal in B, the signal in A lags in phase while its amplitude falls and leads while its amplitude rises. This particular order in amplitude and phase modulations is due to the fact that the neighbor's EOD frequency is lower than the fish's own. If the neighbor's EOD frequency were higher, then a fall in the signal amplitude at A would be paired with a phase lead, and a rise in amplitude would be paired with a phase lag. Following the example in Fig.1, the joint modulations in amplitude and phase can be recorded in a two-dimensional state plane, with the amplitude of the signal at A, $|S|_A$, plotted on the ordinate, and differential phase, H_A-H_B, plotted on the abscissa (lower left). In this case, since the EOD frequency of the neighbor is lower, a circular trajectory, or 'graph', with a clockwise rotation is obtained as the more strongly modulated signal at A is evaluated in reference to the less strongly modulated signal at B. A counterclockwise rotation would be obtained if the neighbor's signal frequency were higher. The sense of rotation thus reflects the sign of the frequency difference, Df, between the neighbor's EOD and the fish's own.

Much as the signal in A can be evaluated in reference to that in B, however, the signal in B can be evaluated symmetrically in reference to that in A (diagram in lower right). This graph is characterized by a smaller amplitude modulation, and, due to the inversion of the differential phase, its sense of rotation is opposite to that obtained for point A. On the basis of the clockwise rotation in A, this area would interpret the sign of Df as negative and 'vote' for a raising of the pacemaker frequency. An opposite vote would emerge from area B. Area A, however, should win in this competition due to its larger amplitude modulation. Note that both areas, A and B, may simultaneously and independently perform similar computations with other, additional reference areas, C,D,E, etc., and that any two areas on the fish's body surface can interact symmetrically in this manner. Contributions to the control of the pacemaker frequency, however, are weaker for interactions between more distant areas, and amplitude modulations in the head region have more powerful effects than those on the more distant trunk.

[Note that the reciprocal evaluation of inputs from two points does not always produce graphs with opposite rotations but only if the vectors of the two interfering currents form similar angles, either acute or obtuse. If, by contrast, fish #1 compared inputs from A and from the opposite side of its body, then both graphs would rotate in the correct sense and induce a rise in pacemaker frequency (5).]

to increase the magnitude of the frequency difference, Df, between the two interfering signals. Promptly and without hunting, Eigenmannia lowers its frequency in response to a signal of slightly higher frequency and raises its frequency in response to a signal of slightly lower frequency. The fish must, therefore, be able to discriminate the sign of Df from the very beginning. Fortunately for the experimenter, the JAR persists even after the fish's EODs have been silenced by injection of curare-like drugs and a sine wave signal of appropriate frequency, magnitude and spatial structure substituted. The medullary pacemaker, which drives each EOD cycle by a single command pulse, is not affected by this treatment, and the fish's efforts to change its own frequency can thus be monitored continually. A multitude of behavioral experiments, reviewed elsewhere (4,5), has revealed a set of computational rules and guided the search for their neuronal implementation. These rules are as follows.

1) The fish can determine the sign of Df without reference to its own pacemaker signal; it thus relies exclusively on afferent information extracted from the interference pattern generated by the superposition of the jamming signal and its own signal, whether the EOD or its substitute.

2) This interference pattern is characterized by modulations in the amplitude and phase, or the timing of the zero-crossings, of the nearly sinusoidal signal which results from the superposition of the two original signals over large areas of the fish's body surface. The depth of these modulations at any point, however, depends upon the local amplitude and orientation of the neighbor's current and thus varies with the location on the body surface. As explained in Fig.2, the temporal and spatial configuration of these amplitude and phase modulations reveals the sign of Df.

3) By sampling amplitude and phase information at many points of its body surface and by comparing phase inputs from many pairs of points, the fish is able to extract the sign of Df. With point A being more heavily affected by the jamming signal than point B, a positive Df implies a rise in the amplitude of the signal at A in conjunction with a phase lag of the signal at A versus that at B, and a drop in amplitude in conjunction with a phase lead. For a negative Df, the opposite pairing holds for amplitude and phase, i.e. a drop in amplitude is paired with a phase lag, and a rise in amplitude with a phase lead. These forms of modulations can be portrayed as rotations in the amplitude-phase plane (Fig.2b).

4) Such evaluations of inputs from pairs of points on the body surface contribute to the JAR in that clockwise rotations raise the frequency of the pacemaker, while counterclockwise rotations lower it. The reciprocal interaction between two points may, however, lead to a graph with opposite rotation (Fig.2b) and will then cause an opposite contribution. Since the graph with the larger amplitude modulation wins in this pairwise competition, the correct behavioral response will result as long as the fish's own signal dominates at most points of the body surface.

5) The JAR is driven by a distributed system of contributions resulting from the evaluations of inputs from pairs of points, and there is no evidence of a central controller or decision maker within this seeming neuronal pandemonium.

The amplitude and phase modulations in the fish's signal induced by the appearance of objects resemble those caused by interference with a neighbor's signal (Figs. 1,2). They differ, however, in that object-induced modulations are more localized than interference-related modulations (4). The fish, therefore, could distinguish these two forms of modulations by using small receptive fields for the detection of objects and large receptive fields for determining the frequency difference between interfering signals. A further separation is achieved through the JAR, which is strongest for frequency differences in the range between 2 and 6 Hz (3). By shifting its own EOD frequency away from that of its neighbor and thus raising the magnitude of the frequency difference beyond 10 to 20 Hz, the modulation rates of beats generated by interference are pushed beyond the spectral range of modulations caused by the motion of natural objects.

The sensory coding of behaviorally relevant stimulus variables

Two types of 'tuberous' electroreceptors, tuned to the fundamental frequency of the individual's EOD, are distributed over the body surface. P-units fire intermittently and raise their rate of firing with an increase in the local amplitude of the signal. T-units fire one spike on each EOD cycle, phase-locked to the zero-crossing of the signal. A comparison of the timing of spikes originating in close temporal succession from T-units in points A and B therefore reveals whether the phase of the signal in A leads or lags in reference to that in B (Fig.3). Both types of receptors project via primary afferent neurons to three somatotopically structured maps of higher-order neurons in the electrosensory lateral line lobe (ELL) of the hindbrain. All three maps receive inputs from each afferent neuron via three separate axonal collaterals.

Central maps of the electrosensory body surface and their computational functions

The parts of the brain involved in the JAR are summarized in Fig.4. The three maps in the electrosensory lateral line lobe (ELL) receiving inputs from tuberous electroreceptors have similar architectures but differ, in a graded and complementary manner, in the spatial and temporal resolution of their receptive fields: the most lateral map shows highest temporal and lowest spatial resolution whereas the most medial map shows lowest temporal and highest spatial resolution (6). Most importantly,
phase and amplitude information are processed separately and are relayed to different targets in a single, though multi-layered, body map in the torus semicircularis of the midbrain.

The spherical cells of the ELL receive inputs from patches of T-type receptors and, within each EOD cycle, fire one spike in response to the nearly synchronous

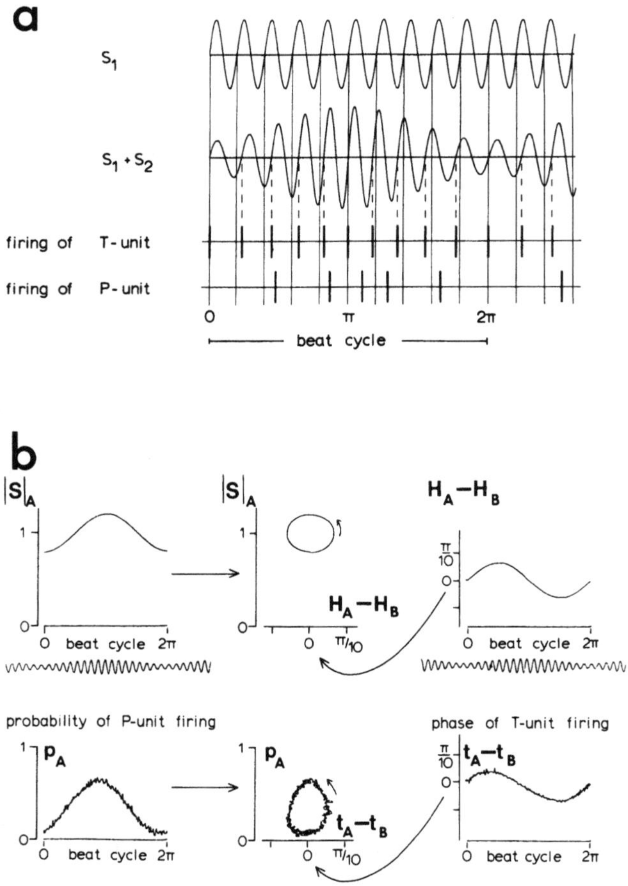

Fig.3: The coding of modulations in instantaneous phase and amplitude by T-type and P-type tuberous electroreceptors. The schematic representation in (a) shows the fish's EOD signal, S_1, in the trace at the top and the interference pattern resulting from addition of a neighbor's signal, S_2, in the trace underneath. This beat pattern is characterized by a modulation of the instantaneous amplitude, or envelope, of the signal and by a modulation of its instantaneous phase, or the timing of its zero-crossings (marked by broken lines), in reference to that of the pure S_1 (marked by continuous lines). The two traces at the bottom indicate the coding of the phase and amplitude modulations by T-unit and P-unit afferents. The phase of the signal is coded by the timing of a single action potential fired by T-unit afferents within each S_1 cycle and at a fixed latency with reference to the timing of the positive zero-crossing. The small difference in the timing of nearly synchronous action potentials of T-units in different parts of the body surface thus reflects the differential phase between the respective signals, and this information is extracted in lamina 6 of the torus semicircularis (see Fig.6). The modulations in the amplitude of the signal are coded by a corresponding modulation in the probability of P-unit afferents firing an action potential within the current S_1 cycle. Fluctuations in the rate of P-unit firing thus indicate corresponding fluctuations in signal amplitude. (b) Patterns of phase and amplitude modulations arcoded by the joint activation of T-units and P-units. Top row: A beat cycle, generated by the interference of the fish's own signal, S_1, with that of a neighbor, S_2, is characterized by a modulation of the instantaneous local amplitude, $|S|_A$ (far left), and by a modulation of the instantaneous differential phase, H_A-H_B (far right), measured in reference to another area, B, less strongly affected by interference with S_2. A plot of $|S|_A$ versus H_A-H_B in a two-dimensional plane (center) yields a circular graph which rotates counterclockwise for positive Dfs, i.e. when the frequency of S_2 is higher than that of S_1. Bottom row: The modulation in amplitude is coded by the probability, p_A, of P-unit firing, in area A, while the modulation in phase is coded by the difference in timing, t_A-t_B, of T-unit firing, in two areas, A and B, with area B being less strongly affected by interference than area A. A plot of p_A versus t_A-t_B yields a circular graph which preserves the sense of rotation of the corresponding graph in the $|S|$,H plane. The data for p_A and t_A-t_B in the lower row are mean values obtained from data collected over 25 beat cycles. For positive Dfs, the values of H and t would be negatives of those shown in this figure, while the associated records of $|S|$ and p would be identical. As a consequence, the circular graphs in the center would rotate in the clockwise direction.

arrival of signals from their afferents. Due to the spatial convergence of their inputs, spherical cells generate spikes that lock to the zero-crossing of the electric signal in the periphery with even less jitter than observed in spikes of individual T-type afferents.

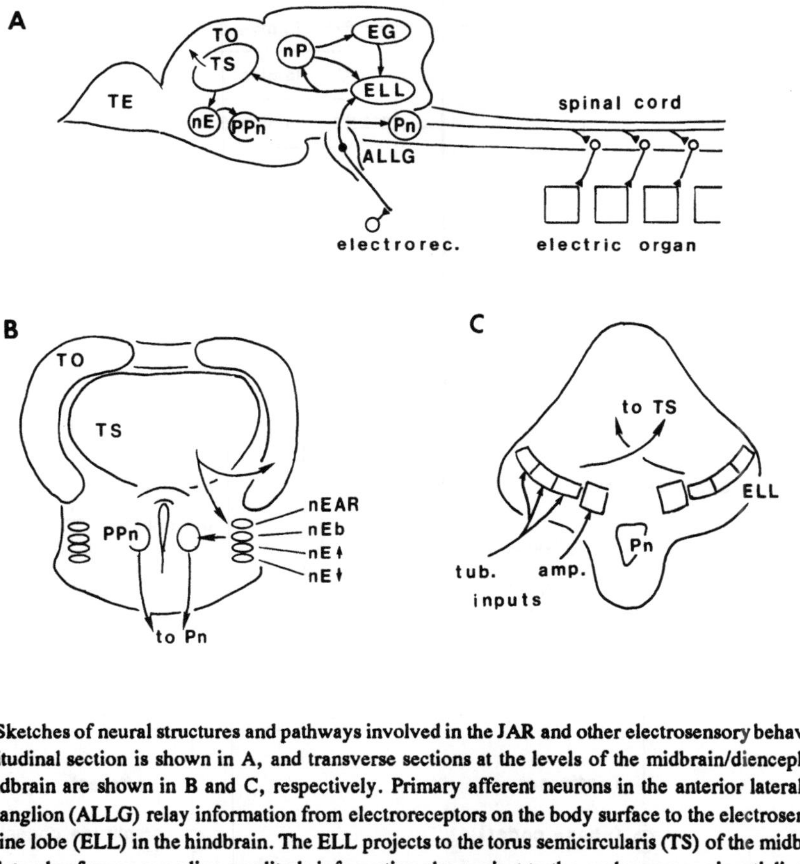

Fig.4: Sketches of neural structures and pathways involved in the JAR and other electrosensory behaviors. A longitudinal section is shown in A, and transverse sections at the levels of the midbrain/diencephalon and hindbrain are shown in B and C, respectively. Primary afferent neurons in the anterior lateral line nerve ganglion (ALLG) relay information from electroreceptors on the body surface to the electrosensory lateral line lobe (ELL) in the hindbrain. The ELL projects to the torus semicircularis (TS) of the midbrain, and collaterals of neurons coding amplitude information also project to the nucleus praeeminentialis (nP). The nP, in turn, is a source of two descending inputs to the ELL, one direct and one indirect via the eminentia granularis (EG) of the cerebellum. The TS projects to the tectum opticum and to the nucleus electrosensorius (nE) of the diencephalon, which appears to drive the prepacemaker nucleus (PPn). The PPn, in turn, induces frequency modulations of the pacemaker nucleus (Pn) in the medulla. The Pn innervates spinal motor neurons which, in turn, innervate the electrocytes of the electric organ. Each discharge of the electric organ is triggered by a single command pulse from the Pn. The TO and the telencephalon (TE) are not required for the JAR. The nE is composed of at least four subnuclei (B) (7), apparently dedicated to different behavioral functions. The transverse section through the hindbrain (C) displays the three maps of the ELL receiving identical input from tuberous electroreceptors (tub.), and the single map receiving input from ampullary electroreceptors (amp.). The latter code low-frequency electric signals and do not participate in the JAR. The Pn is a midline nucleus in the medulla.

P-type afferents excite E-type pyramidal cells of the ELL directly and inhibit I-type pyramidal cells via interneurons. E-units are thus driven by a rise in local signal amplitude, while I-units are driven by a decline in amplitude, much as 'on' and 'off' units in a visual system (Fig.5).

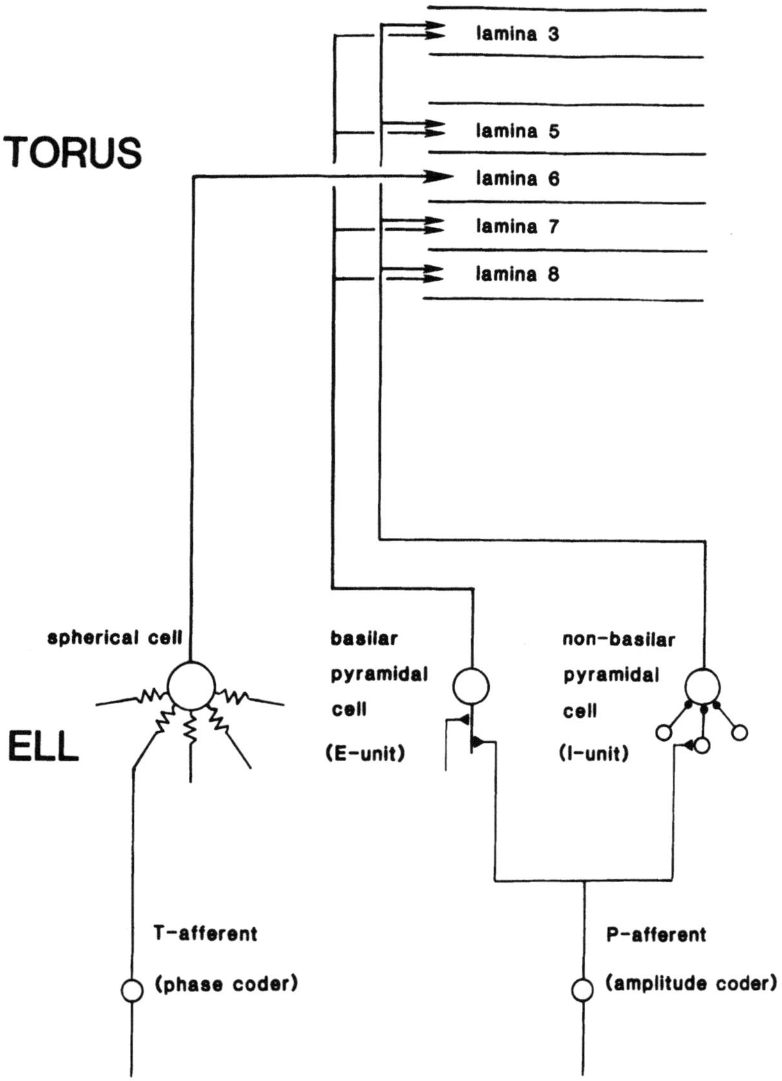

Fig.5: The organization of the electrosensory lateral line lobe (ELL) provides separate substrates for the processing of phase information and amplitude information. Primary afferents from T-type receptors, or phase coders, form electrotonic synapses with spherical cells, and a single spherical cell may receive inputs from several T-afferents originating within a receptive field on the body surface. Primary afferents from P-type receptors, or amplitude coders, form excitatory synapses on the basilar dendrites of basilar pyramidal cells, or E-units, and also excite granule cells which, in turn, inhibit non-basilar pyramidal cells, or I-units. As a consequence of these connections, E-units are excited by a rise and I-units by a fall in stimulus amplitude. Whereas the spherical cells project exclusively to lamina 6 of the torus semicircularis, the pyramidal cells project to various laminae above and below lamina 6. All projections are topographic so that the somatotopic order of electrosensory information is maintained to the level of the torus. Filled triangles indicate excitatory synapses, and filled circles indicate inhibitory synapses. For more details of ELL connectivity see (30,31,33).

The spherical cells of the ELL project exclusively to lamina 6 of the torus, while both types of pyramidal cells project to various laminae above and below. All laminae are organized somatotopically and are aligned in topographic register. A network of neurons within lamina 6 computes phase differences between any two points on the body surface such that a class of 'small cells', located at the representation of point A, discriminates phase advances from phase delays of the signal at point A with reference to signals in a variety of other points on the body surface, with each cell choosing an individual reference point apparently at random (Fig.6). Vertical connections between the laminae of the torus allow for the spatially ordered convergence of amplitude and differential phase information that is necessary for determining the sign of the frequency difference, Df, between the jamming signal and the fish's own signal. So called sign-selective neurons, found in deeper laminae of the torus, discriminate the sign of Df by firing at a higher rate for one sign than for the opposite sign (Fig.7).

The torus projects in topographic order to the optic tectum (Fig.4). Although the tectum contains an abundance of sign-selective neurons, recent ablation studies have shown that it is not required for the JAR (7). The sign-selective neurons of the tectum, instead, appear to serve in the detection of moving objects, which cause modulations in the amplitude and phase of the fish's EOD that are similar to those induced by jamming. As a consequence of their sign-selectivity, these neurons can discriminate the direction of object motion (8).

The emergence of a motor map in the diencephalic nucleus electrosensorius (nE)

A massive projection from the torus to the complex of the nE in the diencephalon apparently abandons the spatial order of electrosensory information preserved up to the level of the torus and leads to the separation of circuits into at least four subnuclei, dedicated to the detection of specific stimulus patterns and the control of corresponding behavioral responses (7,9). Only two of these, the nE↑ and the nE↓, are concerned with the JAR and control the frequency of the medullary pacemaker via the prepacemaker nucleus (PPn) of the diencephalon.

The nE↑ raises the frequency of the medullary pacemaker nucleus when its neurons are excited by application of L-glutamate, and bilateral lesions of this subnucleus eliminate the fish's ability to raise its EOD frequency in the context of the JAR. The nE↓, conversely, lowers the frequency of the medullary pacemaker nucleus, and bilateral lesions of this subnucleus abolish the fish's ability to lower its EOD frequency in the JAR. The nE↑ and nE↓ form mutual, GABA-ergic inhibitions, and their balance can be affected accordingly by local administration of the GABA-receptor blocker bicuculline (9). It appears that the frequency level of the pacemaker is controlled much like the angle of a limb joint, which is determined by the mutually antagonistic influences of flexors and extensors.

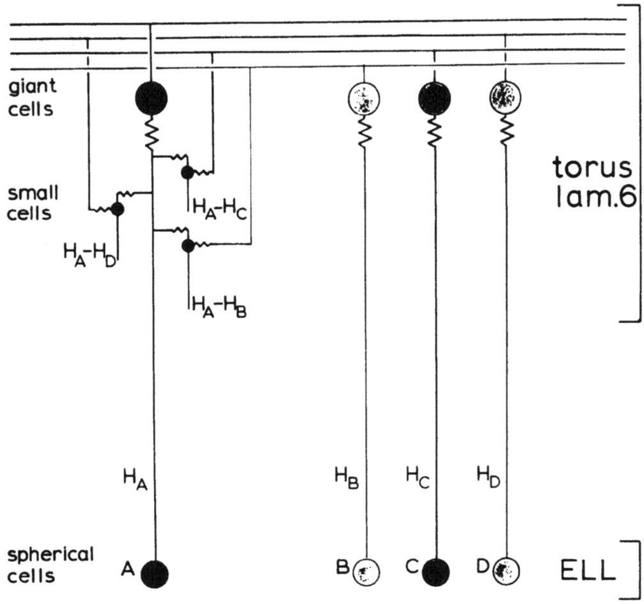

Fig.6: Lamina 6 of the torus semicircularis contains a network for the computation of differential phase between any two points on the body surface. The spherical cells in the ELL (see Fig.5) receive somatotopically ordered information about the phase, i.e. the timing of zero-crossings, of electric signals on the body surface, and they relay this information, H_A, H_B,..., in a topographic projection to lamina 6 of the torus by synapsing electrotonically on the somata of giant cells and the dendrites of nearby small cells. Giant cells, in turn, relay the phase-locked spikes of spherical cells through their large axonal arbors to wide areas of lamina 6, and their axonal collaterals synapse electrotonically on the somata of small cells, which apparently can only accommodate a single large synapse of this kind. Therefore, a small cell is controlled by inputs from two sources: the input at its dendrites is derived from the point on the body surface represented at this location of lamina 6 (point A in the example of this figure), while the input at its soma originates from the point represented at the location of the giant cell soma whose axon collateral forms this contact (point B, for example). Small cells fire intermittently, and their rate of firing is modulated by the differential timing of their two inputs. Small phase advances, for example, may cause an increase in the rate of firing, while small delays may cause a reduction. Giant cells receive inputs from several spherical cells within a receptive field, and only fire in response to a nearly synchronous arrival of spikes from several spherical cells. For clarity, only a single spherical cell is shown to contact a giant cell.

Because the nE↓ lowers the frequency of the pacemaker in the context of the JAR, one might expect that toral neurons excited preferentially by positive Dfs predominantly excite the nE↓, whereas toral neurons excited more by negative Dfs should predominantly excite the nE↑. So far we have been unable to test this hypothesis as we have failed to record intracellularly from neurons in these subnuclei. This is due to the small size of their somata, which rarely exceeds 5-8 μm in diameter, as well as to their sparse distribution.

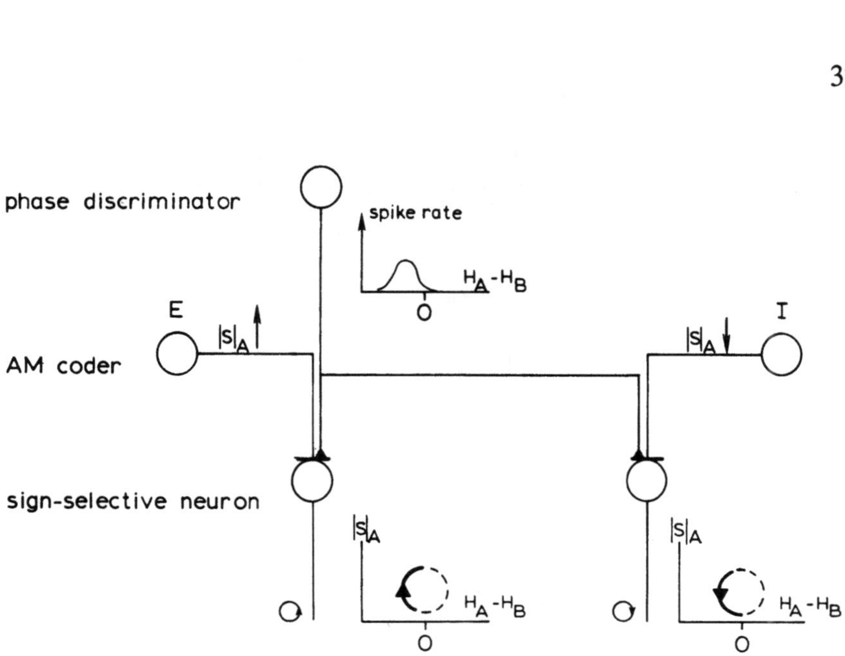

Fig.7: A model for the joint computation of amplitude and phase information by a 'sign-selective' neuron. Information is provided from two areas, A and B, on the body surface. A phase-discriminator neuron which is excited by small negative values of the differential phase, H_A-H_B (top), i.e. small phase leads of area A over B, is assumed to facilitate input from amplitude modulation (AM) coders for area A to sign-selective neurons. Two classes of AM coders, E-neurons and I-neurons, are excited by a rise and by a fall, respectively, in the local stimulus amplitude, $|S|_A$. As a consequence of facilitation, the sign-selective neuron on the left prefers a clockwise rotation centered at phase zero, i.e. a combination of a rise with a phase advance, while the neuron on the right prefers a counterclockwise rotation. Although neurons with these response properties have been recorded and labelled intracellularly, the nature of their synaptic organization is still unknown (35).

In addition to the nE↑ and nE↓, the complex of the nE contains two subnuclei that are involved in the perception and generation of EOD modulations used in aggression and courtship. Since toral neurons commonly send axonal collaterals to more than one subnucleus (11), different subnuclei appear to share afferent information, although they obviously process this information by different rules and contribute to different behaviors.

The diencephalic prepacemaker nucleus (PPn) and its control of the medullary pacemaker nucleus

Although the efferent pathways of the nE have not yet been explored entirely, it appears that the next station in the control of the JAR is a subnucleus of the prepacemaker nucleus (PPn), the PPn-G (7,12, Figs.4,8), which innervates the pacemaker nucleus in the medulla. The small neurons of the PPn-G fire spontaneously and are excited by negative Dfs, i.e. when a neighbor's EOD has a frequency slightly lower than that of the fish's own, and the activation of these neurons, either by an appropriate jamming signal or by the local application of L-glutamate, leads to a gradual, asymptotic rise of the pacemaker frequency. The same neurons are inhibited when a jamming signal is of slightly higher frequency so that the fish lowers its EOD frequency. The resting frequency of the pacemaker thus appears to be maintained by tonic innervation from the PPn-G.

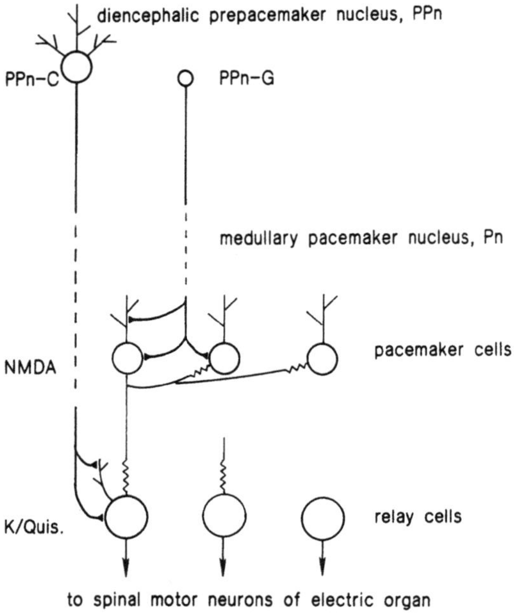

Fig.8: The diencephalic prepacemaker nucleus (PPn) consists of two subunits, the PPn-G and the PPn-C, and innervates the medullary pacemaker nucleus (Pn). While activation of the PPn-G causes a gradual rise of the pacemaker frequency, activation of the PPn-C causes 'chirps', abrupt rises in frequency which may lead to a brief interruption of the pacemaker cycle. The pacemaker nucleus consists of electrotonically coupled pacemaker cells and relay cells. Although the synaptic targets of the PPn-G and PPn-C cannot yet be distinguished, indirect evidence suggests that the PPn-C innervates the relay cells directly. Whereas NMDA receptors mediate smooth rises of the pacemaker frequency, kainate/quisqualate receptors mediate chirps (13).

The medullary pacemaker nucleus is an endogenous oscillator that drives each EOD cycle by a single command pulse (Fig.8) (13). Local application of the NMDA receptor blocker APV (D(-)-2-amino-5-phosphonovaleric acid) reversibly abolishes the fish's ability to raise its pacemaker frequency, either in the context of the JAR or in response to direct stimulation of the PPn-G. In addition, the resting frequency of the pacemaker falls in response to this treatment, supporting the original assumption that it is maintained by tonic activation via the PPn-G.

A second subnucleus of the PPn, the PPn-C, is a cluster of larger, multipolar neurons that also innervate the pacemaker nucleus. Excitation of these cells causes a rapid, brief acceleration of the pacemaker cycle, and a sufficiently strong acceleration will lead to a temporary interruption of the pacemaker altogether (12). These rapid accelerations, also known as 'chirps', are generate in the context of aggression and courtship. The PPn-C neurons innervate the relay cells of the pacemaker nucleus, and their synapses appear to use a kainate-quisqualate receptor, since chirping can be blocked reversibly by local application of GAMS (τ-Dglutamylaminomethyl-sulphonic acid) or CNQX (6-cyano-7-nitroquinoxaline-2,3-dione) to the pacemaker. Most significantly, this block does not affect the performance of the JAR, much as blocking of the JAR by application of APV does not affect the production of chirps (13). The network of the pacemaker nucleus is thus shared by two distinct patterns of behavior.

General principles in the control of the JAR

a) The sharing of networks at the sensory and at the motor level

Neurons at a level as high as that of the torus semicircularis have still rather general response properties and are certainly not dedicated to a single behavioral function. Individual neurons in the torus respond to a variety of modulation patterns in the amplitude and phase of EOD-like signals that can be associated with the motion of objects as well as with the jamming effect of a neighbor's EODs. Through as many as six collaterals, a single neuron of the torus may send its information to a variety of targets for further processing (Metzner, pers. comm., (11)). In addition to the nucleus electrosensorius involved in the control of the JAR, these targets include centers that analyse images of moving objects, play roles in social communication, participate in the coordination of movements, and contribute to a source of descending input to the electrosensory lateral line lobe (ELL) (14,15). These target areas execute specific computations to extract features that are relevant to the control of their respective behavioral function. Whereas, for example, the tectum preserves the spatial order of sensory information for the identification and analysis of objects moving through space, the nE apparently discards spatial information and instead focuses on temporal signal properties that identify specific patterns of social communication.

A similar sharing of pathways is seen in the organization of motor patterns. The medullary pacemaker drives each EOD cycle by a single command pulse, but different inputs to the pacemaker nucleus can modulate its rhythm to generate specific signals either for social communication or for JAR-related frequency shifts. These inputs originate in the diencephalic prepacemaker nucleus which contains distinct populations of neurons dedicated to the generation of specific forms of pacemaker modulation. As a consequence, the stimulation of a small population of neurons, or even of a single neuron, may elicit a complete performance of a natural behavior. The depolarization of a single multipolar neuron in the PPn-C of Eigenmannia, for example, suffices to trigger a weak form of chirping in its pacemaker cycle that normally occurs in the context of aggression and courtship, although a given neuron may not participate in the generation of every single chirp (12,16). It appears to hold across all vertebrates that the diencephalon contains points that are dedicated individually to a specific behavioral function. These points represent nodes upon which sensory pathways that extract specific stimulus features converge and from which motor pathways that orchestrate corresponding patterns of behavior diverge. This form of organization appears to explain why the stimulation of points in the diencephalon routinely elicits the execution of natural behaviors, from fish to mammals.

b) The emergence of 'recognition' units and extreme perceptual acuities

On the way to the diencephalon, individual neurons respond more selectively to specific stimulus features and become increasingly insensitive to stimulus variables that are not relevant to the behavior that these neurons control. Neurons in the torus, for example, may already be able to discriminate whether a neighbor's EOD frequency is slightly higher or lower than that of the fish's own EOD, but their performance strongly depends upon the orientation of the current field associated with the jamming signal (17). Most neurons of the nE and, even more so, the still higher-order neurons of the PPn can discriminate such frequency differences regardless of the neighboring fish's orientation. Moreover, neurons of the nE and of the PPn are far superior to neurons of the torus in their ability to detect very weak jamming signals; their sensitivity often approaches thresholds observed at the behavioral level. These high-order neurons obviously achieve their greater stimulus selectivity and sensitivity through a massive spatial convergence of information from the torus to the nE and PPn (18).

c) The distributed organization of sensory information processing

The neuronal organization of the JAR is characterized by a seeming pandemonium of local computations and by the absence of decision elements that are individually essential for the execution of this behavior. As a consequence, the JAR will persist after extensive loss of sensory inputs or physical damage to structures, such as the ELL and the torus, that are involved in distributed processing of sensory information. The only effect of such losses is a proportional decline in the fish's acuity of detecting weak jamming signals (18), a form of 'graceful degradation'. The convergence of spatially structured, distributed information upon premotor nuclei, however, generates nodes, and the loss of specific nodes, such as the nE↓ or the nE↑, may eliminate entire components of the JAR (9).

Obvious advantages of distributed organizations in sensory processing and a subsequent convergence of information towards higher-order structures are at least twofold. First, they tolerate the local loss of sensors and computational elements, and second, they can afford the use of 'cheap' sensors at the periphery as a sufficiently high acuity of perception can still be obtained centrally by robust mechanisms of averaging and interpolation (19,20). The coding of sensory information in large populations of broadly tuned neurons appears to be a principle of topographic representations in general.

Evolutionary and developmental considerations

The JAR seems to have evolved relatively recently as the genus Sternopygus, which is in the same family as Eigenmannia but shows more ancestral features in its general morphology, lacks the behavior (3,25). Nevertheless, the torus and nE of Sternopygus contain neurons that can discriminate the sign of the frequency difference between an interfering signal and the fish's EOD. One role of these neurons may be to discriminate the motion of objects - as suggested above for the sign-selective cells in the tectum of Eigenmannia. Moreover, this fish can be conditioned to discriminate the sign of a frequency difference between a jamming signal and its own EOD by responding with physical maneuvers (Rose, pers. comm.). With some additional neuronal engineering, it should be possible to use part of the existing circuitry to establish a JAR in Sternopygus (25).

Although the JAR is a relatively simple behavior, the complexity of its neuronal organization might surprise an engineer trained in the efficient design of electronic circuits. And even more surprising would be the obvious imperfections in the design of this hardware (4,5). But this complexity and lack of perfection are to be expected if we assume that the future need for jamming avoidance mechanisms was not anticipated when electrosensory systems began to evolve and that, later in evolution, a JAR had to be implemented by modifying and patching networks adapted to other functions. The multitude of axonal collaterals of the output neurons of the torus, for example, suggests that the hardware for new behaviors evolved through the establishment of additional collaterals and the emergence of new target nuclei possibly by the division of ancestral nuclei. This possibility is supported by the recently described division of the ELL into three maps that appear similar in their cytoarchitecture and are innervated by collaterals of the same primary afferents but differ to some degree in the spatial and temporal properties of their receptive fields (6). In the future they could specialize even further, eventually fulfilling very different functions.

The study of neuronal networks also suggests mechanisms of their ontogeny, and neural networks postulated on the basis of theoretical considerations appear biologically plausible to the extent that they could grow under the guidance of simple and robust rules of self-organization. Such rules have been suggested and explored for the establishment of topographic maps (26,27) as well as for the calibration and scaling of maps required for appropriate matching of sensory and motor representations (see duLac, this issue[22]). Adaptive behavioral responses may develop either on the basis of genetic instructions or on the basis of trial and error learning. Direct genetic determination appears to prevail in organisms, such as many invertebrates, that have a short life span, develop without the assistance of experienced conspecifics and cannot 'afford' to make mistakes. Personal experience and learning through mistakes, on the other hand, may shape behavioral responses in animals growing up in the relatively safe environment of social groups and in all those instances where initial mistakes are not too costly. A general reward mechanism may then teach the individual to build its own response pattern and to shape it according to its particular needs and ecological conditions.

The JAR is a simple behavior not only in view of its motor organization but also with regard to its development. Juvenile fish acquire a correct motor response without any personal experience with jamming signals. Fish raised in isolation from the egg stage to a total length of at least 15mm are able to discriminate on their first exposure whether an interfering signal is slightly higher or lower in frequency than their own EOD and to shift their EOD frequency in the correct direction, i.e. away from the interfering frequency (28). But, even more surprising, some individuals, tested at a slightly earlier stage shift their EOD frequency in the wrong direction (29). These fish, however, correct their JAR within days, often overnight, even though the behavioral test employed does not provide any clues that would reveal the correctness of their response. Possibly the innervation of nuclei, such as the nE↓ and nE↑, follows statistical rules, and occasionally wrong connections may briefly predominate at the beginning. Antagonistic neuron pools could be separated by weeding out synapses that are not activated synchronously with most of their neighbors, using a Hebbian-type mechanism. But genetic information is at least required to assign the activation of one set of neurons to each of the antagonistic pools, to the nE↓ or the nE↑ in the case of the JAR.

Problems and Prospects

Although all behavioral properties of the JAR, such as its computational rules and time constants, can now be explained at the neuronal level, we are still far away from a complete description of all classes of neurons and their connections involved in this behavior. Only some structures, such as the ELL (30,31) and lamina 6 of the torus (32), have been explored at the ultrastructural level to develop detailed and realistic circuit diagrams. Network properties of the ELL have been tested by local pharmacological manipulations of intact structures (33) as well as in slice preparations (34). The most critical element of lamina 6, the small cells that compute phase differences between two inputs, are still too small to be analyzed physiologically, but modelling of their assumed membrane properties has at least suggested plausible ways they may operate (Lytton, pers. comm.).

Most of the neuronal connections participating in the JAR have not yet been examined ultrastructurally and can only be inferred from light microscopic tracing studies and from the spatial distribution of dendrites and axons of intracellularly labelled and physiologically characterized neurons. A comparison of response properties of neurons projecting to a certain area with those of neurons intrinsic to the area suggests certain functional relations between these two classes of neurons and postulates corresponding connections. Following this strategy, a realistic and almost complete circuit diagram for all structures involved in the JAR has been established (Figs.9, 10). Although our goal is to demonstrate connections postulated

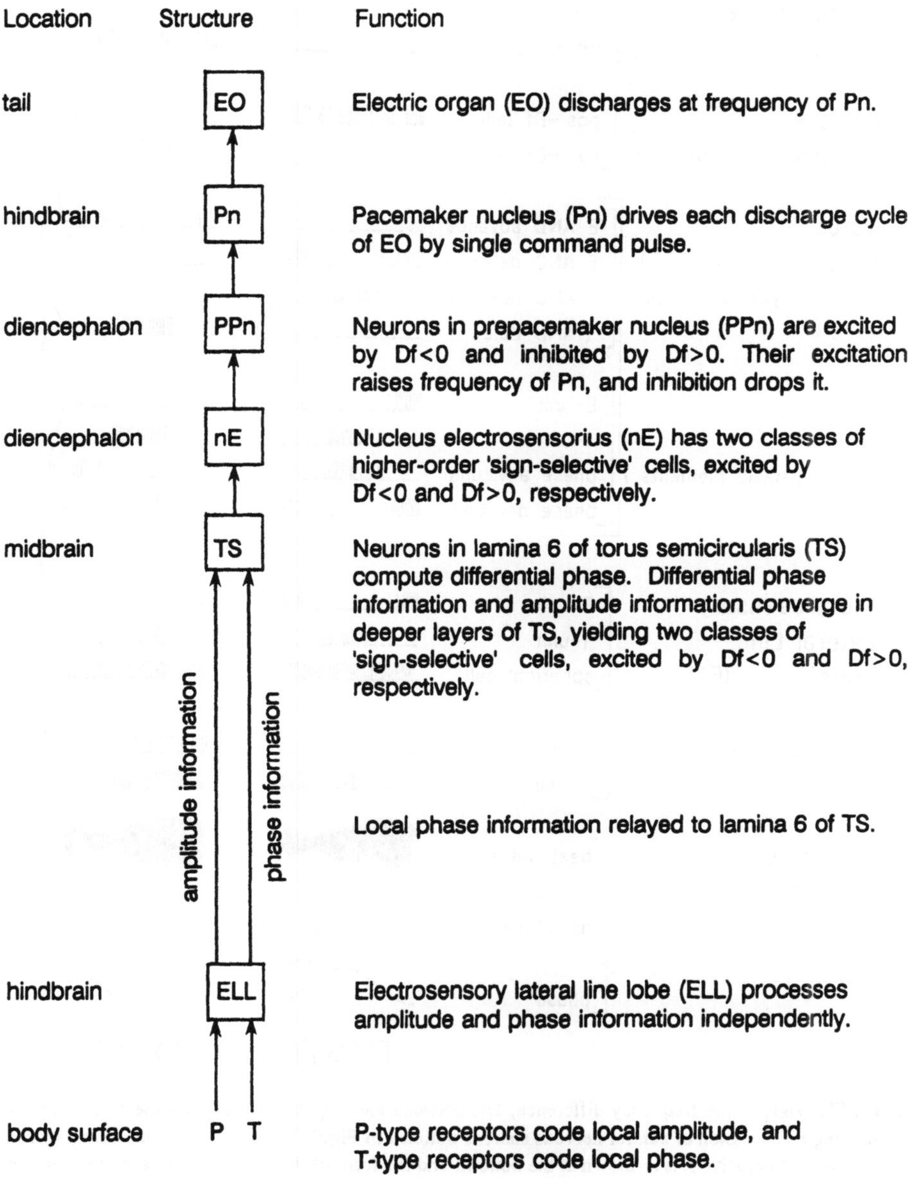

Fig.9: A flow diagram of neuronal structures and their functions in the context of the JAR, showing only elements whose significance for the JAR has been demonstrated experimentally.

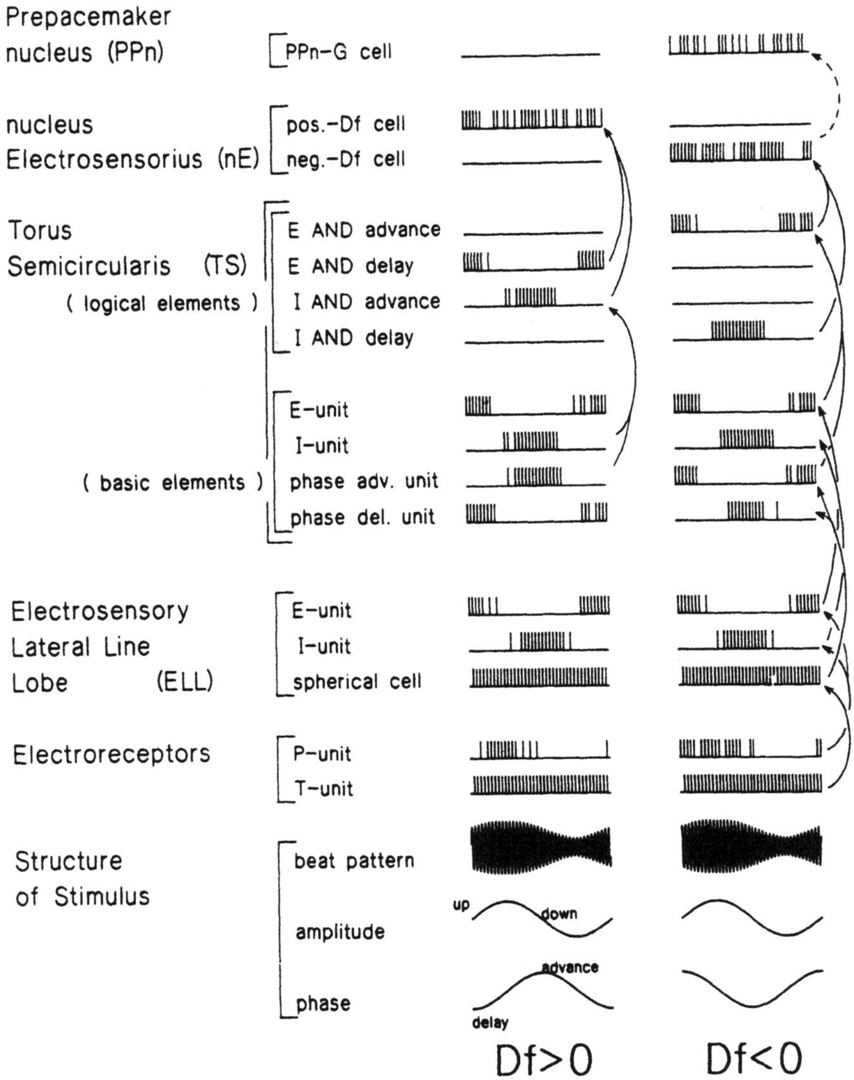

Fig.10: The sign of the frequency difference, Df, between the neighbor's EOD and the fish's own is coded along a succession of central nervous stations outlined in Fig.9. The structure of the beat patterns for positive and negative Dfs is shown at the bottom, and the form of their phase modulations is derived from a comparison of these beat patterns with representations of the fish's own EOD less strongly contaminated by the neighbor's signal (see Fig.2). Curved arrows to the right of symbolic spike records indicate the flow of information among the classes of neurons involved. To minimize clutter, only some connections have been indicated between the 'basic elements' and 'logical elements' of the torus. The assumed structure of the logical elements, also referred to as 'sign-selective' cells, has been discussed in connection with Fig.7. The 'AND' operation indicates a general form of non-linear facilitation between an amplitude-coding input (E- or I-type) and a phase-coding input (phase advance-coding unit or phase delay-coding unit). The nature of the connections from torus to nucleus electrosensorius (nE) is still unknown, and the arrows indicate only a possible way of wiring. Also unknown are the connections from the nE to the prepacemaker subnucleus (PPn-G) which contains cells inhibited by positive Dfs and excited by negative Dfs. The activation of these cells, in turn, raises the frequency of the medullary pacemaker.

on the basis of physiological evidence at the ultrastructural level, the time and resources required for such studies may be justified only in cases where unusual structural designs are expected.

Greatest difficulties have been encountered in the diencephalon where the small size and often sparse distribution of neurons have made intracellular studies almost impossible. Up to the level of the torus, studies have been aided by the fact that the electrosensory system codes temporal information with high fidelity, which, in turn, requires axons with relatively fast conduction. The axons and somata of these neurons, therefore, are sufficiently large for intracellular penetrations. But in the nE of the diencephalon, all necessary computations of sensory information have been performed, and inputs from the torus are pooled in two subnuclei, the nE↓ and nE↑, to lower or raise the frequency of the pacemaker slowly and gradually. Fast conduction, therefore, is no longer needed, and very small neurons, with soma diameters in the range of 5 μm, appear sufficient to establish the link to the PPn and, from there, to the pacemaker.

Acknowledgements: I thank David Sparks, Jennifer Altman and Catherine Carr for most helpful criticism. This work has been supported by grants from NSF, NIMH, and NINCDS

References

(1) Bastian J (1986) Electrolocation: Behavior, anatomy and physiology. In: Bullock TH, Heiligenberg W (eds) Electroreception. Wiley and Sons, New York, pp 577-612

(2) Konishi M, Takahashi TT, Wagner H, Sullivan WE, Carr CE (1988) Neurophysiological and anatomical substrates of sound localization in the owl. In: Edelman GM, Gall WE, Cowan WM (eds) Auditory Function. John Wiley & Sons, New York, 721-745

(3) Bullock TH, Hamstra RH, Scheich H (1972) The jamming avoidance response of high-frequency electric fish, I & II. J Comp Physiol 77: 1-48

(4) Heiligenberg W (1986) Jamming Avoidance Responses, model systems for neuroethology. In: Bullock TH, Heiligenberg W (eds) Electroreception. Wiley and Sons, New York, pp 613-649

(5) Heiligenberg W (1990) Sensory Information Processing in Electric Fish: Computational Rules and Their Neuronal Implementation. MIT Press monograph in: The Computational Neuroscience Series (in preparation).

(6) Shumway C (1989) Multiple electrosensory maps in the medulla of weakly electric gymnotiform fish. I: physiological differences.J Neuroscience 9: 4388-4399

(7) Keller CH, Maler L, Heiligenberg W (1990) Structural and functional organization of a diencephalic sensory-motor interface in the gymnotiform fish, Eigenmannia. J Comp Neurol 293: 347-376

(8) Rose GJ, Heiligenberg W (1986) Neural coding of difference frequencies in the midbrain of the electric fish Eigenmannia: Reading the sense of rotation in an amplitude-phase plane. J Comp Physiol A 158: 613-624

(9) Keller C, Heiligenberg W (1989) From distributed sensory processing to discrete motor representations in the diencephalon of the electric fish, Eigenmannia. J Comp Physiol A 164, 565-576

(10) Hagedorn M, Heiligenberg W (1985) Court and spark: Electric signals in the courtship and mating of gymnotoid electric fish. Anim Behav 33: 254 265

(11) Heiligenberg W (1991) Neural Nets in Electric Fish. In: Computational Neuroscience Series. MIT Press (in press)

(12) Kawasaki M, Maler L, Rose GJ, Heiligenberg W (1988) Anatomical and functional organization of the prepacemaker nucleus in gymnotiform electric fish: The accommodation of two behaviors in one nucleus. J Comp Neurol 276: 113-131

(13) Dye J, Heiligenberg W, Keller CH, Kawasaki M (1989) Different classes of glutamate receptors mediate distinct behaviors in a single brainstem nucleus. Proc Nat Acad Science, USA 86: 8993-8997

(14) Bastian J, Bratton B (1990) Descending control of electroreception. I: Properties of nucleus praeeminentialis neurons projecting indirectly to the electrosensory lateral line lobe. J Neurosci 10: 1226-1240.

(15) Bratton B, Bastian J (1990) Descending control of electroreception. II: Properties of nucleus praeeminentialis neurons projecting directly to the electrosensory lateral line lobe. J Neurosci 10: 1241-1253.

(16) Kawasaki M, Heiligenberg W (1988) Individual prepacemaker neurons can modulate the pacemaker cycle of the gymnotiform electric fish, Eigenmannia. J Comp Physiol A 162: 13-21

(17) Rose GJ, Kawasaki M, Heiligenberg W (1988) 'Recognition units' at the top of a neuronal hierarchy? J Comp Physiol A 162: 759-772

(18) Kawasaki M, Rose GJ, Heiligenberg W (1988) Temporal hyperacuity in single neurons of electric fish. Nature 336: 173-176

(19) Heiligenberg W (1987) Central processing of sensory information in electric fish. J Comp Physiol A 161: 621-631

(20) Baldi P, Heiligenberg W (1988) How sensory maps could enhance resolution through ordered arrangements of broadly tuned receivers. J Biol Cybern 59: 313-318

(21) Sparks D (1990) Neural control of saccadic eye movements in rhesus monkeys. (this issue)

(22) duLac S (1990) The role of visual experience in the registration of sensory and motor maps in the optic tectum of barn owls. (this issue)

(23) Georgopoulos AP, Kettner RE, Schwartz AB (1988) Primate motor cortex and free arm movements to visual targets in three-dimensional space. II. Coding of the direction of movement by a neuronal population. J Neurosci 8: 2928-2937

(24) Suga N (1988) Auditory neuroethology and speech processing: Complex-sound processing by combination-sensitive neurons. In: Auditory Function, eds. G.M. Edelman, W.E. Gall, W.M. Cowan, 679-720, New York: John Wiley & Sons

(25) Rose GJ, Keller C, Heiligenberg W (1987) 'Ancestral' neural mechanisms of electrolocation suggest a substrate for the evolution of the jamming avoidance response. J Comp Physiol A 160: 491-500

(26) Udin SB, Fawcett JW (1988) Formation of topographic maps. Annu Rev Neurosci 11: 289-327

(27) Durbin R, Mitchison G (1990) A dimension reduction framework for understanding cortical maps. Nature 343: 644-647

(28) Hagedorn M, Heiligenberg W, Carr CE (1988) The development of the Jamming Avoidance Response in the weakly electric fish Eigenmannia. Brain Behavior and Evolution 31, 161-169

(29) Viete S (1990) The development of the jamming avoidance response (JAR) in Eigenmannia: Behavioral and anatomical aspects. Diploma Thesis in Biology, University of Tübingen, Germany

(30) Maler L, Sas E, Rogers J (1981) The cytology of the posterior lateral line lobe of high-frequency weakly electric fish (Gymnotoidei): Dendritic differentiation and synaptic specificity in a simple cortex. J Comp Neurol 195: 87-140

(31) Mathieson B, Heiligenberg W, Maler L (1987) Ultrastructural studies of physiologically identified electrosensory afferent synapses in the gymnotiform fish Eigenmannia. J Comp Neurol 255: 526-537

(32) Carr CE, Maler L, Taylor B (1986) A time-comparison circuit in the electric fish midbrain. II. Functional morphology. J Neuroscience 6: 1372-1383

(33) Shumway CA, Maler L (1989) GABA-ergic inhibition shapes temporal and spatial response properties of pyramidal cells in the electrosensory lateral line lobe of gymnotoid fish. J Comp Physiol A 164, 391-407

(34) Mathieson WB, Maler L (1988) Morphological and electrophysiological properties of a novel in vitro preparation: the electrosensory lateral line lobe brain slice. J Comp Physiol A 163, 489-506.

(35) Heiligenberg W, Rose GJ (1986) Gating of sensory information: Joint computations of phase and amplitude data in the midbrain of the electric fish Eigenmannia. J Comp Physiol 159: 311-324

'Small cell' simulations: physiological features of a phase difference detector

William W. Lytton
The Salk Institute
10010 N. Torrey Pines Rd.
La Jolla, CA 92037

Abstract

The 'small cell' is a descriptive designation given an identified cell type in lamina 6 of the torus semicircularis of the weakly electric fish Eigenmannia that behaves as a coincidence detector similar to the cells of nucleus laminaris of the owl. A compartmental model of the small cell with Hodgkin-Huxley channel kinetics was used to explore the origin of the cell's temporal disparity sensitivity. This model was compared with a system level model of the time disparity detecting cell of the nucleus laminaris. The small cell has no identified inhibitory synaptic input yet will fire in response to a single input and cease firing when a second one is added at certain phase differences, a functional inhibitory effect. Simulations showed that such an effect could occur due to the internal inhibition provided by voltage-sensitive potassium channels. An initial subthreshold excitatory postsynaptic potential (EPSP) can thereby increase the threshold for a subsequent EPSP. In contrast to the n. laminaris model, this model shows a abrupt changes in rate at certain time disparities. This characteristic is not seen in the real cell and detracts from the cell's ability to code arbitrary disparities. Obtaining a small cell response that varies smoothly with input will require either adding noise or adding more complex dynamics to the model. These alternatives could be assessed physiologically with current-clamp.

INTRODUCTION

Coincidence detection is a general problem for neural systems that must respond to small phase differences in incoming signals in the auditory domain. The owl and the bat have been extensively studied in this regard but homologous systems are used in all auditory systems (Simmons and Chen, 1989; Takahashi and Konishi, 1986). The electric fish, Eigenmannia, performs the same task in the detection of phase differences in electric fields produced by itself or others of its species (Rose and Heiligenberg, 1986). The 'small cell' appears to be the first cell in a neural hierarchy of cells concerned with determining phase difference. The signals are initially detected by receptors that signal the zero-crossing time of sinusoidal fields impinging on the fish's body. The projections from receptors at two different sites on the body eventually reach the small cell, which does the phase comparison, through two routes (Fig. 1). One route arrives via a single additional synapse onto the 'spherical cell' while the other route arrives via two additional synapses with the spherical cell synapse being followed by a synapse onto the 'giant cell' (Carr et al., 1986b). All synapses are electrotonic or mixed, allowing the spikes at each level to phase lock closely with little jitter. In fact, the variability of the signals arriving at the small cell have been measured and show a standard deviation of about 20 μsec (Carr et al., 1986b).

The physiological characteristics of the small cell have been assessed by studies of the Jamming Avoidance Response (JAR), discussed elsewhere in this volume (Heiligenberg, this volume). Briefly, two body areas are insulated from each other and two artificial signals are presented with a known phase difference (Fig. 1). Eigenmannia can distinguish a 1 μsec time difference between the phase of externally applied sinusoidally varying electrical fields (Kawasaki et al., 1988). The small cell, although it fires in response to phase inputs over 1 msec of time difference, shows a tuning curve with respect to phase difference that give it a sensitivity on the order of tens of microseconds (Heiligenberg and Rose, 1985).

The general design of a coincidence detector is very simple. Two signals arrive in the detector. If they are nearly in-phase they summate producing a large signal. If they are nearly out-of-phase they tend to cancel one another giving a small signal. One can set a threshold to signal the arrival of large signals, thereby signaling that the time disparity between signals is less than that determined by this threshold. This is the basis of this model as well as one previously described for the nucleus laminaris of the owl (Grün et al., 1990). In its simplest form,

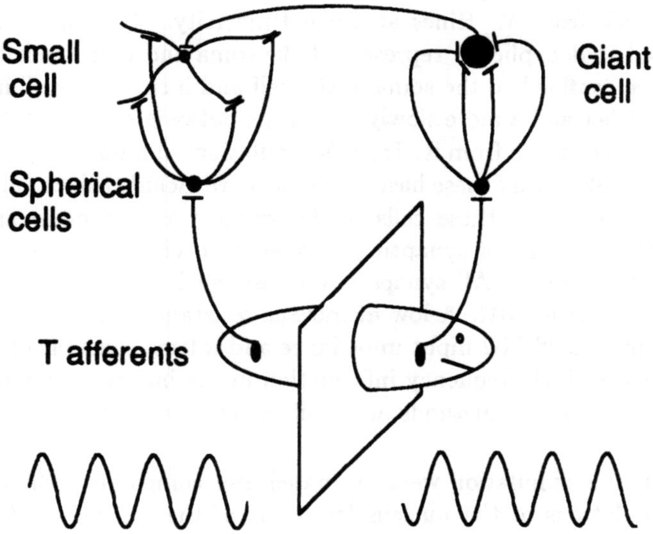

Figure 1: Small cell inputs come from 2 parts of the bodies via 2 different paths as shown. The first path is disynaptic via the spherical cells and the second is trisynaptic via spherical cell and giant cell. In testing the JAR, sinusoidal signals of identical frequency but varying phase are applied to two parts of the fish's body which are insulated from each other. Phase coding tuberous receptors (T afferents) encode the positive zero crossings. All synapses are electrotonic. The small cell receives synaptic input on its dendrites from the spherical cells and on its soma from the giant cell. Several T afferents project to each spherical cell and several spherical cells project to each giant cell; only one is shown in each case for simplicity. It is not known whether the projections onto the small cell dendrites come from one or from several spherical cells.

this type of model will not code arbitrary phase differences since the sharp threshold divides the cell response into two states, firing and not firing. The problem is illustrated by the behavior of the small cell model described in this paper.

METHODS

The small cell was simulated using a modified version of M. Hines CABLE simulator run on a MIPS M/120 computer (Hines, 1989). This simulator uses a Crank-Nicholson integration for voltage and table look-up for voltage-sensitive channel integration. The current version of the simulator is called NEURON and is available from M. Hines at Duke University. A compartment model was produced that explicitly represented the soma and four dendrites. Active channels were confined to the soma of the cell and included a sodium channel, a delayed rectifier and a more slowly activating potassium channel. The former two channels were taken from R. Traub's simulations of hippocampal pyramidal cells (Traub, 1982). Only these basic channels were included since voltage clamp cannot yet be done on these cells to determine what channels are actually present. Two locations of synaptic input were provided: onto the soma and onto the four dendrites. All synapses were assumed to be electrotonic with a resistance of about 50 MΩ. A low membrane resistance was used (330 $\Omega \cdot cm^2$ corresponding to a 26 MΩ input impedance and a time constant of 0.25 msec) to prevent loss of high frequency information in the incoming signals. Further details of the simulation methods are available elsewhere (Lytton, 1991).

Results from this simulation were evaluated and compared with results from previous simulations of the nucleus laminaris of the owl using Mathematica from Wolfram Research Inc.

RESULTS

Manipulation of the firing threshold in a deterministic Hodgkin-Huxley compartment model such as this one would allow the model to be sensitive to an arbitrarily small range of time disparities, down to the round-off errors in the numerical integration. Two aspects of the known physiology prevented taking the model to this extreme. First, the jitter of the synaptic inputs causes some spread in the response. This jitter is known and was included in the model as a Gaussian variation in the timing of synaptic input at each synapse. Second, the cell can be fired by the presence of one input alone. Therefore, a single EPSP must be able to reach threshold. However, two EPSPs of the same size, when arriving out-of-phase, must be subthreshold. Therefore, an added EPSP arriving out-of-phase must somehow functionally inhibit the small cell so that firing does not take place in response to an EPSP that, by itself, would fire the cell. There is no evidence of inhibitory inputs onto this cell so classical inhibitory mechanisms cannot be invoked to explain this functional inhibitory effect (C. Carr, personal communication).

Such paradoxical EPSP inhibition or defacilitation can take place because of the role of active conductances in determining the threshold. The time and voltage dependence of the threshold depends on the time and voltage dependence of the sodium and potassium channels. When total inward current exceeds outward current, positive feedback produces the spike. An EPSP can prevent this by either reducing the inward current by changing the state of the sodium channel or reducing the outward current by changing the potassium channel. In the former case, the initial depolarization partially inactivates sodium channels reducing the inward current in response to the next EPSP. In the latter case, the depolarization partially activates potassium channels, increasing the outward current. In the model, both effects were seen but the effect on the potassium channel predominated. Firing was seen in response to a repetitive EPSP at 300 Hz (Fig. 2A) but no firing appeared when another periodic EPSP at the same rate was added in antiphase (Fig. 2B).

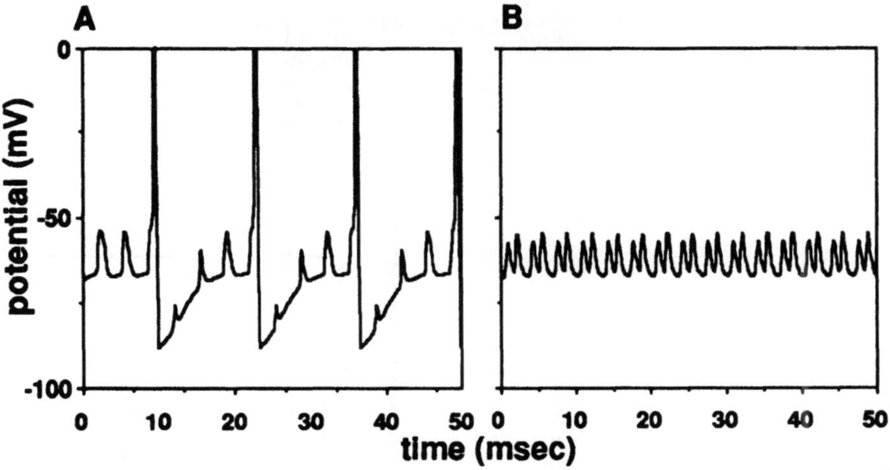

Figure 2: A Small cell model shows repetitive firing with 300 Hz EPSP input into the soma alone. B Identical input onto soma with addition of out-of-phase dendritic EPSPs prevents cell firing.

Having chosen parameters that permit this paradoxical EPSP defacilitation, the cell could then be made very sensitive to a chosen time disparity by choosing the proper threshold (Fig. 3A). This threshold would equal the voltage peak of the compound EPSP produced by adding the dendritic and somatic EPSPs with the chosen time disparity. There is not just a single threshold, however. The small cell will fire upon arrival of an EPSP but will not generally be able to follow at the rate of EPSP arrival, about 300-400 Hz. Instead, it will fire on every second, third or fourth EPSP, taking on firing frequencies that are subharmonics of the input frequency. The threshold for firing at higher frequency is greater since the cell remains in a relatively refractory period. Therefore, as the time disparity between the two EPSPs decreases, the cell will increase firing rate in a quantized manner, going from one subharmonic to the next (Fig. 3B). These quantized firing frequencies can be seen in a histogram of instantaneous frequency at different time disparities (Fig. 3C).

A major difference between the firing pattern of the model and that of the real

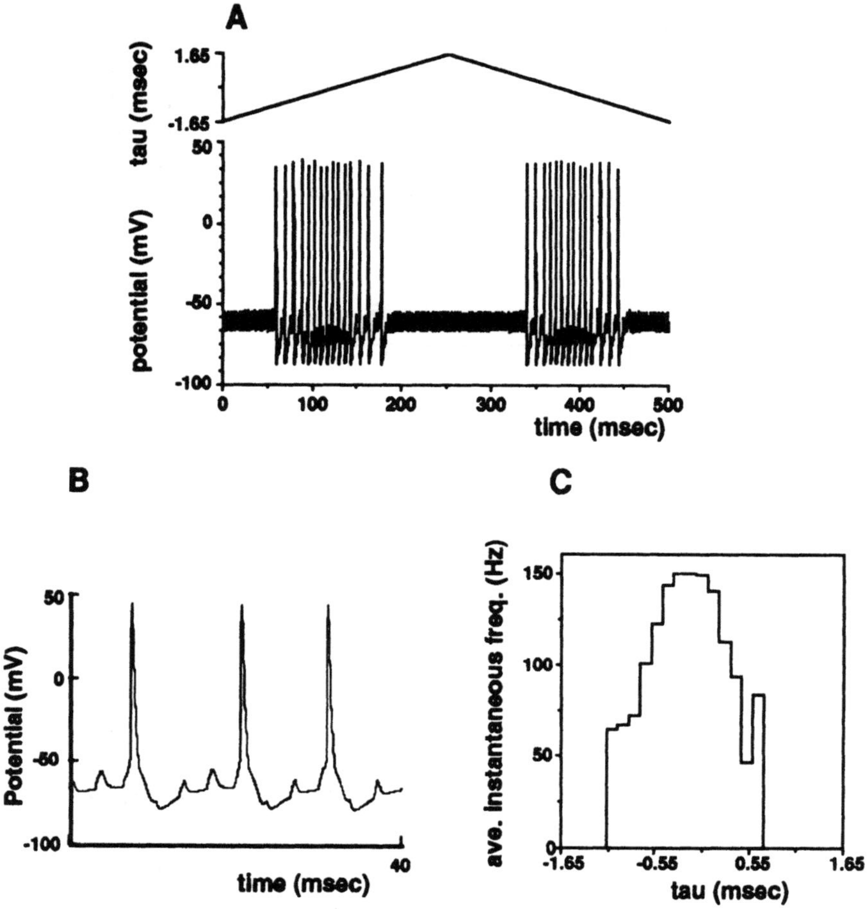

Figure 3: A Small cell fires when dendritic and soma EPSPs are nearly simultaneous. Firing stops when the dendritic EPSP precedes the soma EPSP (negative tau measured from time of onset of soma EPSP) or follows the soma EPSP (positive tau). Frequency of EPSPs is 300 Hz corresponding to a period of 3-1/3 msec. A full cycle of phase difference is seen with tau -1.65 msec to 1.65 msec. B Small segment taken from a simulation similar to that shown in A. As the time disparity decreases, the rate of firing increases. Since firing only occurs on the EPSP, the rate increases incrementally rather than continuously. The small cell goes from firing on every fourth EPSP to every third EPSP. The initial EPSP following the spike is difficult to see since it is buried in the afterhyperpolarization. C Histogram of instantaneous frequency with varying tau. The highest frequency, 150 Hz, occurs when the cell fires on every other EPSP.

cell is this discretization of firing rate. This suggests that the model is missing some source of noise that is present in the real system. The small cell has the same function as the cells of the nucleus laminaris of the owl. A systems model of the latter shows the gradual increase in firing rate with change in time disparity that is lacking in the present model (Grün et al., 1990). Therefore, I analyzed the small cell model using the approach used in the nucleus laminaris model in order to better understand the difference in response pattern.

Several minor differences between the two models became immediately apparent. The n. laminaris cell model assumes that the cell receives a sinusoidal input. This sinusoid cannot be identified with the instantaneous frequency of the presynaptic cell since the frequency of the sinusoid itself (up to 9000 Hz in the owl and about 400 Hz in the fish) is too high to accommodate any frequency coding within a cycle. Therefore, the sinusoid must represent probability of firing at different points in the cycle. The small cell model, on the other hand, directly adds EPSPs. These can be represented as half-wave rectified sinusoids (Fig. 4A).

The signals from the two sources summate to produce voltage variation in the cell. Summation of the two EPSPs will produce a net compound EPSP whose size gradually increases with decreasing time disparity, just as in the n. laminaris model (Fig. 4A). This voltage signal then has to be read out to produce an output signal on the axon of the phase discriminator. In the n. laminaris model, the output is produced by feeding the voltage signal through a sigmoid squashing function

$$f(v) = 1/\left(1 + \exp\left[(v - v_0)/r\right]\right) \tag{1}$$

that represents the probability of firing at different membrane voltages with a location provided by v_0 and a steepness determined by r (Fig. 4B, smooth curve). This permits the voltage signal to be transformed into a frequency code via firing probability. Amplification is achieved by adjusting the location of the squashing function on the abscissa so that the voltage variation of interest occurs over the low-voltage non-linear range. This is accomplished either by adding inhibition to move the signal down to lower voltages or adjusting the threshold to move the sigmoid up to higher voltages. In the small cell model, on the other hand, the relation of voltage to firing does not follow a sigmoid function but instead shows a coarse stair step (Fig. 4B). Therefore the resulting output of the cell is discretized (Fig. 4C). In the figure, probability of firing is shown across time with different time disparity inputs. Although the ordinate is given as a probability, the cell will not fire arbitrarily but will instead fire regularly at the appropriate interval (i.e. on every third cycle if firing probability is given as 1/3).

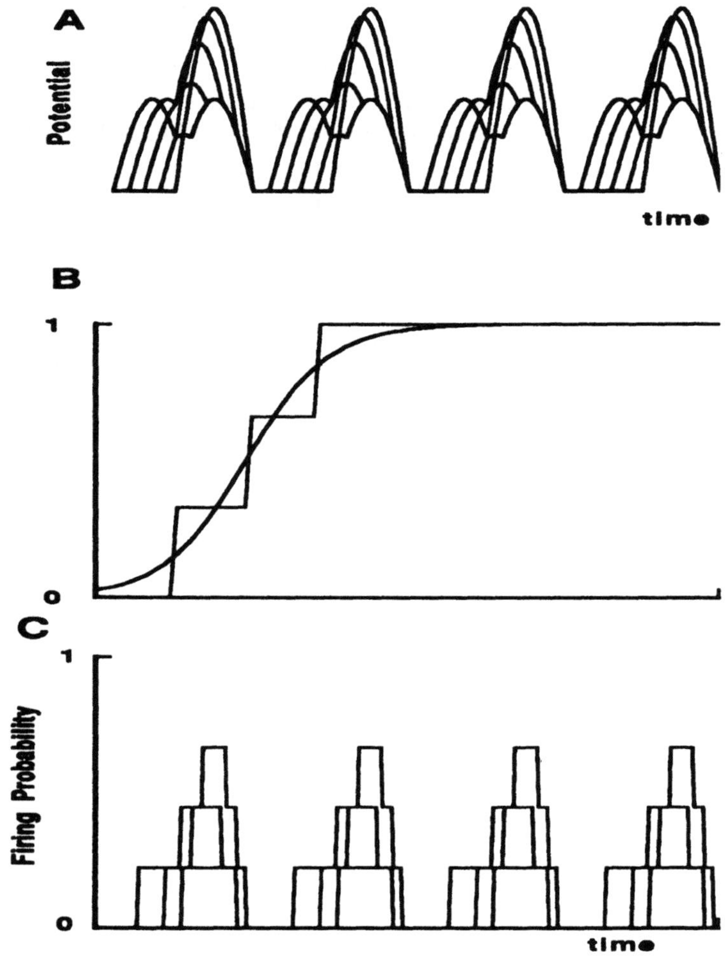

Figure 4: Reinterpretation of small cell model using formalism of Grün *et al.* n. laminaris model. A Cell receives 2 EPSPs with different arrival times. Here the two EPSPs are represented by 2 half-wave rectified sinusoids. With smaller time disparity, a larger compound EPSP results. The original owl model compared non-rectified sinusoids. EPSPs add together depending on their relative arrival time. B. Small cell dynamics provide this stairstep transformation of the incoming signal. This contrasts with the smoother sigmoid transfer function of the original n. laminaris model. C Cell firing tends to be discretized with different rates associated with different thresholds on the stairstep function. The different traces here are firing probabilities associated with 3 of the input patterns illustrated in A. Small cell firing will be regular, the firing probability therefore gives the inverse of the number of cycles between input spikes. A probability of 1/3 means that the cell will fire on every third cycle of input EPSP.

DISCUSSION

Although the general plan of a coincidence detector cell appears clear, many details can vary enormously from one species to another. For example, both the small cell and n. laminaris cell show firing to a single sided input with increased firing when a second input is added in phase and decreased firing when a second out-of-phase input is added. In n. laminaris there is evidence of abundant inhibitory innervation (Carr et al., 1989), while the small cell lacks inhibitory input. Therefore, under certain conditions, the small cell shows reduced firing rate in the setting of increased net excitatory input. The properties of active channels provide a potential explanation for this paradox. Active channels possess a memory by virtue of their time constants. For this reason, the firing of a nerve cell depends on its history. An EPSP causes an increase in threshold as outward currents are activated and inward currents are inactivated. This change in threshold is a minor version of the relative refractory period that occurs following a spike.

Proper balancing of several time constants was necessary to produce this paradoxical EPSP defacilitation. First, the membrane time constant must be short enough to permit the individual EPSPs to be clearly localized in time. Too long a time constant will cause EPSP effects to overlap, making it impossible to distinguish time differences. Second, the time constant of the slow potassium channel must be long enough to provide the memory of the prior EPSP. Third, the spiking mechanism must itself be fast so as not to produce voltages that interfere with the interpretation of signals in the next cycle.

The threshold must be placed relatively low in order for firing to occur in response to a single input. Yet the height of the cell threshold is the primary determinant of the cell's tuning curve for time disparity. The higher the threshold can be placed, the narrower the tuning curve can be. With a narrower tuning curve, the cell's coding would seem to be more precise. However, there are disadvantages to such precision. In many cases, a "coarse-coding" cell that responds over a broader range may be preferable (Baldi and Heiligenberg, 1988). A narrowly tuned cell is of no use when time disparities outside of its small range are presented. Therefore, in most cases it contributes nothing to the interpretation of the signal. A coarse-coding cell responds over a broader range and thereby carries information more robustly. The achievement of ever greater precision may not be an evolutionary goal for cells of this type.

The small cell model showed a tendency to shift abruptly from one firing rate to another as its thresholds were reached. Multiple thresholds were seen because of the nature of the input. There was a relatively high threshold for firing on every other input cycle, a somewhat lower threshold for firing on every third and a still lower one for firing on every fourth. These multiple thresholds gave the cell response function a stairstep appearance. These quantized jumps in firing rate reduce the amount of information the cell can carry since a single firing rate lies on a plateau that represents a range of time disparities. Again, a more useful encoding strategy is the coarse-coding of time disparities in a cell that shows a gradual alteration in firing rate with changes in input. This second element of coarse-coding is complementary to the first, since a very narrowly tuned cell will necessarily have a relatively abrupt change in firing rate over a short range.

A major difference between the small cell and the n. laminaris cell lies in the nature of the synaptic input. While the small cell generally has about five tightly coupled electrical synapse (Carr et al., 1986b), the n. laminaris cell has about 100 chemical synapses coming into a single cell (Carr and Konishi, 1990). On the one hand the chemical synapse can be expected to show greater transmission noise and also produce a low pass filter effect that will obscure the precise timing of the signal. On the other hand, the multiple inputs should produce an average signal with reduced variance (Carr et al., 1986b; Lytton, 1991). Overall, the incoming signal may be expected to be far noisier than the signal arriving at the small cell (Spence and Pearson, this volume). This added noise could help broaden the signal response represented by the relatively shallow sigmoid response function used in the Grün et al. model.

What noise could be added to the small cell model to allow it to show a more gradual change in firing rate in response to input time disparity? The most obvious choice is channel noise in the threshold mechanism. Nossal and Lecar (1971) have studied the variation in firing threshold as a function of the variance in unitary conductance of sodium channels in the Hodgkin-Huxley model. Their results, based on the assumption that the cell only has a fast sodium and delayed rectifier channel, suggests that the threshold is related to channel opening variance by the following formula:

$$P\,(\text{fire}|I) = \frac{1}{2}\left[1 + \text{erf}\left(\frac{1}{R}\cdot\frac{I - I_\theta}{I_\theta}\right)\right] \tag{2}$$

where P is the probability of firing, I is the injected current, I_θ is the voltage of the threshold and R is the variance due to the single channel statistics. Experiments done on the frog node of Ranvier show that variance of sodium channel conductance is approximately 1% of the average conductance. Taking this value for R, figure 5 shows that the probability transfer function for a given input is extremely sharp. The needed variability is lacking. The firing mechanism is also subject to smaller amounts of shot noise, thermal noise and $1/f$ noise but these are probably not significant (Stevens, 1972).

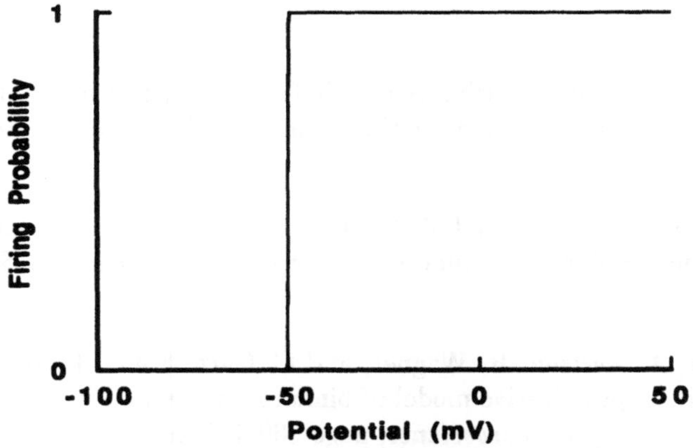

Figure 5: Expected variation in firing threshold due to probabilistic opening of individual sodium channel molecules. Channel noise (R) of 1% and threshold (I_θ) of -50 mV are used. The threshold function is extremely sharp.

The large number of sodium channels needed to produce an action potential greatly reduces the variability in threshold due to the opening of individual channels. Perhaps a mix of conductances could be found that would enhance threshold variability by making the threshold primarily dependent on a single current carried by a relatively small number of channels. Alternatively, the needed variation in firing could be provided in a deterministic system that shows intermediate firing patterns allowing it to switch back and forth between different subharmonics despite constant input. Such a chaotic system might be difficult to distinguish from a stochastic system, since its strong dependence on history would be impossible to trace. However, current-clamp might be able to distinguish these two possibilities since one can then establish identical initial conditions from one test to the next. As current-clamp study of these cells is becoming possible, this variability can be directly assessed.

References

[1] P. Baldi and W. Heiligenberg. How sensory maps could enhance resolution through ordered arrangements of broadly tuned receivers. *Biol.Cyber.*, 59:313–318, 1988.

[2] C.E. Carr, I. Fujita, and M. Konishi. Distribution of gabaergic neurons and terminals in the auditory system of the barn owl. *J. Comp. Neurol.*, 286:190–207, 1989.

[3] C.E. Carr, W. Heiligenberg, and G. Rose. A time-comparison circuit in the electric fish midbrain. I. behaviour and physiology. *J.Neurosci.*, 6:107–119,

1986a.

[4] C.E. Carr and M. Konishi. A circuit for detection of interaural time differences in the brain stem of the barn owl. *J. Neurosci.*, 10:3227–3246, 1990.

[5] C.E. Carr, L. Maler, and B. Taylor. A time-comparison circuit in the electric fish midbrain. II. functional morphology. *J.Neurosci.*, 6:1372–1383, 1986b.

[6] S. Grün, A. Aertsen, H. Wagner, and C. Carr. Sound localization in the barn owl: a quantitative model of binaural interaction in the nucleus laminaris. In *Soc. for Neuroscience*, page 360.6, 1990.

[7] W. Heiligenberg. Central processing of sensory information in electric fish. *J.Comp.Physiol.A*, 161:621–631, 1987.

[8] W. Heiligenberg and G. Rose. Phase and amplitude computations in the midbrain of an electric fish: Intracellular studies of neurons participating in the jamming avoidance response of eigenmannia. *J.Neurosci.*, 5:515–531, 1985.

[9] M.L. Hines. A program for simulation of nerve equations with branching geometries. *Int.J.Biomed.Comp.*, 24:55–68, 1989.

[10] A.L. Hodgkin and A.F. Huxley. A quantitative description of membrane current and its application to conduction and excitation in nerve. *J.Physiol.*, 117:500–544, 1952.

[11] M. Kawasaki, G. Rose, and W. Heiligenberg. Temporal hyperacuity in single neurons of electric fish. *Nature*, 336:173–176, 1988.

[12] H. Lecar and R. Nossal. Theory of threshold fluctuation in nerves I,II. *Biophys. J.*, 11:1048–1084, 1971.

[13] W.W. Lytton. Simulations of a phase comparing neuron of the electric fish eigenmannia. *J.Comp.Physiol.A*, 169:117–125, 1991.

[14] R.J. MacGregor. *Neural and Brain Modeling*. Academic Press, Inc., San Diego, 1987.

[15] G. Rose and W. Heiligenberg. Structure and function of electrosensory neurons in the torus semicirvularis of eigenmannia: morphological correlates of phase and amplitude sensitivity. *J.Neurosci.*, 5:2269–2280, 1985.

[16] G. Rose and W. Heiligenberg. Limits of phase and amplitude sensitivity in the torus semicircularis of eigenmannia. *J.Comp.Physiol.A*, 159:813–822, 1986.

[17] I. Segev, J.W. Fleshman, and R.E. Burke. Compartmental models of complex neurons. In C. Koch and I. Segev, editors, *Methods in neuronal modeling*, pages 63–96. MIT Press, Cambridge, 1989.

[18] J. Simmons and L. Chen. The acoustic basis for target discrimination by fm echo locating bats. *J.Acoust.Soc.Am.*, 86:1333–1350, 1989.

[19] C.F. Stevens. Inferences about membrane properties from electrical noise measurements. *Biophys. J.*, 12:1028–1047, 1972.

[20] T. Takahashi and M. Konishi. Selectivity for interaural time difference in the owl's midbrain. *J.Neurosci.*, 6:3413–3422, 1986.

[21] R. Traub. Simulation of intrinsic bursting in CA3 hippocampal neurons. *Neuroscience*, 7:1233–1242, 1982.

Compartmental modeling of macular primary neuron branch processes

Thomas C. Chimento[1], David G. Doshay[2], Muriel D. Ross[1],
[1]NASA-Ames Research Center, Moffett Field, CA 94035
[2]Sterling Software, Palo Alto, CA 94303

INTRODUCTION: Compartmental models of electrotonic (passive) voltage spread in neural structures depend upon the morphology of the structures being modeled and the electrical characteristics of the cell. The primary morphological parameters are the length and diameter of each compartment. The electrical characteristics include membrane resistance and capacitance, cytoplasmic resistance, and conductance and time course of membrane channels. The accuracy of these model parameters determines how well the model can reflect the actual operation of the functioning neural structure.

In most compartmental models only a rough approximation of morphology is available. With precise measurements of the length and diameter of neural processes we can constrain one set of computational variables to test the importance of morphology on voltage transfer characteristics. Because the primary afferents in the vestibular system have an assortment of collateral-like processes, each of which can contribute to the net depolarization at the impulse initiation zone, it is critical to understand the contribution of each process.

The goal of this research is to determine whether applying ultrastructural morphometry to compartmental modeling yields significant differences between voltage contributions of processes with different morphology. This short-term objective is viewed as a necessary first step in the development of a full three-dimensional (3-D) model of a working vestibular macula. Specifically, other investigators have built vestibular models based upon the assumption that the primary afferent receives an equivalent depolarization from each process on the neuron branch [1]. The simulations reported here demonstrate that morphology plays a significant role in determining the effect of each process on the neuron branch.

METHODS: This paper focuses on compartmental models based on measurements obtained from 3-D reconstructions of actual vestibular neuron afferent processes derived from serial section electron microscopy of rat maculas. Tissues were obtained and prepared for transmission electron microscopy as detailed elsewhere [2]. Micrographs of serial sections were used to produce 3-D images of the terminal endings. Details of the reconstruction method are provided elsewhere [3, 4]. Briefly, terminal and receptive fields were traced onto acetate sheets from montages of photographed sections. These were digitized into a PC and reassembled as filled slice reconstructions using software provided by Kinnamon [5]. Data were transferred to a Silicon Graphics IRIS 4D/ 210 VGX graphics workstation for reconstruction as color-coded, shaded or transparent solids in 3-D using software developed in our lab (ROSS2.3 [6]) .

Within the macula processes emanate from neuron branches and calyces. For simulation purposes all processes were located at the center of a uniform neuron branch 25 μm in diameter and 1118 μm long. This length is two times λ, the electrotonic length. Only simulations of processes on neuron branches are presented here. Numerical routines for two-dimensional current spread, needed to model the calyx properly, are under development. The simulations reported in this paper were done with NEURON [7].

Diameter and length measurements of the processes were used to construct schematic representations for analysis. The measurements were taken where the taper rate of the process changed. This procedure results in model input comprised of sections, each section being a cylinder or a frustrum (truncated cone). Figure 1 is an illustration of two neuron branches with processes. The figures are shown to scale, except that the length used for the neuron branch in the simulation is 100 times longer than shown.

Frustrums are approximated by a series of stacked cylindrical isopotential compartments varying linearly in diameter, as shown in Figure 2. All process segments are modeled with 30 compartments, and the neuron branch is modeled with 200 compartments. The required number of compartments was arrived at empirically, increasing the number of compartments until a finer spatial mesh did not effect the results of the computations. Even with this moderately large number of compartments and a time step of 1/1000 τ(the time constant of the membrane), a single simulation run took less than two minutes on a Sun SPARC-station 4-370.

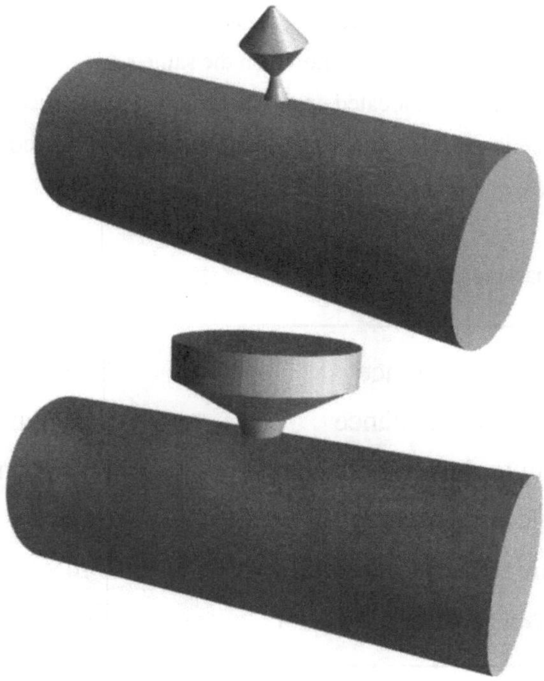

Figure 1. 3-D illustration of two neuron branches with processes.

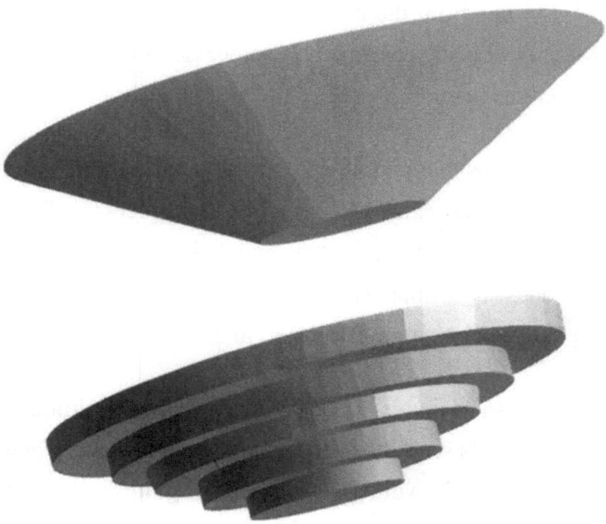

Figure 2. One section of a process. Frustrums are modeled as isopotential stacked cylinders. For clarity, only 5 of 30 cylinders are shown here.

We present simulations of four processes, each shown as the insets in Figures 3 A through D. These schematics are all shown on the same scale. A synapse, modeled as a time-varying conductance, is located on the distal end of each process. The time course of the conductance is modeled as a standard alpha function. All model parameters are listed in Table 1.

Table 1. Model parameters.

membrane resistance	5000 Ω cm^2
membrane capacitance	1 μF/cm^2
axial (cytoplasmic) resistance	100 Ω cm
synaptic conductance	5 μS
synaptic reversal potential	0 mV
membrane resting potential	-65 mV

Simulated voltage is measured at four sites: 1) at the distal end of the compartment containing the synapse, 2) at the base of the process, 3) on the neuron branch process at 33.5 μm and 4) 145 μm from the center of the base of the process. These lengths approximate the range of distances separating processes and the impulse initiation zone in the macular afferents being modeled.

RESULTS: In some cases maximum depolarization varies greatly between processes with roughly similar shapes (Fig. 3A, 3B). The two processes illustrated are funnel shaped, their distal ends have approximately equal diameters and the lengths from their proximal to distal ends are similar. The peak amplitude of the depolarization is greater, and the time to peak depolarization is shorter, in 3A than in 3B.

Although the processes in Figures 3C and 3D have rather dissimilar morphology, the voltages measured at the base of the process and along the neuron branch are indistinguishable from each other. The voltages, however, are quite different from those shown by processes illustrated in Figures 3A and 3B. From these simulations it becomes clear that morphological details can have profound effects on the magnitude of voltage swings in the neuron branch.

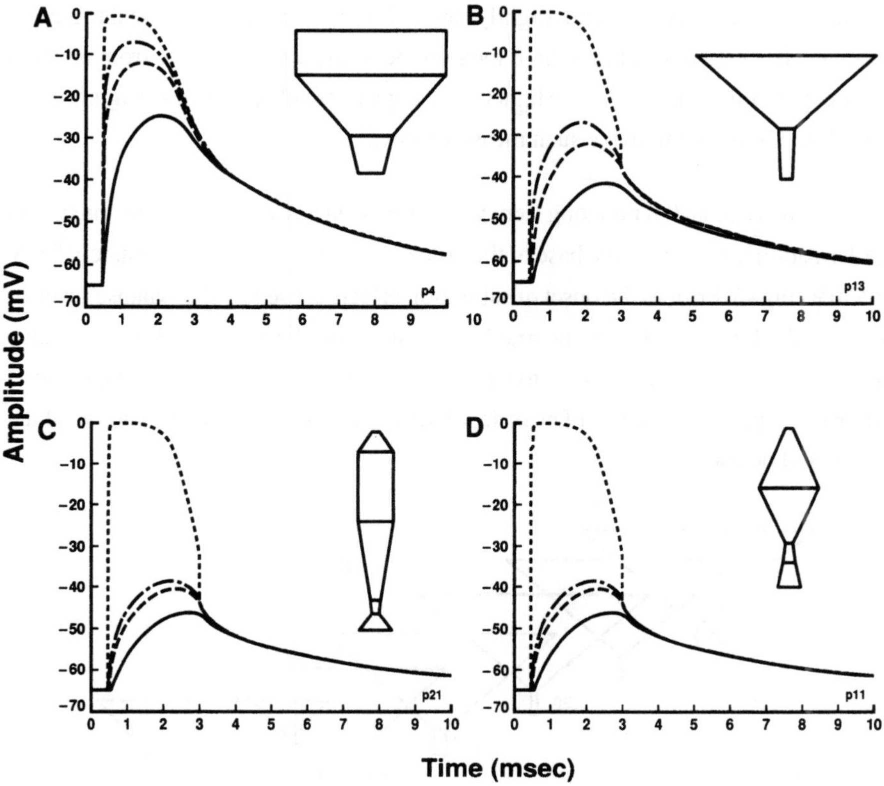

Figure 3. Simulations of four processes. Lines represent voltage at the tip of the process (small dashed line) , the base of the process (alternating dashed line), and two locations along the neruon branch (large dashed and solid lines). See text for details. Insets are diagrams of the processes being modeled.

The above results led us to investigate the particular morphological features that dominate in determining the depolarization at the base of a process. It is apparent from the equations for membrane and cytoplasmic resistance that the diameter of a neural element is of greater consequence than the length in determining the total resistance of the process. In the steady state (constant current) the voltage drop over the length of the process is easily determined by Ohm's Law and the resistance. However, the time-dependent case is not as trivial. Therefore, we examined the effect of length and diameter by first varying the diameter of one section of a process (the stem-like constriction) while keeping all other parameters constant and then varying just the length of the same process stem while holding its diameter constant. With this methodology we investigated just one row and one column of the full martix of possible stem sizes. Results of the full matrix will be reported separately. The lengths and diameters of the processes we examined in our models extend slightly beyond those measured in macular neuron processes. It must be noted that these measurements are available only from electron microscopic examination, because light microscopy does not provide the magnification or resolution required to make such measurements.

As the diameter of an 0.8 mm long stem was varied from 0.07 to 1.6 μm, the maximum depolarization measured at the base of the process increased from approximately 5 mV to 65 mV (Fig 4). Prior to the onset of saturation effects caused by the synapse reversal potential, the linear portion of the graph has a slope of 90 mV/mm. This large value makes clear that the system is sensitive to minute changes in diameter of the narrowest portion of the process. Latency of peak depolarization also changes significantly, which is discussed below.

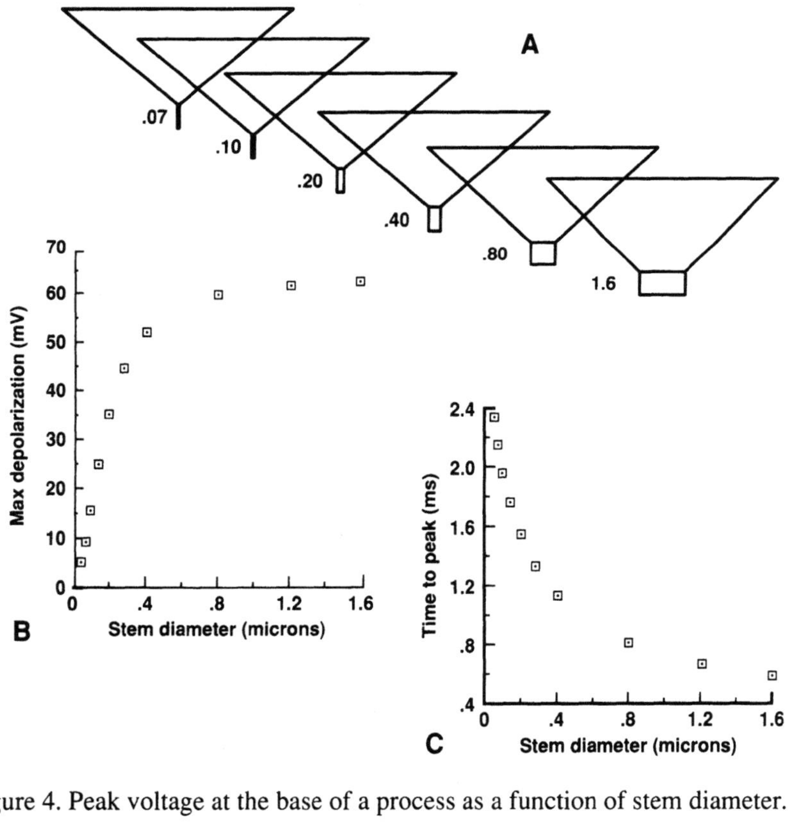

Figure 4. Peak voltage at the base of a process as a function of stem diameter.

Increasing the length of a 0.2 μm diameter stem from 0.5 μm to 16 μm decreased the maximum depolarization from just over 40 mV to nearly 5 mV (Fig 5). Obviously, there is not a linear relationship between stem length and either maximum depolarization or time to peak. This should be expected because the *percentage* change in size is the relevant parameter, not the absolute change in length. To view this more clearly we plot the independent parameter (length) on a log scale in Figure 6. In these graphs we plot the data for lengths of the stem as well as the data for the stem diameters to compare the relative importance of each parameter.

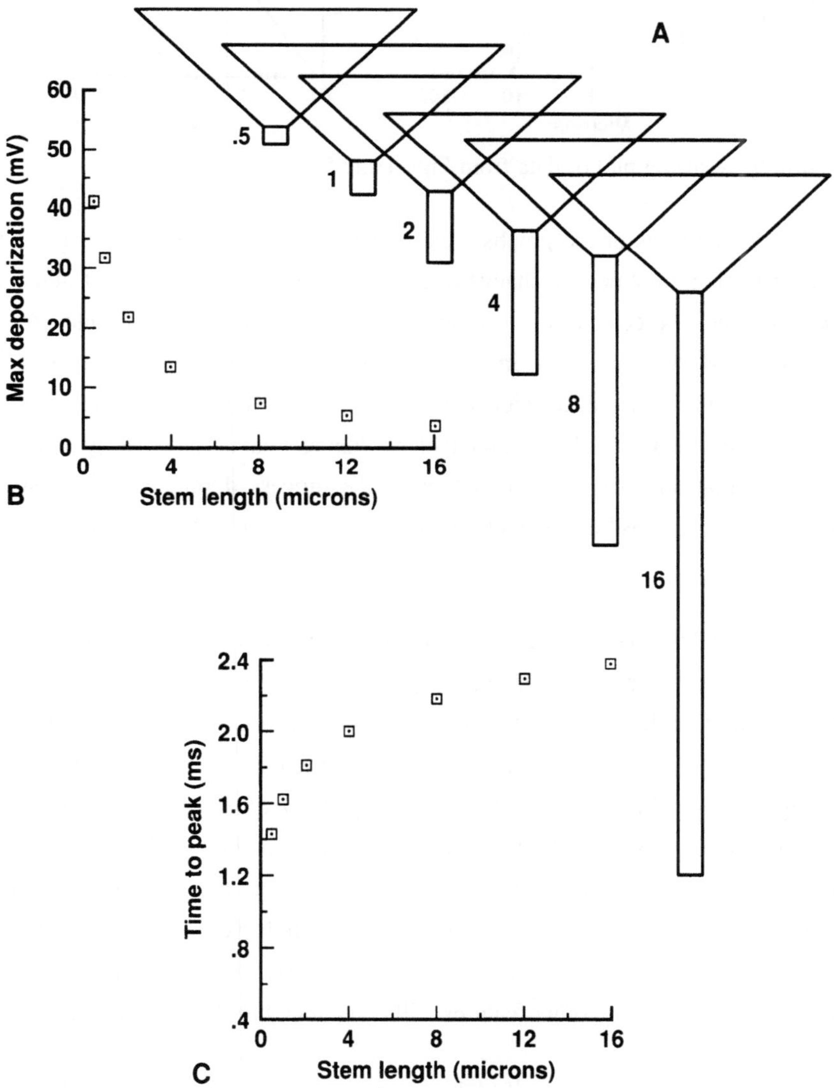

Figure 5. Peak voltage at the base of a process as a function of stem length.

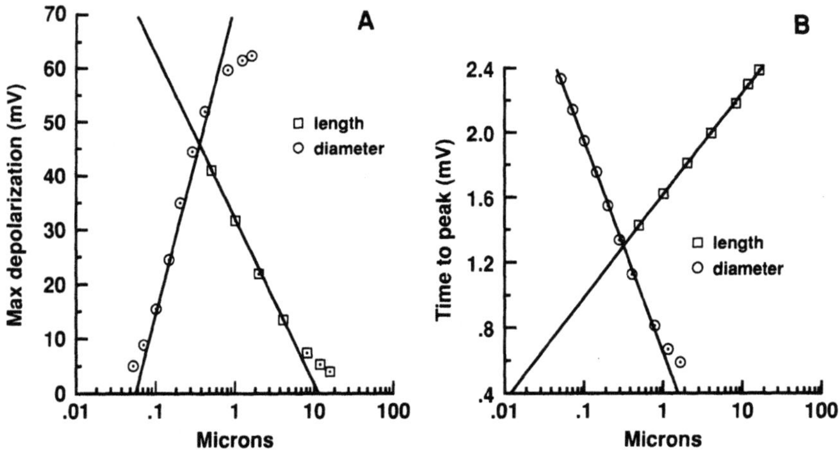

Figure 6. Semi-log plot of data from Figs 4 and 5.

The linear portions of the graphs yield a slope of -32 mV/decade change in length compared to 64 mV/decade change in diameter (Fig 6A). This factor of minus two is exactly what is expected from the steady state analysis of the axial resistance. As shown in Figure 6B, the change in time to peak is 0.75 ms/decade for changes in length compared with -1.7 ms/decade for changes in diameter. Note the very linear relationship for the length data, as well as for most of the diameter data. The linearity of these plots indicates that any narrow constriction in the process dominates other aspects of morphology (such as head diameter or length) in determining voltage transfer characteristics

CONCLUSIONS: The depolarization of the base of a process is very sensitive to small changes in the diameter of the stem, and not quite as sensitive to changes in stem length. Thus, morphologically accurate measurements of processes with afferent input are necessary for realistic compartmental models of passive current flow and the resultant voltage changes in the vestibular macula. Specifically, the voltage contribution of processes along a neuron branch varies considerably in both amplitude and latency as a result of changes in size that can be resolved only by electron microscopy. Other key features only available from electron microscopy include the size, location and polarization of synapses, as well as the presence of cytoplasm-occluding structures such as mitocondria. These and other features will be included in future simulations

Our results have implications for understanding the functional units of the vestibular endorgan. The voltage changes transmitted to the neuron branches of the primary afferent by a variety of morphologically distinct collateral-like processes contribute to the depolarization of the impulse initiation zone, and influence the resultant firing pattern. Furthermore, bouton endings with small diameter constrictions may be sites of local processing in which long term voltage dependent changes in postsynaptic dynamics are altered independently of other boutons on the same afferent.

REFERENCES:

1. Baird, R.A., G. Desmadryl, C. Fernandez, and J.M. Goldberg, *The vestibular nerve of the chinchilla. II. Relation between afferent response properties and peripheral innervation patterns in the semicircular canals.* J Neurophys, 1988. **60**: p. 182-203.
2. Ross, M.D., A*natomic evidence for peripheral neural processing in mammalian graviceptors.* Aviat. Space Environ. Med., 1985. **56**: p. 338-43.
3. Ross, M.D., L. Cutler, G. Meyer, T. Lam, and P. Vaziri, 3-*D components of a biological neural network visualized in computer generated imagery. I. Macular receptive field organization.* Acta Otolaryngol., 1990. **109**: p. 83-92.
4. Ross, M.D., G. Meyer, T. Lam, L. Cutler, and P. Vaziri, 3-*D components of a biological neural network visualized in computer generated imagery. II. Macular neural network organization.* Acta Otolaryngol., 1990. **109**: p. 235-244.
5. Kinnamon, J.C., S.J. Young, and T.J. Sherman, *3DEd, Recon. and Show.* 1986, Laboratory for High Voltage Electron Microscopy. University of Colorado, Boulder.
6. Ross, M., R. Cheng, and T. Lam, R*OSS (Reconstruction of Serial Sections) version CARMA, Computer Aided Reconstruction of Macular Accelerometers.* 1989, Moffett Field, CA.
7. Hines, M., A *program for simulation of nerve equations with branching geometries.* Int. J. Bio-Med. Comput., 1989. **24**: p. 55-68.

Modeling of Chaotic Dynamics in the Olfactory System and Application to Pattern Recognition

Koji Shimoide*, Michael C. Greenspon, Walter J. Freeman

Dept of Molecular and Cell Biology

U.C. Berkeley, CA 94720

*Asahikasei Microsystems Co., Ltd. Atsugi LSI Plant
Atsugi, Kanagawa 243-02 JAPAN

Abstract

Previous research has suggested the possible widespread relevance of chaotic dynamics to the functioning of higher-ordered neural systems. Using a multilayer model of primary and secondary olfactory structures, we have demonstrated the generation of such chaotic activity. We are presently examining the characteristics of these dynamics and exploring the implications for pattern recognition applications.

Structure of KIII Model

Our present model (Figure 1, from Yao and Freeman, 1990) is derived from the anatomical features of the olfactory system and consists of a distributed layer of coupled olfactory bulb (OB) oscillators linked via feedforward and feedback pathways to lumped oscillators with incommensurate frequencies representing the anterior olfactory nucleus (AON) and the prepyriform cortex (PC). Each oscillator simulates the local population dynamics of interacting pools of excitatory and inhibitory cells (circles.) Activity of these neuron pools is modeled by sets of simultaneous second-order nonlinear differential equations numerically integrated using a fourth-order Runge-Kutta method. (The equations are shown for the mitral (M1) and granule (G1) cells in the OB layer.)

Figure 1. A connection diagram for the KIII model with a distributed OB layer (N units), a lumped AON layer, and a lumped PC layer (From Yao and Freeman, 1990).

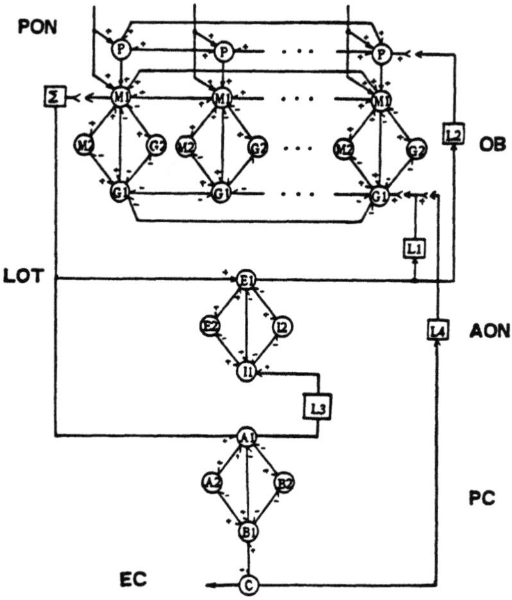

$$\frac{d^2X_{M1}[i]}{dt^2} + A\frac{dX_{M1}[i]}{dt} + BX_{M1}[i] = \tag{1}$$

$$K_{mm}Q(X_{M2}[i]) - K_{gm}(Q(X_{G1}[i]) + Q(X_{G2}[i])) + K_{pm}Q(X_p[i]) + \frac{1}{N-1}\sum_{j=1\, j\neq i}^{N} K_{mm}[i,j]Q(X_{M1}[j])\,) + K_m Input[i]$$

$$\frac{d^2X_{G1}[i]}{dt^2} + A\frac{dX_{G1}[i]}{dt} + BX_{G1}[i] = \tag{2}$$

$$-K_{gg}Q(X_{G2}[i]) + K_{mg}(Q(X_{M1}[i]) + Q(X_{M2}[i]))$$

$$- \frac{1}{N-1} \sum_{j=1\ j\neq i}^{N} Kgglink[i,j]Q(X_{G1}[j])$$

$$+ \frac{K1}{(Ts1-Te1)} \sum_{k=t-Ts1}^{t-Te1} Q(X_{E1}(k)) + \frac{K4}{(Ts4-Te4)} \sum_{k=t-Ts4}^{t-Te4} Q(X_C(k))$$

where i = 1, 2, ..., N; A = a + b ; B = ab; K1, K2, K3, and K4 are feedback gains.

Output representing the averaged activity of each pool of cells is transformed by a nonlinear axonal activation function, $Q(v) = Q_m[1 - e^{-(e^v-1)/Q_m}]$; a and b represent open loop rate constants (Freeman, 1975). Transformed output is globally distributed via symmetrically strengthened intralayer synaptic connections. Spatial activity patterns are presented as input to the model via a distributed set of lines simulating primary olfactory nerve projections.

Dynamics of KIII Model

Chaotic dynamics arises in the KIII model mainly due to differences in the time delay constants and dissonant characteristic frequencies in the feedback interactions between the bulb and the lumped AON and cortical layers. The dynamics of the model shows sensitivity to the values of the internal parameters; for example, an increase in the intralayer excitatory synaptic coupling strength (K_{mm}) causes a shift in the period of the resulting attractors and a growth of the amplitude (Figure 2).

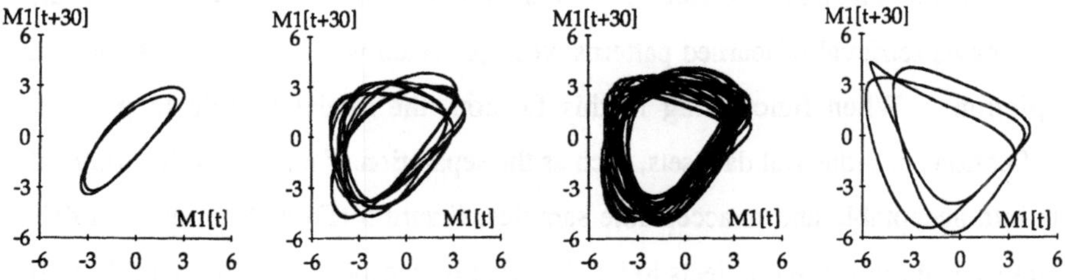

(a) $K_{mm,hi}$=0.5 (b) $K_{mm,hi}$=1.0 (c) $K_{mm,hi}$=1.25 (d) $K_{mm,hi}$=2.0

Figure 2. Graphs show the shift in the period of attractors and the trajectory of the output amplitude of a mitral cell in the KIII model which is plotted against itself delayed 30 msec. Increasing the value $K_{mm,hi}$ corresponds to the progress of learning.

Presentation of a spatial input pattern that is superimposed on a step input forces the system to bifurcate from a basal attractor to another attractor characteristic of the input pattern (Figure 3). Further, we have demonstrated the ability of the system to jump between attractors rapidly and without returning to an intervening basal state (Figure 4). We consider these features advantageous for use of the system in pattern recognition applications.

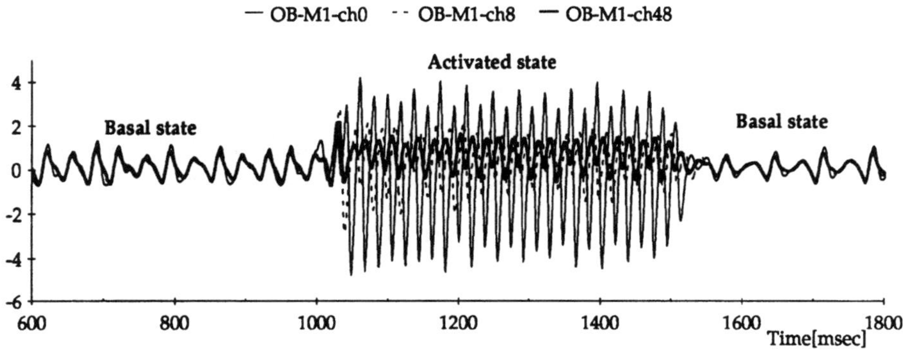

Figure 3. Dynamics of mitral cells in the KIII model. Not having input during time periods 600-1000 and 1500-1800msec, the system shows basal activities.

Pattern Recognition by KIII Model

When the input to the model is a spatial pattern representing a class that was previously learned from examples by Hebbian modification of excitatory synapses in the bulb, presentation forces the system to bifurcate to an attractor characteristic of that input class (Yao and Freeman, 1990). Functioning as an associative memory, the model shows retrieval of learned patterns when provided with partial inputs (pattern completion.) When functioning in this fashion, the model is able to perform classifications of industrial data sets, such as the separation of ultrasound imagings of parts into acceptable and unacceptable samples. Figure 5 (From Yao, et al., 1991) illustrates examples of the objects to be classified by the system. Each input data set includes 64 binarized values and is given to the distributed primary layers.

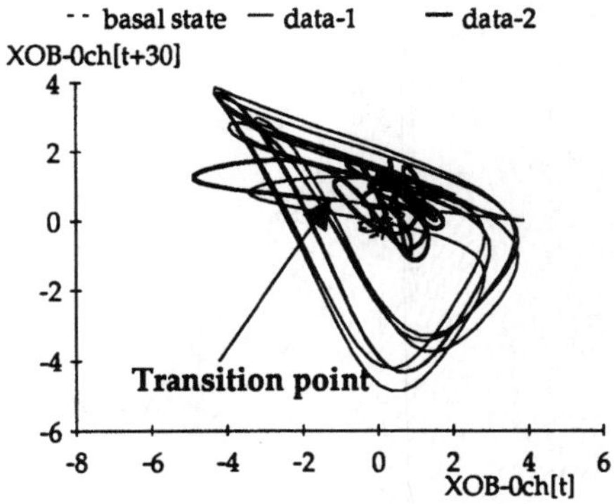

Figure 4. Given two different input data patterns, the system jumps rapidly between two attractors. Neither of the attractors are similar to the basal state.

Figure 5. Samples of the recognition objects (From Yao, et al., 1991).

Figure 6 shows an example of the dynamics of mitral cells in the OB layer of a 64 channel KIII model. The outputs of three different channels are shown in Figure 7. Each channel represents one of the channels belonging to a specified category; acceptable samples, unacceptable samples and neither. Six different input patterns are given at 200msec intervals representing three acceptable samples and three unacceptable samples. Given no input data during two time periods at 0-200 and 1400-1600msec, the system shows basal state activity. Attractors of each channel are shown in Figure 8, where differences in features and amplitudes can be observed. The pattern recognition capacity of the KIII model is now being investigated.

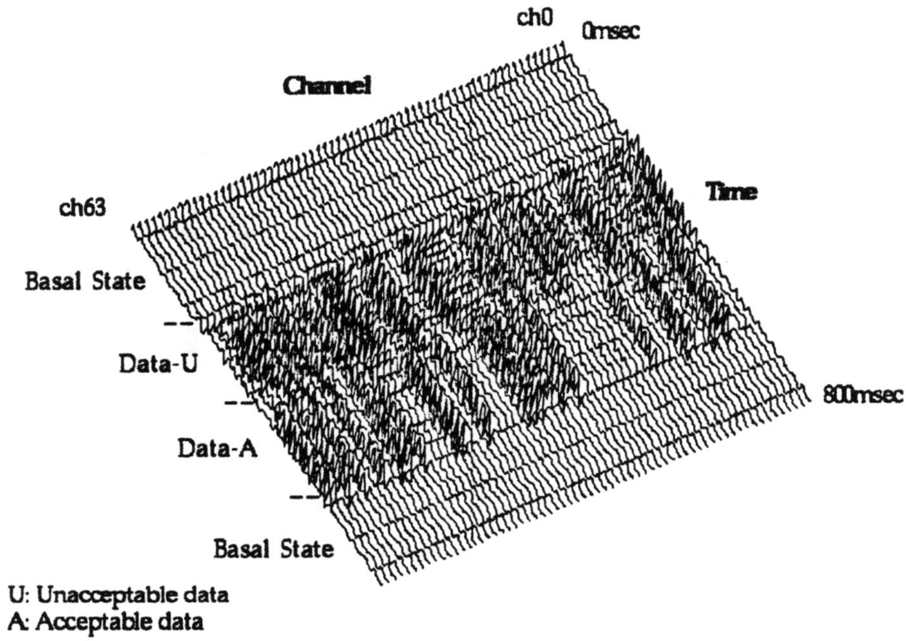

Figure 6. The dynamics of mitral cells in the 64 channel KIII model.

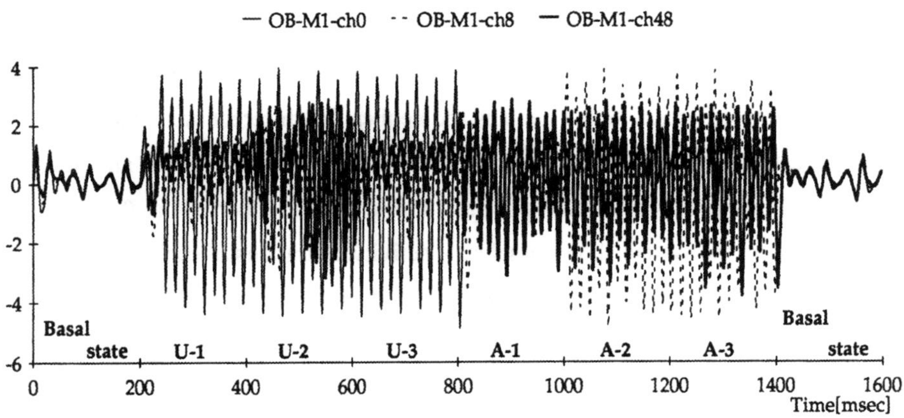

Figure 7. Dynamics of three different channels. Inputs are three unacceptable and three acceptable samples from an industrial data set.

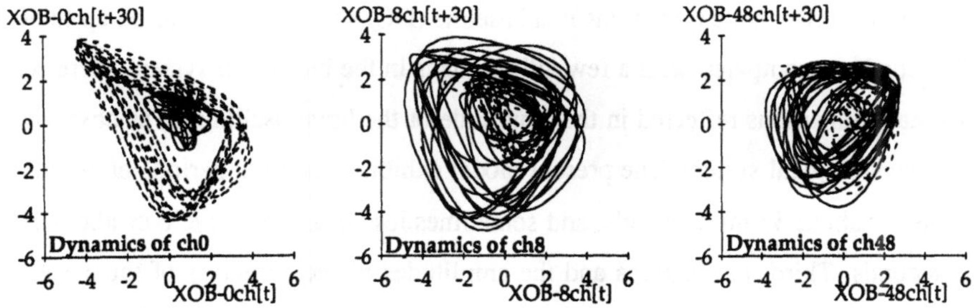

(a) Dynamics of ch0 **(b) Dynamics of ch8** **(c) Dynamics of ch48**

Figure 8. Each graph illustrates the dynamics of one of the channels belonging to specified category; acceptable samples, unacceptable samples and neither. Differences are observed in features and amplitudes. (solid line: acceptable data, dotted line: unacceptable data; each line represents the results of three data sets.)

Distributed Software Simulator

Our simulations are conducted using a dynamic modeling system implemented as a graphical front-end to a computational kernel; the two parts communicate via a network remote call interface. This arrangement provides us with a freedom to independently evolve the computational mechanism and the control system and to compare various computational implementations. Presently we are using a Cray-based kernel and are exploring various novel parallel architectures including the multi-DSP Ring Array Processor. The front-end provides standardized formatting and tracking of experimental data sets; visualization is accomplished using commercially available software.

Discussion

We have demonstrated some of the significant characteristics of a distributed olfactory system model, the KIII model. First, the system shows ranges of dynamic stability with varying internal parameters while maintaining sensitivity to relative changes in these parameters. This means the system is effectively sensitive to a learning process which allows the system to easily categorize the input data. Second, the system shows the ability to jump rapidly between attractors without returning to an intervening basal state. Our previous work (Yao, et al., 1991) showed the system's capability to reflect state transitions in a baseline (DC) shift, which required up to 50 milliseconds as compared with a few milliseconds in the biological system. Here we show state transitions reflected in the amplitude of the burst oscillation, as observed in the physiological system. The present model exhibits transition periods of at most 2 cycles or about 35 milliseconds, and sometimes less than a half cycle or about 10 milliseconds. Third, the features and the amplitudes of the attractors of the system show differences depending on the class of input data. This result leads to the possible effective usage of the KIII model for pattern recognition.

Still, however, questions remain as to the role of chaotic dynamics in the olfactory system itself. Such dynamics is observed both in activated states and the basal state. The importance of the broad power spectrum of the basal activity, which may let the system remain in a semi-activated or "restless" state and may facilitate the transition to patterned attractors has been previously suggested (Skarda and Freeman, 1987). In other words, chaos may precondition the system to accept novel inputs. We are also wondering if chaotic dynamics gives an advantage to the capability of pattern recognition. As chaos naturally involves many micro state fluctuations, this characteristic may assist in the classification of obscure input data. Further investigations are proceeding.

References

Freeman, W. J. *Mass Action in the Nervous System.* (1975) Academic Press.

Skarda, C. A. and Freeman, W. J. *How brains make chaos in order to make sense of the world.* Brain and Behavioral Science, **10** (1987): 161-195.

Yao, Y. and Freeman, W. J. *Model of Biological Pattern Recognition with Spatially Chaotic Dynamics.* Neural Networks **3** (1990): 153-170.

Yao, Y., Freeman, W. J., Burke, B. and Yang, Q. *Pattern Recognition by a Distributed Neural Network: An Industrial Application.* Neural Networks **4** (1991): 103-121.

PART III

MOTOR SYSTEMS

SECTION 6

SECTION 7

SECTION 6

CENTRAL PATTERN GENERATORS (CPG'S)

The stomatogastric ganglion of crustacea has proven to be a fertile ground for investigation of central pattern generators. Whereas these CPG's were once thought of as fixed hardwired networks, recent research has revealed that they are dynamic both at the level of the single neuron and the network.

Eve Marder et al. use a compartmental model of the Lateral Gastric neuron of the stomatogastric ganglion to show that the existence of chemosensitive spike initiating zones, at some distance from the normal spike initiating zone, allow the neuron to function in several modes of operation.

Computational Implications of a Serotonin-Sensitive Region of Axonal Membrane On a Dual Function Motor Neuron

Eve Marder, James M. Weimann,
Thomas B. Kepler, and L.F. Abbott
Center for Complex Systems
Brandeis University
Waltham, MA 02254

Abstract

Experiments on the Lateral Gastric (LG) motor neuron of the crab stomatogastric ganglion demonstrate a serotonin-sensitive region of the axonal membrane. This serotonin-sensitive region becomes a spike initiation site, and propagates action potentials antidromically into the integrating regions of the stomatogastric ganglion, where they do not ordinarily evoke transmitter release. A compartmental model of the neuron which includes a serotonin-sensitive slow inward current on the axon successfully simulates the physiological data. A mathematical study of axonal membranes that include slow currents provides insight into the mechanisms by which spike initiation zones can travel as waves down axons that contain slow currents.

The Lateral Gastric Neuron

The Lateral Gastric (LG) neuron of the stomatogastric ganglion (STG) of the crab, *Cancer borealis*, innervates several muscles that participate in movements of the gastric mill (Weimann et al., 1991). Additionally, the LG neuron is an important component of the pattern generating networks of the stomatogastric ganglion, where the synaptic connections it makes and receives play significant roles in shaping the motor patterns of the stomatogastric ganglion (Weimann, 1992). These dual roles of the LG neuron allow the investigator to study simultaneously its role in behavior (by analyzing its action on muscles) and its role in shaping network function (by analyzing its central connections).

The structure of the crab LG neuron is shown in Figure 1. Note that this neuron, like all STG neurons, is monopolar. The LG soma leads to a large diameter major neurite, from which branch many small processes. On these neurons the input and output synapses are made on these small processes (King, 1976a,b). The soma and the neuropilar processes of STG neurons do not display the ability to generate rapid action potentials, but action potentials are initiated near the region where the axon exits the neuropil, and propagate out to the neuromuscular junctions along several centimeters of nerve (Figure 2).

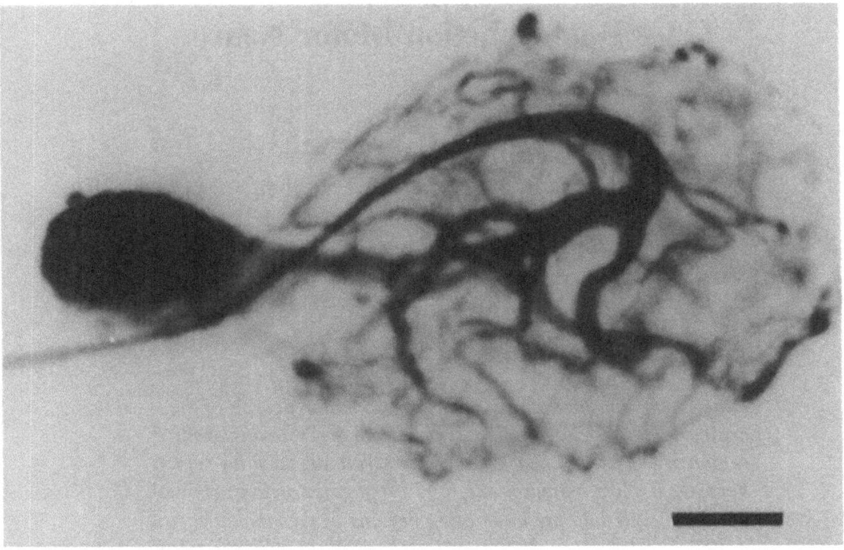

Figure 1. Lucifer yellow fill of a LG neuron from the crab, *Cancer borealis*. Note the soma, large diameter neurite, and numerous small processes that ramify throughout the neuropil. The axon leaves the ganglion (towards the left) to carry action potentials to the muscles. Scale bar, 100μm.

The LG Neuron Shows Multiple Spike Initiation Zones

In preparations in which the STG was left attached to the muscles innervated by the LG neuron, long depolarizations of the LG soma triggered bouts of action potentials that followed the depolarizing pulse and were initiated at a significant distance from the STG, along the peripheral axon (Meyrand et al., 1992). In Figure 3 it can be seen that during the depolarization of the LG soma, action potentials were initiated between electrode 2 and the soma. Subsequent to the depolarization, action potentials were initiated at a site between electrodes 3 and 4, from which they propagated antidromically towards the STG and orthodromically towards the muscles

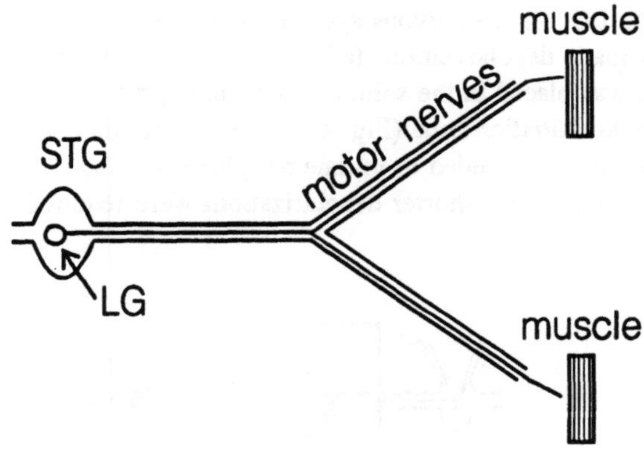

Figure 2. Schematic diagram of the LG neuron, showing its general structure.

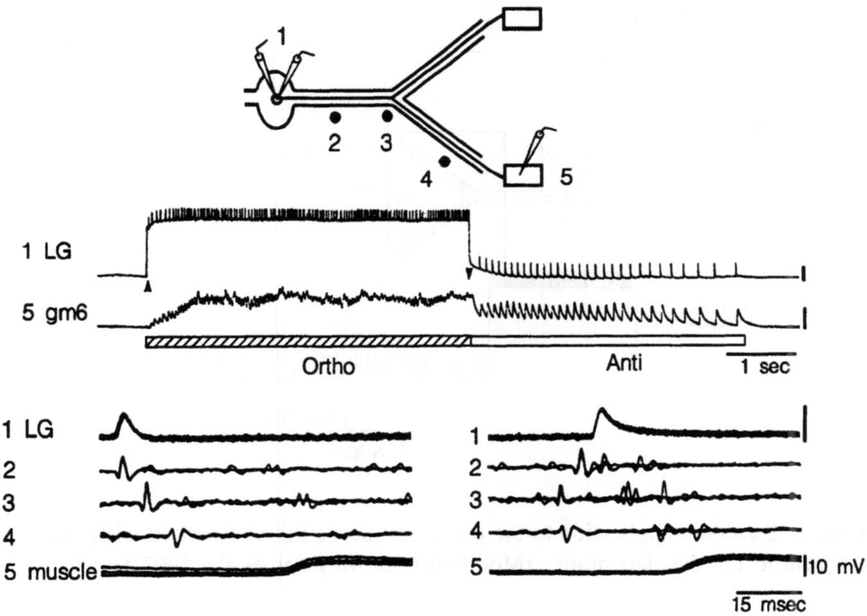

Figure 3. The schematic shows the position of the recording electrodes. 1 and 5 are intracellular electrodes in the LG soma and muscle, respectively. 2, 3 and 4 are extracellular recording electrodes on the motor nerve at the positions indicated. The LG was depolarized at the up arrow head. Current was removed at the down arrowhead. Bottom left panel shows the action potentials during depolarization of the LG (ortho) and after the depolarization (anti). (Modified from Meyrand et al., 1992.)

innervated by the LG. If the nervous system was entirely removed from its muscles, long somatic depolarizations failed to evoke peripheral spike initiation unless serotonin was placed in the solution surrounding the nerve in the region of the peripheral spike initiation zone (Fig. 4). In low serotonin concentrations (10^{-7}M), long depolarizations were needed to initiate peripheral spiking; as the serotonin concentration was increased, shorter depolarizations were required (Meyrand et al., 1992).

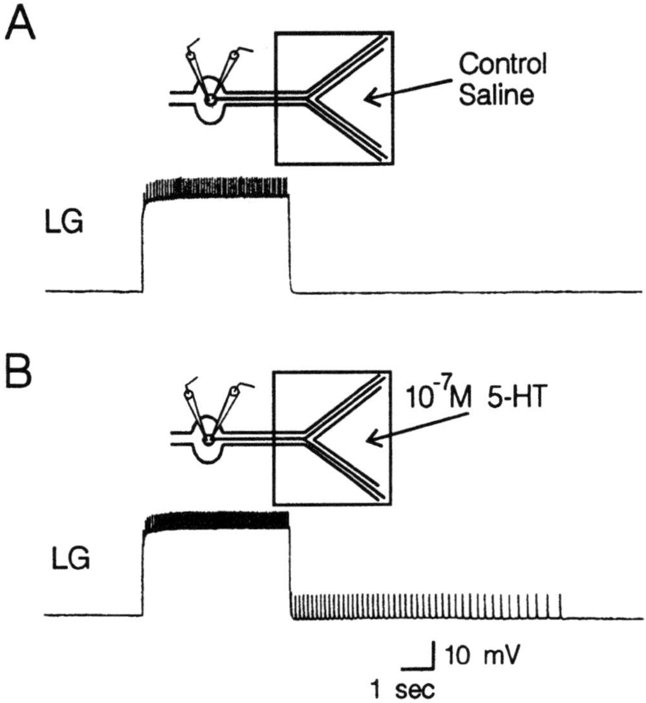

Figure. 4. Serotonin application to the peripheral nerve initiates peripheral spiking subsequent to LG depolarization. (Modified from Meyrand et al., 1992.)

These data suggested to us the possibility that a serotonin and depolarization sensitive slow inward current might be present in the axonal membrane and that this could lead to spontaneously generated peripheral spiking. This possibility is explored in the modeling studies described below.

Slow Inward Current on Axonal Membrane Can Lead to A Site of Spike Initiation

Kepler and Marder (1992) studied the properties of an axon with a slow inward current in addition to the more classical Na^+ and K^+ conductances described in Hodgkin and Huxley (1952). The Na^+ and K^+ currents were modeled as described in Connor et al (1977). An additional slow, voltage-dependent current was added to the model. For simplicity these channels open and close at the same rate, and the reversal potential for this current is taken to be very positive. Therefore, this current can be thought of as a simplified version of a voltage-dependent Ca^{++} current, or a slow voltage-dependent Na^{+} current.

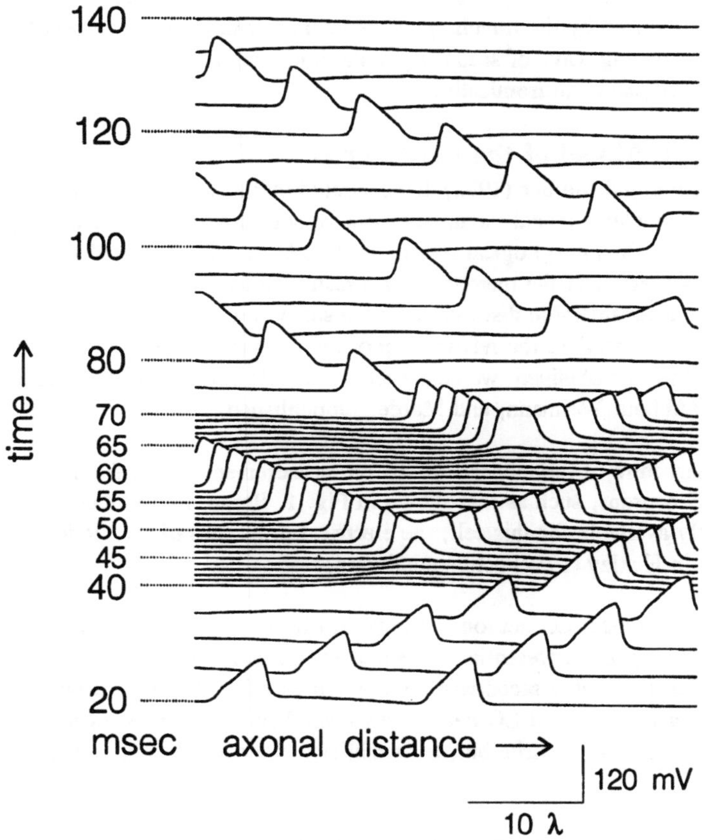

Figure 5. A traveling zone of spike initiation in an axon with a slow inward current. Each horizontal trace is a plot of membrane potential as a function of distance along the axon. Successive times are shown in the upward direction, at the intervals indicated. Two pulses were initiated at the left of the axon. The slow current activated by these depolarizations brings the axon to threshold behind the second pulse, leading to a third action potential, which then propagates in both directions. The site of "spontaneous" spike initiation travels down the axon behind the first two pulses. (Modified from Kepler and Marder, 1992.)

In this model, when the amplitude of the slow current is large, depolarization of the model sufficient to elicit action potentials can result in activity that outlasts the depolarization. At even higher values of the slow current, depolarization of the membrane to elicit action potentials can lead to a permanent plateau state in which the axon remains spontaneously active, until reset by a hyperpolarizing pulse (Kepler and Marder, 1992).

An interesting case is seen when the slow current is not large enough to sustain indefinite plateauing in response to a depolarization, and is illustrated in Figure 5. Here, two spikes were initiated by depolarization of one end of the axon. The depolarizations produced by the traveling spikes interact with the slow current to produce a new, traveling wave of secondary spike initiation that propagates spikes both orthodromically and antidromically.

Compartmental Model of the LG Neuron

The study of Kepler and Marder (1992) demonstrated that a slow inward current distributed over the whole axonal membrane could result in a secondary site of spike initiation. However, the physiological results on the LG neuron suggested that the serotonin-sensitive region of the membrane was restricted to an area of the nerve, and that this was a considerable distance from the soma, and the presumed usual site of spike initiation. To explore the relationship of the geometry of the LG neuron to the physiological results obtained, we constructed a simple compartmental model of the LG neuron (Abbott, Weimann, and Marder, unpublished).

The geometric structure of the model is based on measurements taken from Lucifer yellow filled LG neurons, such as that shown in Fig. 1. Because no voltage clamp data are currently available for this cell, the electrical properties of the model are based on Hodgkin-Huxley (1952) active conductances.

The structure of the model LG neuron is shown in Fig. 6. The model neuron contains two large regions representing the soma and primary neurite. These are connected to an axon and to a secondary process representing the LG neuron's presynaptic terminals. The real LG neuron has several secondary processes emerging from the primary neurite but, for simplicity, we consider only a single process.

The axon and secondary process are represented by multi-compartmental cables. The soma is modelled as a 150 μm diameter sphere. The primary neurite is 30 μm in diameter and 100 μm long. However, because the primary neurite in the real LG neuron has large, irregular side branches (see Fig. 1), we have used a total surface area for the primary neurite 3 times larger than the area of this cylinder. Depending on the position of its soma in the STG, the LG neuron may have a cylindrical neck with a diameter of about 10 μm and a length of up to 100 μm between the soma and primary neurite. Because the entire soma and primary neurite region is electrotonically compact (even with such a neck), and because the neck contributes a negligible amount to the total input impedance and capacitance of the neuron, we have ignored it in the model. The total membrane surface area for the soma and primary neurite is 100,000 μm^2. The axon of the model LG neuron is 10 μm in diameter and its length in the simulations is 1.5 cm. The secondary process has a

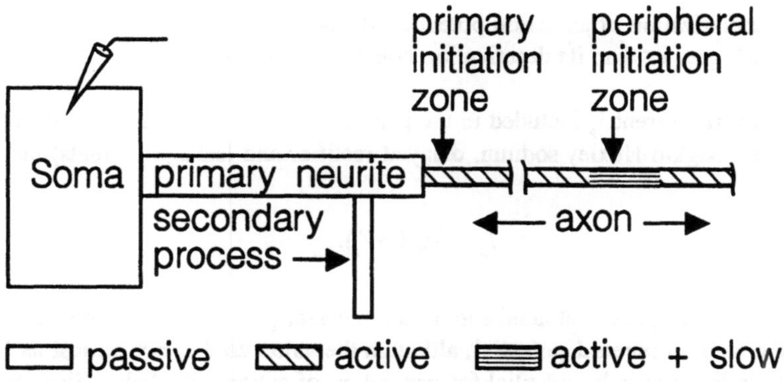

Figure 6. Schematic of model LG neuron.

diameter of 1 μm and a length of 250 μm.

The resting potential of the model neuron is -65 mV. The membrane capacitance is 1 $\mu F/cm^2$ and the resistivity of the intracellular fluid is 50 Ω-cm. The membrane conductances of the model neuron are different in different regions of the neuron. The passive soma and primary neurite have a voltage-independent membrane resistance of 10 KΩ-cm^2 to provide a total input impedance of 10 MΩ which matches typical measured values. The secondary process is also passive with a voltage-independent membrane resistance of 1.5 KΩ-cm^2. This matches the membrane resistance of the model axon when it is at the resting potential.

In the model, the axon has transient sodium and delayed rectifier conductances in addition to a passive leakage conductance, similar to those of Hodgkin and Huxley (1952) except for two minor modifications. Because the action potentials generated in the primary spike initiation zone of the LG neuron ride on top of a prolonged depolarization, we increased the threshold potential for de-inactivation of the sodium conductance so that the sodium current does not remain refractory during this depolarization. In addition, the maximal conductance of the sodium current has been reduced from 120 to 80 mS/cm^2. The same sodium, potassium and leakage conductances are present in all of the active compartments of the cable representing the axon.

Peripheral Generation of Action Potentials

Antidromic action potentials are generated in a single compartment of the model neuron located midway on the axon (labeled peripheral initiation zone in Fig. 6). To produce spiking in this region we have added to this single compartment a slow conductance with properties described above. This work suggests that the mechanism that generates spiking in the peripheral region involves a serotonin-dependent inward current that is activated by action potentials passing through the peripheral zone and that deactivates slowly. Since the activation variable reacts to the brief depolarization of an action potential, the activation rate for this current must be much greater than its deactivation rate of several seconds.

The slow inward current I_p included in the peripheral spike initiation zone (which also has the Hodgkin-Huxley sodium, delayed-rectifier and leakage currents) takes the form

$$I_p = gz(V-V_r). \tag{1}$$

where V is the membrane potential and V_r the reversal potential. We have set V_r equal to the sodium reversal potential, although the value used is not critical as long as it is above the threshold potential for generation of action potentials. Since both serotonin and depolarization are required to activate this current, we have written the conductance as the product of two factors g and z. The factor g depends on the serotonin concentration in a graded manner increasing with increased serotonin concentration. The factor z is voltage-dependent. We will assume for simplicity that under normal conditions z is not near its saturation limit.

Since peripheral spiking is initiated by action potentials, z must increase rapidly when an action potential passes through the peripheral spike initiation zone. Since this is such a rapid process, we model it by incrementing z by an amount ϵ every time an action potential passes through the peripheral zone,

$$z \rightarrow z+\epsilon. \tag{2}$$

When an action potential is not present, z decays slowly back to zero deactivating the current. The decay of z is modelled by the equation

$$T\frac{dz}{dt} = -z. \tag{3}$$

The parameter T is the time constant for the decay of z. The spontaneous peripheral spiking reported by Meyrand et al. (1992) typically lasts for 3 to 5 seconds suggesting that T is of order a few seconds. We have used a value of 3 seconds.

Figure 7 shows a simulation of spiking in the LG model. A pulse of current is injected into the soma of the model neuron (third trace) to depolarize it. A similar, though somewhat attenuated depolarization appears at the end of the secondary process (second trace). Conduction from the soma to the end of the secondary process is passive. The depolarization causes a series of action potentials to be produced in the primary spike initiation region at the proximal end of the axon. These are seen in the soma and secondary process traces as highly attenuated spikes riding on the current-induced depolarization. The soma attenuation matches that seen in the real LG neuron quite well (Meyrand et al., 1992). The orthodromic propagation of action potentials down the axon can be seen by examining the three lower traces in Fig. 7 which give the potential in the primary spike initiation zone (Zpri) at the proximal base of the axon, the peripheral initiation zone (Zper) in the middle of the axon, and the distal end of the axon (DAx).

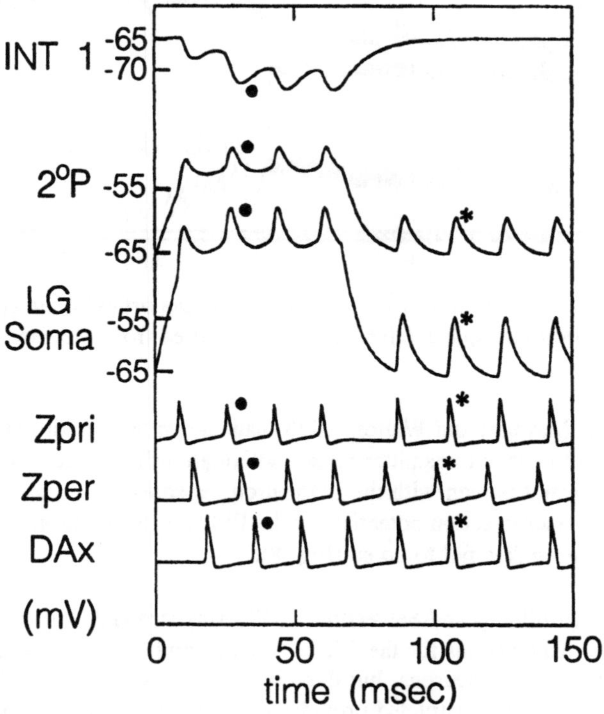

Figure 7. Depolarization of the model LG neuron for 50 msec evokes centrally initiated spikes (●) during the depolarization. Subsequent to the depolarization, four peripherally initiated spikes (*) are seen.

In the simulation shown in Figure 7, the current injection into the soma is long enough to activate the slow inward current in the peripheral spike initiation zone to the point where peripheral spiking begins. Examination of the voltage traces for the different regions of the axon shows that these action potentials (*) originate in the peripheral zone and from that point propagate both orthodromically to the distal part of the axon and antidromically back to the soma. In the soma and secondary process they appear as highly attenuated spikes rising from the resting potential rather than from a prolonged depolarization. Notice that the shape of the antidromic action potentials as seen in the soma is different from that of the orthodromic spikes. This difference is seen in the real LG neuron.

Antidromic Action Potentials Do Not Evoke Transmitter Release

Previous work on other stomatogastric ganglion neurons (Mulloney and

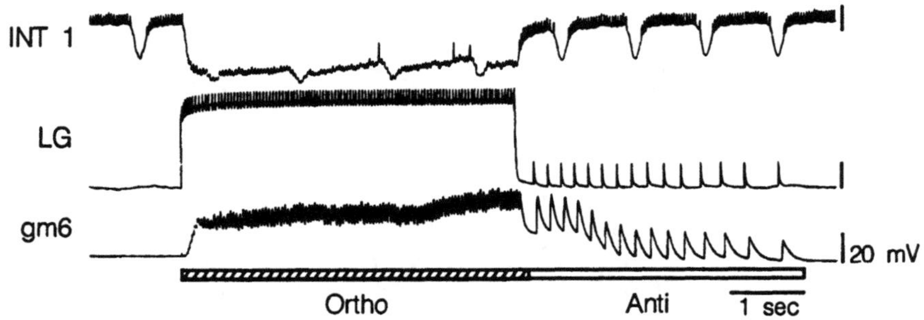

Figure 8. Intracellular recording from Int1 shows that the orthodromic spikes evoke IPSPs, but the antidromic spikes fail to do so. (Modified from Meyrand et al., 1992).

Selverston, 1972; Maynard and Walton, 1975) had shown that antidromic action potentials can fail to evoke transmitter release although orthodromic action potentials do so. The same result is seen with the LG neuron, as shown in Figure 8. Orthodromic LG neuron action potentials evoke IPSPs in Interneuron 1 (INT 1), although antidromic spikes fail to do so (Fig. 8).

Previous authors (Mulloney and Selverston, 1972; Maynard and Walton, 1975) had suggested several explanations for the failure of antidromic action potentials to evoke transmitter release in STG neurons, but these were highly speculative. The compartmental model we constructed provides a simple, and extremely plausible explanation for this phenomenon.

To examine synaptic transmission of orthodromic and antidromic spikes from the LG neuron, we include in the model a passive postsynaptic neuron taken to represent INT 1. The model INT 1 has a capacitance of $1\mu F/cm^2$ and a voltage-independent membrane resistance of $10K\Omega\text{-}cm^2$. The synaptic current in INT 1 has a reversal potential of -75mV and a conductance proportional to the amount of transmitter released by the model LG neuron.

Transmitter release from STG neurons is a graded function of membrane potential (Graubard, 1978; Graubard et al., 1983). Below a threshold potential, no transmitter is released and above the threshold, release increases with depolarization. For our model LG we have used -52mV as the threshold value for transmitter release. This value is taken from Meyrand et al. (1992) which shows the threshold effect followed by a linear rise in the IPSP seen in INT 1 as a function of LG depolarization. In the model, the amount of transmitter released is a linear function of membrane potential above the threshold.

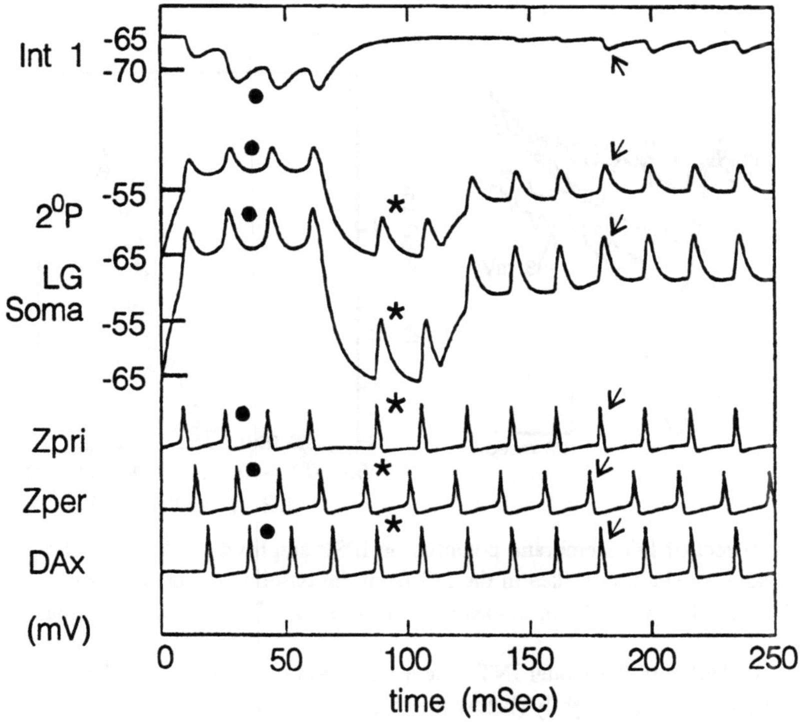

Figure 9. The model LG was depolarized for a brief period, to elicit centrally generated spikes (●) which evoke IPSPs in Int1. After the depolarization, peripherally generated spikes (*) failed to evoke IPSPs in Int1. However, when the model LG was depolarized again by a small amount, the action potentials riding on this depolarization (↓) evoked IPSPs in Int1. The amplitude of the IPSPs increased as a graded function of LG membrane potential.

As in the physiological experiments shown in Figure 8, in the model, orthodromic spikes elicited during the depolarization of the LG neuron evoke IPSPs in the model INT 1 neuron (Figure 7, 9). However, no corresponding IPSPs are evoked by the antidromic spikes. This is because the antidromic spikes are highly attenuated and, unlike the orthodromic spikes, they do not ride on top of a steady depolarization. As a result, they do not reach the threshold of -52mV required for synaptic transmitter release and so they do not evoke IPSPs. This suggests that if the LG soma were depolarized slightly during antidromic action potential generation, the antidromic spikes would ride on sufficient baseline depolarization to be above the threshold for transmitter release and therefore evoke IPSPs.

Figure 9 shows the result of depolarizing the soma of the model LG neuron during peripheral spiking. Orthodromic spikes were first produced by a fairly large somatic depolarization. These orthodromic spikes initiate peripheral spiking that

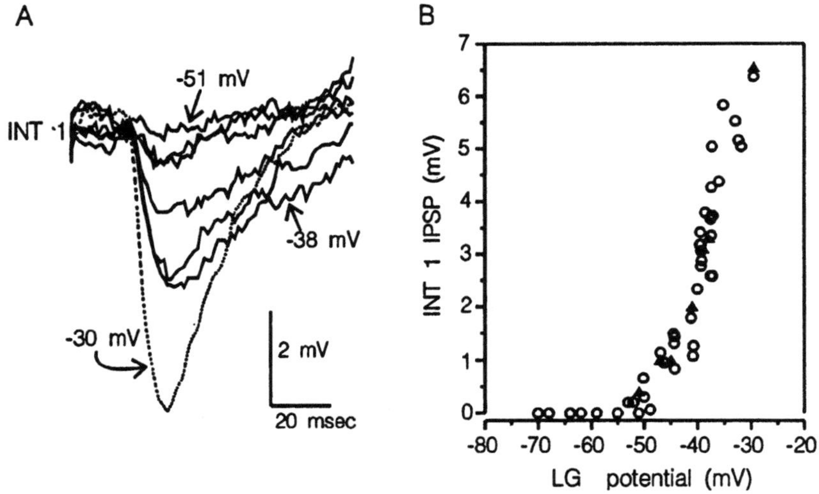

Figure 10. Effect of LG membrane potential on IPSP amplitude in Int1. A. IPSPs in Int1 evoked by eliciting spikes in the LG from the baseline membrane potentials indicated. B. Plot Int1 IPSP amplitudes as a function of LG membrane potential.

fails to evoke IPSPs in the model INT 1 neuron. However, when the LG soma is subsequently depolarized slightly (below the threshold for activation of the primary spiking zone) small IPSPs can be seen in the INT 1 trace. When the LG soma is then depolarized slightly further, the IPSPs in INT 1 get larger.

These simulations predicted that if the biological LG neuron were depolarized during peripherally generated action potentials, that these would then produce IPSPs in Int1. Figure 10 shows the results of experiments done to test this idea. Figure 10A shows an intracellular recording from Int1. The motor nerve was stimulated to evoke LG action potentials, and the membrane potential of the LG neuron was changed by current injection into the soma. As the baseline membrane potential of the LG neuron was depolarized, the amplitude of the resultant IPSP (Fig. 10B) increased.

Conclusions

The serotonin sensitivity of the peripheral nerve of the LG neuron is quite remarkable. Immunocytochemical results (V. Kilman and E. Marder, unpublished results) show serotonin-like immunoreactive terminals in the region of the LG nerve that shows serotonin-sensitive spike initiation. The source of these terminals is not clear, but they may be associated with serotonin-containing sensory neurons (Katz et al., 1989).

What functional role do these peripherally initiated spikes play? When the peripherally-initiated spikes enter the ganglion subsequent to a LG burst, and when the LG neuron is relatively hyperpolarized, they fail to evoke transmitter release. Thus, under these conditions, the peripheral spikes will influence muscle tension, but not pattern generation.

However, should the peripherally generated spikes arrive coincident with another depolarization, either synaptic or intrinsically generated, then the peripherally generated spikes will influence muscle tension as well as influence the central pattern-generating circuitry. Thus, inputs that influence the LG neuron's baseline membrane potential "gate" the transmitter release from its terminals.

Thus, the existence of chemosensitive spike initiation zones at a distance from the normal site of spike initiation allows this neuron to function in several modes of operation.

Acknowledgments

We thank Ms. Joan McCarthy for help with the manuscript and Mr. Michael O'Neil for help with figure preparation. This work was supported by NS17813 and T32 NS07292 from the National Institutes of Health, MH46742 from the National Institutes of Mental Health.

References

Connor JA, Walter D, McKown R (1977) Neural repetitive firing. Modifications of the Hodgkin-Huxley axon suggested by experimental results from crustacean axons. Biophys J 18:81-102.

Graubard K (1978) Synaptic transmission without action potentials: Input-output properties of a nonspiking presynaptic neuron. J. Neurophysiol. 41:1014-1025.

Graubard K, Raper JA, Hartline DK (1983) Graded synaptic transmission between identified spiking neurons. J. Neurophysiol. 50:508-521.

Hodgkin AL, Huxley AF (1952) A quantitative description of membrane current and its application to conduction and excitation in nerve. J Physiol 117:500-544.

Katz PS, Eigg MH, Harris-Warrick RM (1989) Serotonergic/cholinergic muscle receptor cells in the crab stomatogastric nervous system. I. Identification and characterization of the gastropyloric receptor cells. J. Neurophysiol. 62:558-570.

Kepler TB, Marder E (1992) Spatial and temporal dynamics of spike initiation on axons with slow inward currents. (submitted for publication).

King DG (1976a) Organization of crustacean neuropil. I. Patterns of synaptic connections in lobster stomatogastric ganglion. J. Neurocytology 5:207-237.

King DG (1976b) Organization of crustacean neuropil. II. Distribution of synaptic contacts on identified motor neurons in lobster stomatogastric ganglion. J. Neurocytology 5:239-266.

Maynard DM, Walton KD (1975) Effects of maintained depolarization of presynaptic neurons on inhibitory transmission in lobster neuropil. J. Comp. Physiol. 97:215-243.

Meyrand P, Weimann JM, Marder E (1992) Multiple axonal spike initiation zones in a motor neuron: serotonin activation. J Neuroscience, in press.

Mulloney B, Selverston AI (1972) Antidromic action potentials fail to demonstrate known interactions among neurons. Science 177:69-72.

Weimann JM (1992) Multiple task processing in neural networks: numerous central pattern generators in the stomatogastric nervous system of the crab, *Cancer borealis*. PhD Thesis, Biology Department, Brandeis University.

Weimann JM, Meyrand P, Marder E (1991) Neurons that form multiple pattern generators: Identification and multiple activity patterns of gastric/pyloric neurons in the crab stomatogastric system. J Neurophysiol 65:111-122.

SECTION 7

CORTEX, CEREBELLUM AND SPINAL CORD

Charles Wilson uses an model of a representative neostriatal spiny neuron to illustrate that the presence of various voltage-gated channels increases the complexity of computation performed by an individual cell.

Thomas Anastasio reviews control theory approaches to the vestibulo-ocular reflex (VOR) and presents a dynamic neural network model that extends the description of the VOR to the level of individual neurons.

Simon Giszter et al. present results suggesting that the spinal cord possesses a small number of movement primitives, and that these primitives may be combined to produce adjustable behaviors.

Edward Blum et al. present two mathematical models of a simplified cerebellar network necessary for the learning and memory of the rabbit nictitating membrane response in classical conditioning. Computer simulation results of these two models show reasonable agreement with experimental test data from Thompson's laboratory.

Nonlinear Synaptic Integration in Neostriatal Spiny Neurons

Charles. J. Wilson
Department of Anatomy and Neurobiology
University of Tennessee, Memphis
Memphis, TN 38163

Abstract

The presence of a variety of voltage-gated channels which are activated at subthreshold membrane potentials increases the complexity of the computations that can be expected from individual neurons. This is especially so for neurons receiving large numbers of relatively weak synaptic inputs, so that many spatially distributed inputs are required to excite each cell. In a model of one such neuron, the spiny projection cell of the mammalian neostriatum, inclusion of a simple and very common membrane nonlinearity, the time-independent inwardly rectifying potassium conductance, caused spatially distributed weak synapses to summate cooperatively, while having no effect on the integration of spatially clustered synaptic input.

Voltage-Gated Ion Channels and Synaptic Integration

The development of new voltage clamp recording techniques has transformed our view of the electrical properties of neurons in the mammalian central nervous system. Findings made using these methods have shown that along with their specialized connectional arrangements and morphological specializations, neurons are specialized in their electrical characteristics in ways that suit them to perform their unique functional roles. These electrical specializations consist of combinations of ionic channels that confer nonlinear properties to the membranes of the cells. While some channels are only activated over the range of membrane potentials associated with action potentials, others are active at or near the resting membrane potential of neurons. These latter contribute to the generation of the resting membrane potential, and can be expected to react to deviations of the membrane potential caused by subthreshold synaptic inputs. Thust there may be no membrane potential range over which it is accurate to consider the neuron to be a linear system. The assumption that neurons can be considered to be linear for small disturbances of the membrane potential is deeply embedded in both the formal theory and the folklore of synaptic integration. Neuronal modeling without the assumption of linearity is frightening, because nonlinear systems have almost unlimited potential for variety. Moreover, neurons often receive very large numbers (tens of thousands) of synaptic inputs. Systems with such large numbers of inputs are difficult to understand, even when they are linear. But the validity of the superposition principle for linear systems allows some confidence in extrapolating empirical data and theoretical results obtained from manipulation of a small number of inputs to the more realistic case in which many inputs are active. When neurons cannot be considered to be linear in the subthreshold voltage range, such extrapolation is not reliable, and the scale of synaptic inputs presents a much more imposing barrier to understanding synaptic integration.

One possible avenue of escape from this dilemma is computer simulation of realistic models of neurons containing accurate representations of both the ionic channels of interest and the natural distribution of synaptic inputs. By adding accurate models of individual ion channels to simulations that include realistic patterns of inputs, the computational implications of individual ionic channels can be explored under conditions close to those expected *in vivo*. Although not a complete description of the potential of the ion channel to influence the behavior of the neuron, such simulations can reveal aspects of their action under certain well defined circumstances. This method could not pretend to the kind of generality expected from linear systems analysis, and to be useful it must be combined with accurate experimental data on particular neurons and particular patterns of synaptic input.

The Morphology of Neostriatal Cells

Spiny projection neurons are the most common cell type in the mammalian neostriatum, accounting for between 75 and 95% of cells in various species (DiFiglia at al., 1976). The internal organization of the neostriatum consists principly of two tissue compartments, projecting to different target structures, but with approximately the same cellular consituents and organization (for review see Wilson, 1990). Afferents to the neostriatum, arising mainly from the cerebral cortex, thalamus, amygdala, and substantia nigra, make synapses with the spiny projection cells in both compartments. Most of the afferents contact dendritic spines of the projection cells, and form synaptic connections employing excitatory amino acids as neurotransmitters. The dendrites of the neostriatal neurons project in all directions from the soma and arborize in an approximately spherical field surrounding the cell. An example showing the dendritic field of a neostriatal spiny cell is shown in Figure 1.

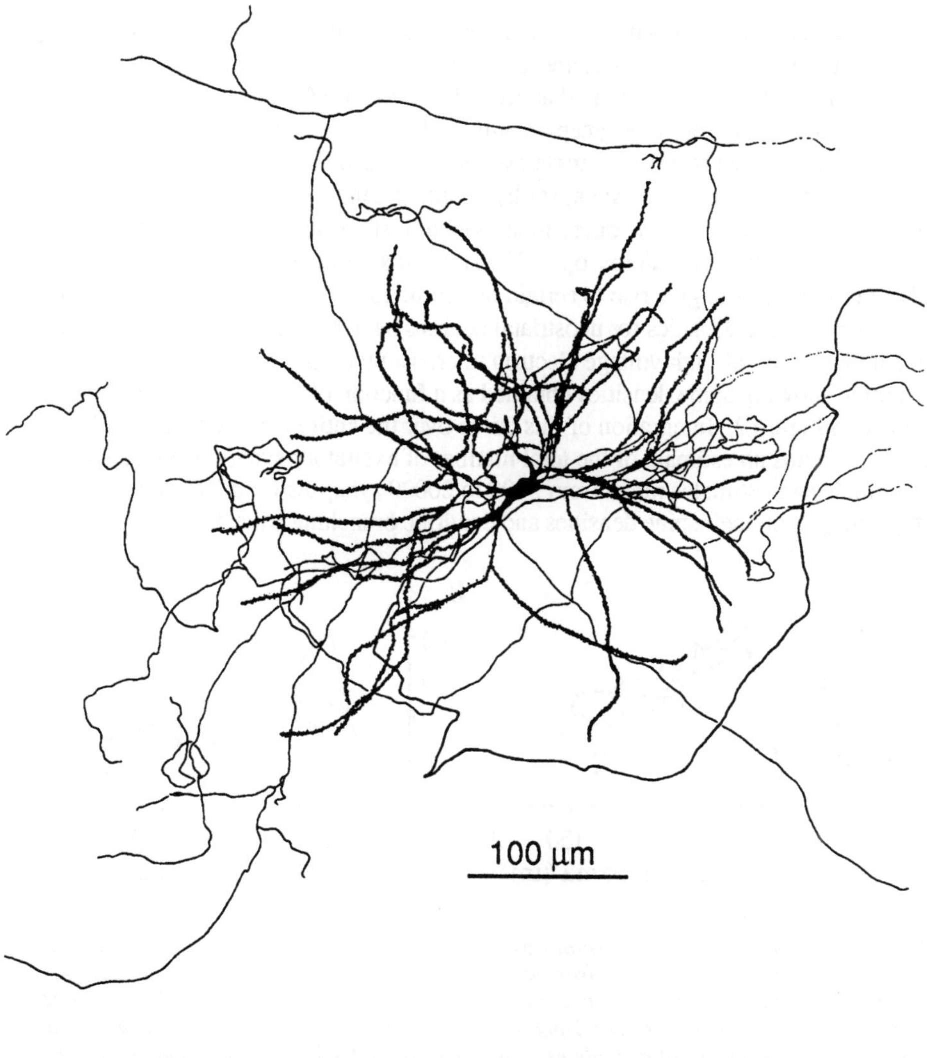

Figure 1. A reconstruction of the dendritic tree of a spiny projection neuron from serial sections. The neuron was injected intracellularly with horseradish peroxidase and drawn from serial sections using a drawing tube. The dendrites branch once or twice, to form 20-40 dendritic tips. The soma and proximal dendrites are free of spines, but the secondary and tertiary dendrites are densely covered with an investment of dendritic spines that continues to their terminations. The fine aspiny processes are local axon collaterals of the neuron. From Kitai and Wilson, 1982.

There is apparently no specialization among dendrites, and no preference for synaptic inputs to end in a layered fashion on the dendrites. Thus all the synaptic inputs have approximately the same distribution on the dendrites (Kemp and Powell, 1971). This is in contrast to structures like the cerebral cortex and hippocampus, where inputs generally end in a laminar pattern and so different inputs contact the neuron at different distances from the soma. Individual afferent axons form extended arborizations in the neostriatum, generally much larger than the dendritic field of any spiny neuron. Synaptic contacts are sparsely distributed within the axonal arborizations. It is thus assumed that each afferent neuron contacts many spiny neurons, and forms very few synapses with each spiny cell (Wilson, 1990). Because each dendritic spine receives a single excitatory synaptic contact, it is possible to estimate the number of such synapses by counting dendritic spines. The density of dendritic spines has been measured in individually-stained spiny neurons by observation in the light microscope. The small size of the spines, and the obscuring effect of the dendrite give rise to certain inaccuracies in these measurements. Counts of dendritic spine densities for neostriatal neurons have been corrected using the more accurate method of high voltage electron microscopy (Wilson et al., 1983). An example showing spine densities obtained as a function of distance from the soma is shown in Figure 2. Integration of this curve over the entire dendritic field of the neurons yields an estimate of the total number of excitatory synapses on a spiny neuron. These estimates range from 5000 to 20000 synapses for different neurons, depending upon their spine densities and dendritic branching patterns.

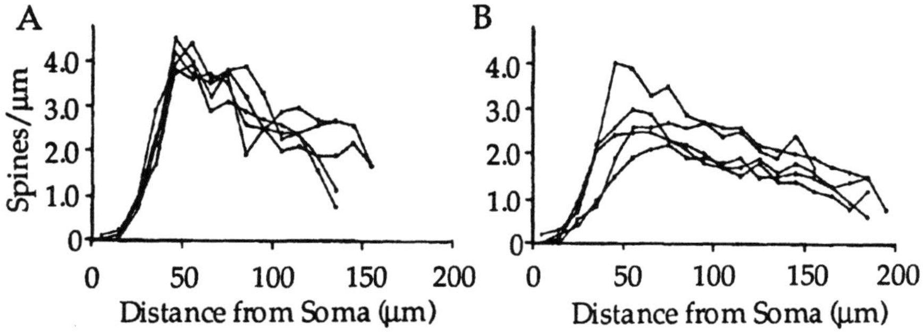

Figure 2. Density of dendritic spines as a function of distance from the soma. A. Spine density for 5 dendrites arising from a single neuron. B. Average spine density for 5 different neostriatal neurons, including the cell shown in A. Each curve in B is an average of results from several dendrites arising from a single neuron. The spine density distribution on the various dendrites of a single neuron is very reproducible. While some neurons have much higher spine densities than others, the shape of the spine density distribution is similar among cells. From Wilson, et al., 1983.

The shapes of dendritic spines of neostriatal neurons have been described from serial section electron microscopic reconstructions, and from high voltage electron microscopic studies of intracellularly stained neurons. These descriptions have included measurements of the surface areas of dendritic spines, as well as their linear dimensions. These studies have shown that surface areas of dendritic spines in the neostriatum vary from about 0.5 to 3.0 μm^2, with a mean of about 1.5 μm^2. The dimensions of the spine head and stalk vary independently, as do the stalk length and cross-sectional diameter. Stalk lengths vary among spines from about 0.5 to about 2.5 µm, and stalk diameters from less than 0.1 µm to about 0.4 µm. Spine heads vary between 0.2 and 1.0 µm, and the surface area of the spine head is linearly related to the area occupied by the synaptic specialization. Spines of all shapes and sizes are distributed over the dendritic tree, with no preference for particular kinds of spines to be found proximally or distally in the dendrites. This is indicated by the absence of a correlation between spine shape parameters and the diameter of the dendrite from which the spines arise (Wilson et al., 1983).

A Model of the Neostriatal Spiny Neuron

The morphological data on the neostriatal spiny neuron were used to construct a compartmental model intended to represent this class of neurons (Wilson, 1991). The model was not intended to reflect the morphology of any one single neuron, but rather the features shared by all spiny neurons. The representative cell had a 15 μm diameter spherical soma and 6 identical dendritic trees, each of which branched once at a distance of 10 μm from the soma, and a second at 30 μm from the origin of the dendritic tree. The primary dendrites had diameters of 2.25 μm. The secondary dendrites had initial diameters of 1.0 μm and tapered slightly (to 0.96 μm) over their 20 μm length. The tertiary dendrites were 190 μm long and tapered from 0.63 μm at their origin to 0.29 μm at their termination. These values were obtained from high voltage electron microscopic measurements, and, like all the morphological parameters of the model, were average values from previous anatomical studies. Dendritic spines were distributed using a tabular form of the averaged values from the curves shown in Fig. 2. Spines were divided into two groups, those that would be synaptically active during the simulation and those that would not. Synaptically silent spines were represented only by their surface area contribution to the compartment on which they were placed. The surface area of each of the silent spines was fixed at 1.46 μm^2, the average value obtained from the electron microscopic studies described above. Synaptically active spines were represented more variably, and could be adjusted in size over the natural range of values for neostriatal spiny cells. They were represented as 4 compartments, one representing the spine head, and three to represent the spine stalk. Dendrites were divided into 10 μm compartments, yielding 81 compartments for each dendritic tree. The number of active spines was varied, but in some simulations was as high as 2000, which required 8000 additional compartments.

Computer simulations were performed using the Saber simulator (Carnevale et al., 1990). Compartments were represented by a membrane capacitance (based on 1.0 $\mu F/cm^2$) and a voltage-dependent resistance consisting of a leak resistance and a time-independent inward (anomalous) rectification. The anomalous rectification was a potassium conductance based on the model of Hagiwara and Takahashi (1974). The voltage-dependence of the conductance, and the current which arises from it, are shown in Figure 3.

Because the conductance attributable to the inward rectifier approaches zero as the cell is depolarized, it was necessary to add an additional conductance to prevent the membrane potential from growing without limit upon depolarization. There is little evidence for a significant leak conductance in neostriatal neurons. Rather, the input resistance of the cells in the membrane potential range near threshold is probably dominated by a slowly-inactivating voltage-dependent potassium conductance which is activated by depolarization (Surmeier et al., 1991). To prevent the effects of this conductances from obscuring the effect of the inward rectifier, it was not included in the model, but rather a leak conductance was used. The magnitude of the leak conductance was selected to match the input resistance of the spiny neuron when measured close to the threshold for action potential generation (about -45 mV).

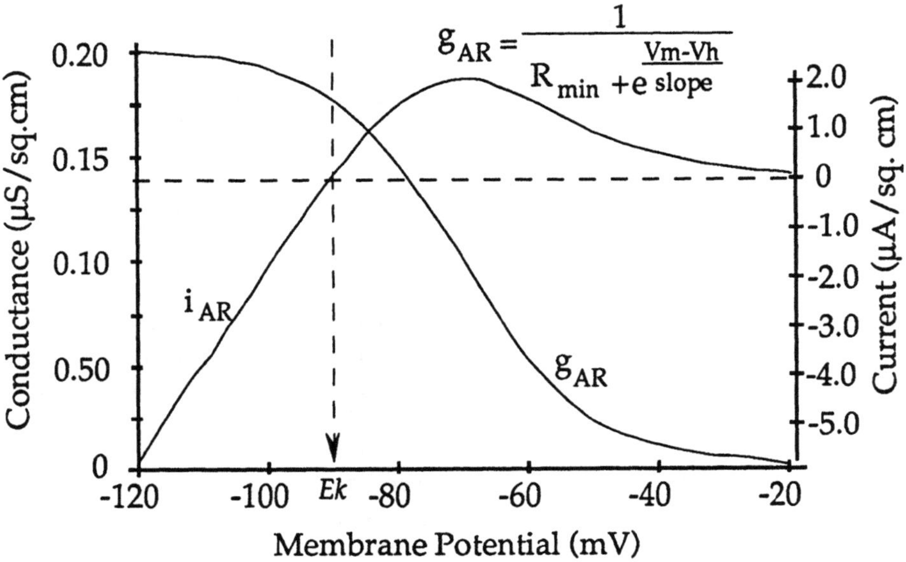

Figure 3. The voltage-sensitivity of the inward rectifier conductance (gAR) as represented in the model, and the current that results from it (iAR). The potassium equilibrium potential is indicated by the vertical dotted line (Ek). The minimum resistivity (Rm) is the inverse of the maximal conductance when all channels are open. Vm is the membrane potential, and Vh and slope are parameters determining the half activation voltage and the slope of the activation curve respectively. From Wilson, 1991.

The Effect of the Classical Synaptic Nonlinearity

Synapses are intrinsically nonlinear, and their nonlinear behavior has profound consequences for neurons receiving large numbers of synaptic inputs. Because the driving force for any synaptic input is decreased by the membrane potential changes produced by other inputs of the same type, excitatory synaptic inputs sum in a less than linear fashion. Because the reversal potential for most excitatory synaptic potentials is far from the resting potential, this nonlinearity is small when the synaptic potentials are small. The effect becomes important when the local depolarization due to synaptic input exceeds a few millivolts. Because it is strictly dependent upon the local membrane potential, the synaptic nonlinearity affects synaptic summation of nearby synapses much more than distant ones. It also is much more significant for synaptic sites with high input resistances, such as fine dendritic branches, where local synaptic potentials may be very large. One such site is the dendritic spine, the high input resistance of which has been the subject of intense theoretical discussion. Previous studies have concentrated on the effect of the axospinous location of single excitatory inputs or very small numbers of excitatory inputs, and have shown that the local synaptic potential in the spine may be much greater than that in the dendrite to which it is attached (see references in Wilson, 1984). This large EPSP amplitude, combined with the classical synaptic nonlinearity, can result in a reduced synaptic current. But aside from reducing the EPSP amplitude, it has little effect on the linearity of synaptic integration in a neuron receiving many axospinous synaptic inputs. This is because there is very little voltage decrement from the dendrite to the dendritic spine (Shepherd and Brayton, 1979). Thus, dendritic depolarization decreases the driving force for axospinous excitatory synapses as effectively as it does for axodendritic synaptic inputs. To the extent that an axospinous synapse can depolarize a dendrite, its effects will summate nonlinearly with axodendritic and other axospinous inputs.

A much larger effect on synaptic integration is exercised by the distribution of inputs on the dendritic tree. For the model spiny neuron, clustered synaptic input was compared with uniformly distributed excitatory synaptic input. An example is shown in Figure 4. For the simulations shown there, synaptic input was placed on a single dendrite, about half-way between the soma and dendritic tip. Synaptic inputs were placed either on dendritic spines or on the dendritic shafts in the same location (curves marked *axodendritic* or *axospinous*).

Synaptic inputs were synchronous identical synaptic conductance changes with a reversal potential 80 mV positive to the resting potential, a time course described by an alpha-function with alpha=100 with a peak conductance of 1 nS. Postsynaptic spines were 3.0 μm in length and 0.1 μm in cross-section, to maximize the current attenuating effect of the dendritic spines. That this was effective can be verified in the lower panel in Fig. 4, which shows the contribution made by a single synapse as a function of the number of synchronously-activated synapses. The leftmost point in this graph represents the effect of a single synaptic event. It shows that under the conditions of the simulation, the effectiveness of a single synapse was attenuated by the spine to about half of that expected for an axodendritic synapse. However, as the number of activated synapses increased, the difference between axospinous and axodendritic synapses decreased rapidly. When 20 synapses were active, the current attenuating effect of the spine had all but disappeared. The reason for this is apparent from the peak EPSP amplitudes occurring locally in the dendrite at the level of synaptic input and in the soma, as shown in the upper panel of Fig. 4. As the number of synaptic inputs increased, the local dendritic EPSP rapidly approached the reversal potential for the excitatory synapse. This caused the dramatic decrease in synaptic effectiveness seen in the lower panel, and also the reduction in the effect of dendritic spines. The dendritic spine's current attenuating effect is due to the large amplitude of EPSPs in the spine head. Thus the occurrence of large EPSPs in the dendrite during synchronous activation of several synapses caused current attenuation due to the classical synaptic nonlinearity to be about the same in the dendrite and spine.

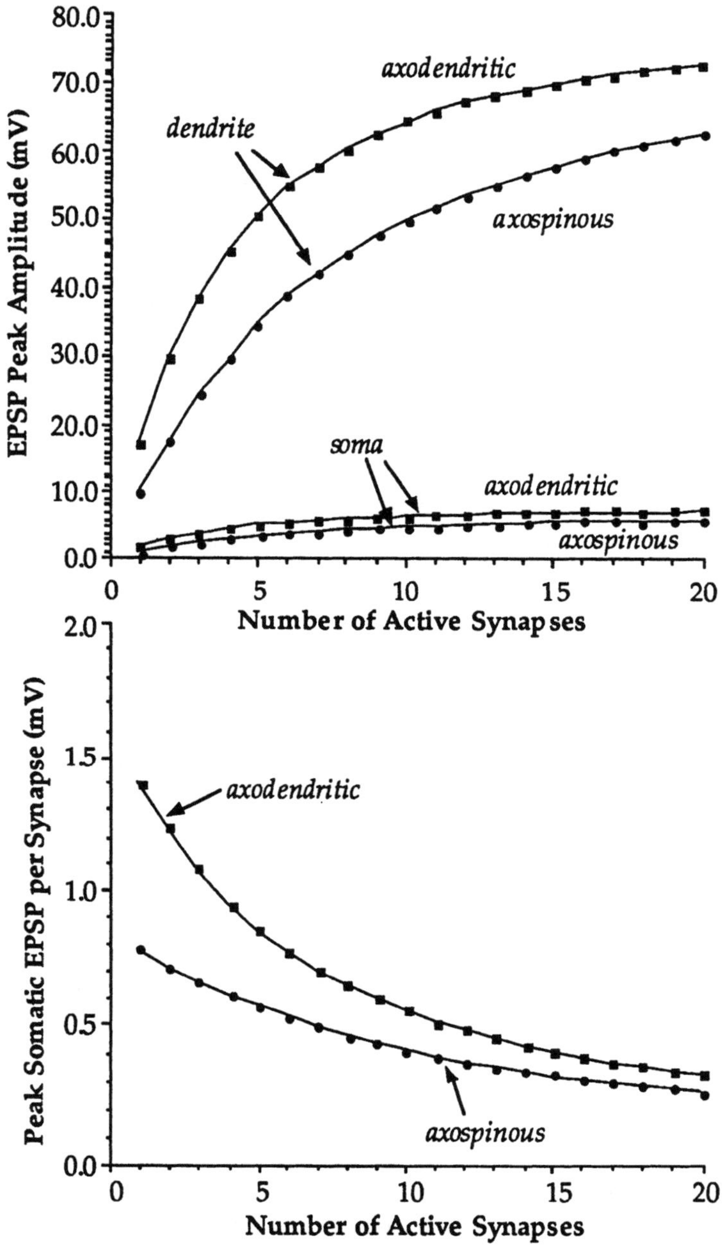

Figure 4. Effectiveness of clustered synaptic input. From Wilson, 1991.

It is important that large amplitude local EPSPs occur with relatively low densities of co-activated synaptic inputs. The giant local EPSPs and greatly decreased synaptic effectiveness seen with 20 inputs only represented about 25% of the total innervation density of the neuron in the region of the dendritic field used in the simulation. Similar simulations were performed using inputs whose latencies were normally distributed, with standard deviations varying from 1 to 5 ms. Although the loss of synaptic effectiveness during coactivation was mitigated somewhat by dispersion of synaptic latencies, temporally overlapping but asynchronous synaptic inputs engaged the synaptic nonlinearity over nearly the same range of synaptic densities. Thus, spatially clustered excitatory synaptic input was especially ineffective at activating the neurons. At relatively low synaptic densities the local synaptic potential approached its maximal value, limiting the charge that could be applied to the neuron by the clustered synaptic activation. This limited charge was able to effect only a small depolarization in the soma due to the load presented by the rest of the dendritic tree (upper panel in Fig. 4, curves labeled soma). Of course, this result depends upon the absence of voltage-gated inward currents in the dendrites that could boost synaptic potentials, but such channels were not included in the simulations shown in Fig. 4.

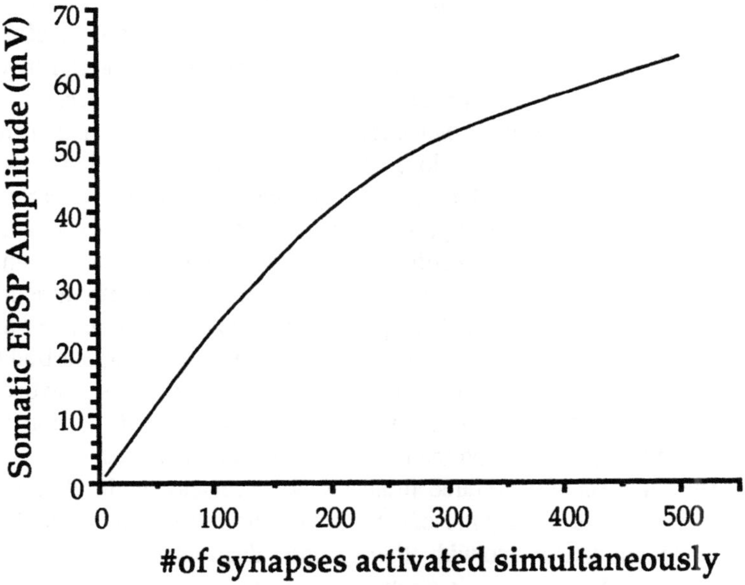

Figure 5. Distributed synaptic activation of the spiny neuron. From Wilson, 1991.

In the absence of voltage-activated dendritic booster currents, a low density of inputs dispersed on the dendritic tree was much more effective in depolarizing the soma. Because coactivated synapses were separated on the dendritic tree, the local synaptic potentials were kept small enough so that synapses acted relatively independently. The effect of the classical nonlinearity was thus minimized. The total amount of charge that could be delivered to the neuron wass greatly increased, increasing the maximal depolarization that could be achieved at the soma to cover the entire dynamic range of the synaptic mechanism (80 mV in the simulations used here). This is illustrated in Figure 5.

Effect of Time-Independent Anomalous Rectification

Addition of the inward rectifier to the model of the spiny neuron altered the responses to synaptic input in three ways, all predictable consequences of the dependence of membrane resistivity upon membrane potential. Firstly, the local input resistance of synaptic sites was dependent upon the membrane potential. Thus the amplitude of synaptic potentials was not as simple a function of the local geometry and the synaptic conductance change as in the passive case. Small conductance changes occurring in isolation on large diameter dendrites saw very low input resistances due to the high conductance of the anomalous rectification. However, a much more common result was the opposite one, in which one or two synaptic inputs was enough to drive the local membrane potential into the range in which the inward rectifier conductance was mostly turned off and the local membrane resistivity approached the maximal value (determined by the leak resistance). This was especially true for dendritic spines, where nearly any synaptic conductance change could drive the spine membrane to the high resistance state. A second effect of the inward rectifier was to make the electrotonic length of the dendrite sensitive to membrane potential. The increased conductance of the membrane when hyperpolarized caused the electrotonic length of the dendrites to be much greater than that due to the leak conductance alone. Thus as the dendritic membrane close to a synaptic input was depolarized, the synapse became electrotonically nearer the soma, and also nearer other synaptic inputs, with which it might interact via the classical nonlinearity or the anomalous rectifier. The third effect was to alter the time course of synaptic potentials depending upon the membrane potential. Because the membrane resistivity was dependent upon voltage, the time constant was as well. Excitatory synaptic potentials superimposed on a relatively hyperpolarized baseline (due to the absence of other inputs) suffered a shortened time course due to decreased membrane time constant. This decreased the degree of latency dispersion that could be tolerated without loss of summation of the synaptic potentials.

The net effect of these three influences was a decrease in the effectiveness of excitatory synapses occurring in isolation on a hyperpolarized background, compared to synaptic inputs coactivated with other synapses distributed along the dendrites. The influence of this alteration in synaptic effectiveness depended upon the spatial pattern of synaptic input to the cell. In simulations of clustered synaptic input, anomalous rectification had little effect on the results. This was because the portion of the neuron depolarized by the clustered synaptic input was very small. The dendritic membrane in the region of the synaptic input was very depolarized and so was driven to its high resistance state. Thus these synapses saw a very high input resistance, which increased the current attenuating effect of the classical synaptic nonlinearity. There was a slight decrease in electrotonic length of the activated dendrite due to the depolarization, increasing the effectiveness of the inputs in depolarizing the soma, but this influence decremented rapidly with distance along the unstimulated region of dendrite lying between the synaptic inputs and the soma. When synaptic input was distributed uniformly along the dendrite, the inward rectifier's effects were maximized. Under these conditions, the neuron was depolarized in a more homogeneous way, and there were no extremes of depolarization anywhere along the neuron (except in the dendritic spines). Increasing the number of synaptic inputs caused gradual increases in the level of depolarization, with resulting decreases in dendritic electrotonic length, increases in time constant, and increasingly better charge transfer throughout the dendritic tree.

This increase in synaptic effectiveness counteracted the effect of the classical synaptic nonlinearity when the number of activated synapses was low. As a result, the relationship between synaptic effectiveness and total number of activated synapses was greatly altered by anomalous rectification when combined with a uniform distribution of synaptic input. This is shown in Figure 6. That curve shows the average synaptic effectiveness (peak EPSP amplitude at the soma per active synapse) as a function of the total number of synapses activated. At low densities of synaptic input, the effectiveness of individual synapses was decreased by the presence of the fast anomalous rectifier, due to the increased electrotonic length of the dendrites. The synaptic nonlinearity played a relatively minor role with such low numbers of

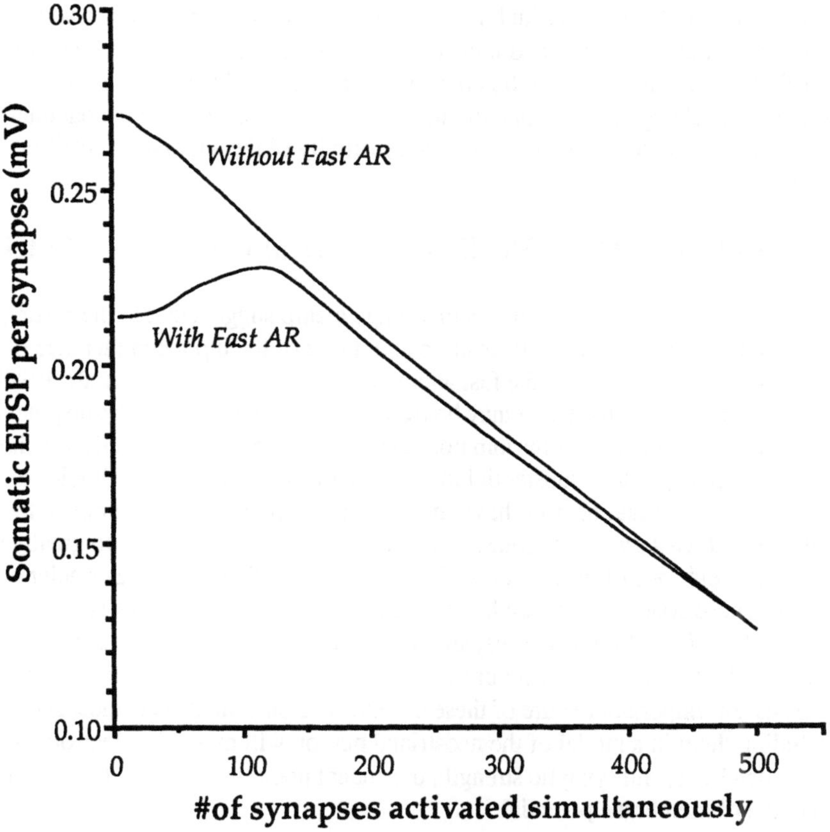

Figure 6. Effect of anomalous rectification on synaptic effectiveness. From Wilson, 1991.

activated synapses anyway, but the reduced input resistance due to the anomalous rectifier reduced its effect still further by decreasing the local (dendritic) EPSP amplitude. Increasing the density of activated synapses in the model without anomalous rectification slowly engaged the classical synaptic nonlinearity and resulted in a monotonic decrease in synaptic strength. In the model containing anomalous rectification, however, increasing the number of active synapses actually increased synaptic effectiveness, due to an increase in membrane resistance and an accompanying decrease in dendritic electrotonic length. This increased over a range of membrane potentials determined by the voltage-sensitivity of anomalous rectification. As the conductance turned off, the membrane resistivity continued to increase toward the maximal value and the dendritic field became increasingly compact, but the input resistance of the dendrites likewise increased, and EPSP amplitudes in the dendrite gradually became large enough to engage the synaptic nonlinearity. Eventually, this caused synaptic strength to begin to decline. The result was a biphasic effect of synaptic density on synaptic strength, as shown in Fig. 6. In the simulation shown in Fig. 6, the peak synaptic effectiveness occurred with about 130 simultaneous excitatory inputs, and this caused a depolarization of about 45 mV in the soma (which is about the depolarization required to activate an action potential). The preferred number of activated synapses depended greatly on the distribution of synapses (Fig. 6 is for a uniform distribution), the maximal conductance change per synaptic input, dendritic spine shapes, and the parameters Vh and slope, which determined the voltage sensitivity of the anomalous rectifier conductance.

Prospects for More Realistic Neostriatal Neuron Models

Studies of neostriatal neurons in voltage clamp suggest that at the resting membrane potential seen in slices up to about half of the input conductance of the neurons is accounted for by the fast anomalous rectifier (Uchimura et al., 1989). Thus the model presented here may be reasonably accurate for membrane potentials close to the potassium equilibrium potential. Even relatively ineffective excitatory synaptic inputs produce substantial local depolarizations, however. Studies of dissociated neostriatal neurons have shown that moderate depolarizations of the cell can evoke at least two other potassium conductances, which will act to greatly reduce the input resistance of the neuron and limit synaptic effectiveness (Surmeier et al., 1991). These conductances can be engaged at membrane potentials that are subthreshold for action potentials, and thus they must be included in any model of the cell which can claim to account for generation of the firing patterns of the neurons. An important feature of these currents is their time dependence, and including them in a model of the neostriatal neuron will make the electrotonic length of the dendrites, and synaptic strength, dependent upon the recent history (ca. 1-2 seconds) of membrane potential.

References

Carnevale, N.T., Woolf, T.B. and Shepherd, G.M. (1990) Neuron simulations with SABER. J. Neurosci. Meth. 33:135-148.

DiFiglia, M., Pasik, P., and Pasik, T. (1976). A Golgi study of neuronal types in the neostriatum of monkeys. Brain Res., 114, 245-256.

Hagiwara, S. and Takahashi, K. (1974). The anomalous rectification and cation selectivity of a starfish egg cell. J. Membrane Biol., 18, 61-80.

Kemp, J.M. and Powell, T.P.S. (1971) The termination of fibres from the cerebral cortex and thalamus upon dendritic spines in the caudate nucleus: A study with the Golgi method. Phil. Trans. R. Soc. Lond. (Biol.) 262:429-439.

Kitai, S.T. and Wilson, C.J. (1982) Intracellular labelling of neurons in mammalian brain. In: Cytochemical Techniques in Neurobiology. S. Palay and V. Chan-Palay (eds) Alan R. Liss, New York, pp.533-549.

Shepherd, G.M. and Brayton, R.K. (1979) Computer simulation of a dendrodendritic synaptic cirduit for self- and lateral-inhibition in the olfactory bulb. Brain Res. 175:377-382.

Surmeier, D.J., Stefani, A., Foehring, R.C. and Kitai, S.T. (1991) Developmental regulation of a slowly-inactivating potassium conductance in rat neostriatal neurons. Neurosci. Lett., 122, 41-46.

Uchimura, N., Cherubini, E. and North, R.A. (1989) Inward rectification in rat nucleus accumbens neurons. J. Neurophysiol., 62:1280-1286/

Wilson, C.J. (1984). Passive cable properties of dendritic spines and spiny neurons. J. Neurosci. 4, 281-297.

Wilson, C.J. (1990) Basal ganglia. In: The Synaptic Organization of the Brain, 3rd edition, G.M. Shepherd (ed.) Oxford Univ. Press, New York, pp.279-316.

Wilson, C.J. (1991) Dendritic morphology, inward rectification and the functional properties of neostriatal neurons. In: Single Neuron Computation, McKenna, T., Davis, J. and Zornetzer, S.F. (eds.), Academic Press, (in press).

Wilson, C.J., Groves, P.M., Kitai, S.T. and Linder, J.C. (1983b). Three-dimensional structure of dendritic spines in the rat neostriatum. J. Neurosci., 3, 383-398.

Modeling Vestibulo-Ocular Reflex Dynamics: From Classical Analysis to Neural Networks

Thomas J. Anastasio
University of Illinois
Beckman Institute
405 North Mathews Avenue
Urbana, Illinois 61801

Abstract

The vestibulo-ocular reflex, which stabilizes the eyes during head rotation, is an elegant sensorimotor system that has long attracted the attention of theoretical neuroscientists. Previous models, based on classical analysis and control systems theory, provided precise descriptions of overall vestibulo-ocular reflex dynamics, but indicated little about the nature of the neurons that mediate this response. More recently, dynamic neural network models have extended the description to the neural level, providing new insights into the mechanisms of sensorimotor signal processing in the vestibulo-ocular reflex.

Introduction

The central aim of sensorimotor neurophysiology is to elucidate the mechanisms by which the nervous system processes sensory inputs to produce motor outputs. The vestibulo-ocular reflex (VOR) is one of the most well-studied examples of a sensorimotor transformation in the nervous system. A large amount of data has accumulated concerning the behavior of VOR and the anatomy and physiology of the neurons that mediate it. This data has paved the way for analytical models that precisely described the function of VOR, and answered the question: What does the VOR do? Although the analytical models suggested possible neural mechanisms, they failed to account for many of the properties of VOR neurons. They were therefore unable to answer the neurophysiologically more interesting question: How does the VOR do it?

Recently, the VOR has been modeled using dynamic neural networks. These models incorporate some of the features of real neurons, such as nonlinearity and complexity in interconnections, that are ubiquitous neurophysiologically but intractable analytically. The dynamic VOR neural network models can account for many of the properties of VOR neurons, and provide insight into neural mechanisms that might actually be used by the brain in producing the VOR transformation.

The function of the VOR is to stabilize vision by producing eye rotations that counterbalance head rotations. It is mediated by interneurons in the vestibular nuclei (VN) that relay head rotational velocity signals from vestibular, semicircular canal afferents to the motoneurons of the extra-ocular muscles (Wilson and Melvill Jones 1979). Far from being a simple relay, however, the vestibular signal is modified by VN and other neurons before it is transmitted to the motoneurons. These modifications include mathematical integration of the velocity signal into a position signal, the insertion of pulses, and the addition of other oculomotor commands (Robinson 1975).

Modification of the vestibular head rotational velocity signal also includes a process known as velocity storage (Raphan et al. 1979). For the horizontal VOR (which operates about a vertical axis of rotation), velocity storage is a relatively simple form of signal processing in which the time constant of the canal afferent signal is prolonged before it is used as an eye movement command. This allows the VOR to have a longer time constant than the canal signal that drives it. Velocity storage in the VOR serves as the focus of the models described in this chapter. By concentrating on one mechanism, a clear (if simplified) view of signal processing in the VOR can be obtained.

The purpose of this chapter is to briefly describe previous VOR models that utilized classical analysis and control systems theory, and to present new findings from dynamic neural networks in the context of this previous work. It is hoped that in the comparison between analytical models and neural networks, something about the nature of signal processing by neurons can be understood.

Classical Analysis of Canal Hydrodynamics

Head rotations are transduced by the semicircular canal receptors. Essentially, the canal receptors are fluid-filled tori, carved out of the bone of the skull, with a membrane spanning the lumen at one end (Wilson and Melvill Jones 1979). The fluid within the canal is known as endolymph, while the membrane is called the cupula. Head rotation causes relative movement between the canal and the en-

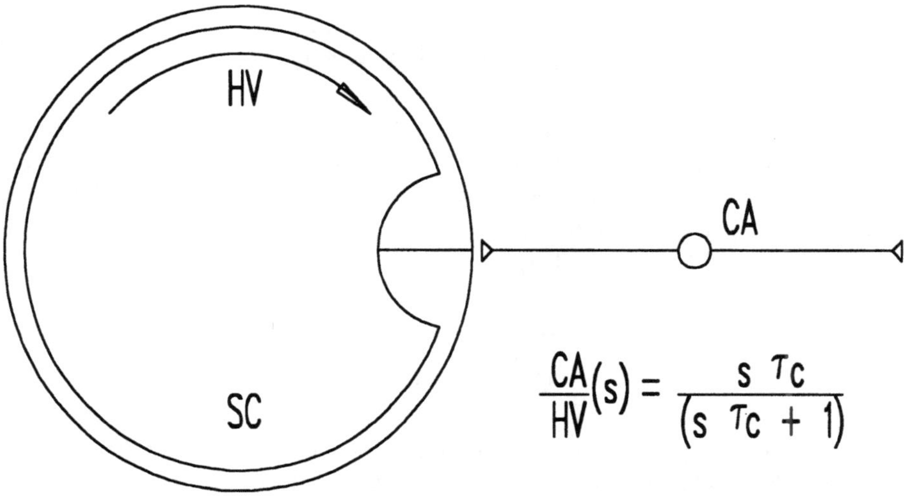

Figure 1. Arrangement between semicircular canal receptors and canal afferent neurons. The system response can be described adequately as a first-order lag. *SC*, semicircular canal; *HV*, head rotational velocity; *CA*, canal afferent, and the head rotational velocity signal it carries; τ_C, canal signal time constant; *s*, Laplace operator.

dolymph, which in turn causes a displacement of the cupula. Canal afferent neurons receive synaptic input from hair cell receptors that are sensitive to cupular displacement. The arrangement between canal receptors and sensory afferents is schematized in Figure 1.

Theoretical studies of the vestibular system began with classical analysis of semicircular canal hydrodynamics (Steinhausen 1933; Mayne 1950; Wilson and Melvill Jones 1979). If some force rotates the head, then the canal will experience an angular acceleration, but the endolymph will tend to stay in place due to its inertia. The inertial force causes the endolymph to flow in the accelerating canal. This flow will be opposed by forces due to endolymph viscosity and cupular elasticity. Thus, canal hydrodynamics are governed by the following second-order differential equation:

$$I(\ddot{q} - \ddot{\theta}) = B\dot{\theta} + K\theta\,, \qquad (1)$$

where q is the angular position of the canal, θ is the relative angular position between the endolymph and the canal, and the single and double dots designate the first and second time derivatives, respectively (θ is also equal to cupular displacement). I, B, and K are the coefficients of endolymph inertia, endolymph viscosity, and cupular elasticity, respectively. This equation can be solved using the method of Laplace transforms to give:

$$\frac{\theta}{\ddot{q}}(s) = \frac{\tau_s}{(s\tau_s + 1)} \frac{\tau_l}{(s\tau_l + 1)}, \tag{2}$$

where τ_S and τ_l are the so called short and long time constants. The value of τ_S is so short that it can be ignored for practical purposes (Wilson and Melvill Jones 1979). This leaves a simplified, first-order lag equation for canal dynamics, wherein the τ_S term has been removed and τ_l (approximately equal to *B/K*) is renamed the canal time constant, τ_C:

$$\frac{\theta}{\dot{q}}(s) = g\frac{s\tau_c}{(s\tau_c + 1)}. \tag{3}$$

As described by equation (3), canal dynamics produce an imperfect (leaky) integration of head rotational acceleration, where cupular deflection is proportional to head rotational velocity (ibid.). For this reason, equation (3) relates cupular displacement (θ) to head rotational velocity (*q dot*), and the right hand side is multiplied by *s*, the Laplace operator. To simplify the following discussion, constants of proportionality (*g*) are assumed to be equal to one. In the time domain, the canal response to a head rotational velocity step (acceleration impulse) is:

$$\theta(t) = e^{-t/\tau_c}. \tag{4}$$

The theoretical analysis of the semicircular canal served as the basis for analyses of the response dynamics of VOR and the neurons that mediate it. Single unit recordings showed that the baseline firing rates (spontaneous rates, SRs) of semicircular canal sensory afferent neurons are modulated by head rotation, and that the dynamics of these modulations can be adequately described as a first-order lag with a time constant of about five seconds (Fernandez and Goldberg 1971; Büttner and Waespe 1981). The afferents have other response components as well. While the tonic response component appears to be related to cupular position as predicted in theoretical studies, some units have a phasic component that may be related to cupular velocity (Fernandez and Goldberg 1971). Also, canal afferents can exhibit neural adaptation (ibid.). Although these other response features probably influence the VOR, their time constants (about 0.05 and 80 seconds for cupular velocity sensitivity and adaptation, respectively) are far enough away from the dominant canal time constant (five seconds) that they can be disregarded in order to simplify the following discussion. The equation (transfer function) relating the firing rate modulations of canal afferents (*CA*) to head velocity (*HV* = *q dot*) is given in Figure 1, where the canal time constant (τ_C) equals five seconds.

Control Theory Models of Velocity Storage in the VOR

Like the canal afferents, the VOR also exhibits primarily first-order dynamics (Buettner et al. 1981; Paige 1983). However, the time constant of the VOR (τ_V) is about 20 seconds long, approximately four times longer than that of the canal signal that drives it (τ_C). VOR dynamics can be described by:

$$\frac{EV}{HV}(s) = \frac{s\tau_v}{(s\tau_v + 1)}, \tag{5}$$

where *EV* and *HV* are eye and head rotational velocity, respectively. In the time domain, the responses of the VOR to head rotational acceleration impulses and steps are given in equations (6) and (7):

$$EV(t) = e^{-t/\tau_v}, \tag{6}$$

$$EV(t) = (1 - e^{-t/\tau_v}). \tag{7}$$

Equations (3, 4) and (5, 6) illustrate the signal processing problem that must be solved by the brainstem circuitry of the VOR. Neural elements must transform the first-order canal afferent response with a time constant of five seconds (τ_C), into the first-order VOR response with a time constant of 20 seconds (τ_V). Two control theory models have been put forward to account for this phenomenon (Raphan et al. 1979; Robinson 1981). Simplified versions of these models are schematized in Figure 2.

In realizing the VOR transformation, it would not be sufficient to simply follow the canal afferents with another first-order lag. This could prolong the time constant, but the VOR would then lack the sharp onset that it needs to counterbalance rapid head movements. In the original model (Figure 2A), Raphan and coworkers (1979) suggested that a velocity storage element, consisting of a leaky integrator (first-order lag), existed in parallel with a direct path from the canal afferents to the motoneurons. By appropriately weighting the outputs of both the direct and integrated paths at the motoneurons, the correct VOR transformation could be achieved.

As an alternative, Robinson (1981) suggested that the same thing could be accomplished with only one path (Figure 2B). Feedback through a leaky integrator would produce a lag compensator (Milsum 1966) that would simply swap the VOR time constant for the canal time constant, without affecting the first-order nature of the response. The form of this lag compensation, in Laplace notation, is given at

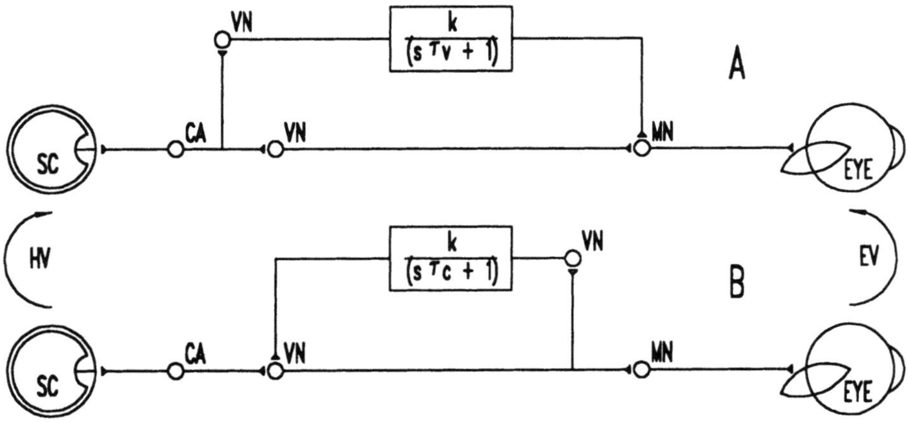

$$\frac{s\,\tau_C}{(s\,\tau_C+1)} \times \frac{\tau_V\,(s\,\tau_C+1)}{\tau_C\,(s\,\tau_V+1)} = \frac{s\,\tau_V}{(s\,\tau_V+1)} = \frac{EV}{HV}(s)$$

Figure 2. Control theory models of velocity storage in the VOR. These models are simplified versions of: A, Raphan et al. 1979; B, Robinson 1981. The VOR transformation is represented in Laplace notation at the bottom. *EV,* eye rotational velocity; *VN,* vestibular nuclei neurons; *MN,* extra-ocular motoneurons; τ_V, VOR time constant; k, gain of the leaky integrators (first-order lags); other abbreviations an in Figure 1.

the bottom of Figure 2. Evaluating the positive feedback loop in Robinson's model (Figure 2B) gives (Robinson 1981):

$$\frac{1}{1-\dfrac{k}{(s\tau_c+1)}} = \frac{1}{(1-k)}\frac{(s\tau_c+1)}{\left[s\left(\dfrac{\tau_c}{1-k}\right)+1\right]} \equiv \frac{\tau_v}{\tau_c}\frac{(s\tau_c+1)}{(s\tau_v+1)}. \qquad (8)$$

Remembering that τ_C and τ_V are approximately equal to five and 20 seconds, the required lag compensator will be formed if the strength of the feedback path (k) is equal to 0.75 ($1 - \tau_C/\tau_V$). It is assumed that the time constant of the feedback element is the same as the canal time constant, τ_C.

By applying this type of analysis to Raphan's model (Figure 2A), it can be shown that the two control theory models compute equivalent transformations, depending upon the value of k. Here it is assumed that the time constant of the integrating path is equal to the VOR time constant, τ_V. Evaluating the feedforward loop in Raphan's model (Figure 2A) gives:

$$1+\frac{k}{(s\tau_v+1)}=(k+1)\frac{\left[s\left(\dfrac{\tau_v}{k+1}\right)+1\right]}{(s\tau_v+1)}\equiv\frac{\tau_v}{\tau_c}\frac{(s\tau_c+1)}{(s\tau_v+1)}. \quad (9)$$

If k equals three (τ_V/τ_C - 1), then the feedforward loop in Raphan's model will also form a lag compensator that will cancel the canal transfer function. This will result in a simple first-order VOR transfer function with a time constant equal to τ_V, as shown in equation (5) and Figure 2. In the original model, Raphan and coworkers set the equivalent of k to a value greater than three. This added more of the leaky integrator component, which allowed them to reproduce the slight plateau in the VOR response that is observed under certain circumstances (see below). The point of the analysis, however, is to demonstrate the fact that the two control theory models rely on different mechanisms but are computationally equivalent.

Implications of Control Theory Models for Neural Processing

These models formally describe signal processing in the VOR, as they were intended to do. They both transform the canal afferent head velocity signal into the motoneuron eye velocity command, and they both do so via an intermediate processing stage corresponding to the vestibular nuclei (VN). It is instructive to see if there is any correspondence between the intermediate representations in these models and the properties of real VN neurons.

The responses to two impulse head rotational accelerations, one to each side in the horizontal plane, are shown for each of the elements in the control theory models in Figure 3. The time steps, each equaling five seconds, and arbitrary response size scale are chosen to match the results from neural network simulations to be presented below. The canal afferent signal and motoneuron command are shown in Figure 3A (dashed and dotted, respectively). The motoneuron response is equal but opposite to the canal signal, indicating that the VOR will produce perfectly counterbalancing eye movements. (It is understood that, in the bilaterally symmetric VOR, VN neurons on one side effectively invert their signals by sending them to motoneurons on the contralateral side.) The time constants are five and 20 seconds for the afferents and motoneurons, respectively, reflecting velocity storage. The motoneuron response is arbitrarily delayed from the afferent response, simply to match input-to-output delay in the neural networks.

In both models, the VOR transformation is accomplished by combining the canal response with a leaky integrated (low-pass filtered) signal. In Raphan's model, this summation takes place at the level of the motoneurons (Figure 2A). The signal in the direct VN neuron path of Raphan's model (Figure 3B, dashed) is the same as the canal afferent signal (Figure 3A, dashed). The output of the integrated

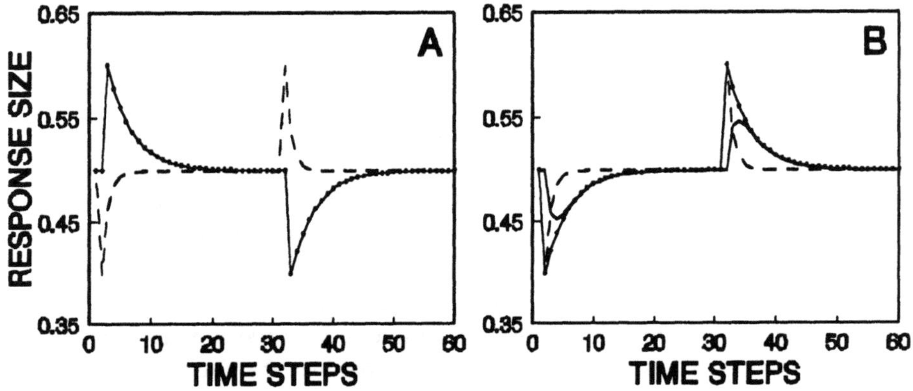

Figure 3. Acceleration impulse responses of the elements in control theory models of the VOR. A, canal afferent, dashed; motoneuron, dotted. B, direct path VN neurons from Raphan's and Robinson's models, dashed and dotted, respectively; leaky integrated path VN neurons from both Raphan's and Robinson's models, solid (leaky integrated path is feedforward in Raphan's model and feedback in Robinson's model).

VN path is also shown in Figure 3B (solid). It is simply a low-pass filtered version of the canal signal. The direct and integrated contributions add-up to give the required motoneuron command.

In Robinson's model (Figure 2B), the VOR transformation is accomplished at the level of the VN neurons, by feedback through a leaky integrator. The response of the forward path VN neuron in Robinson's model is shown in Figure 3B (dotted). It has the same form as the motoneuron signal (Figure 3A, dotted), but with reversed polarity. This signal is the sum of the canal afferent signal (Figure 3A and B, dashed), and the integrated signal from the feedback path (Figure 3B, solid). The latter has the same form as that of the integrated path in Raphan's model.

What implications do these models have concerning the properties of VN neurons? Both models suggest that some VN neurons should have smoothed responses on the time scale of the VOR, corresponding to integrated path neurons. Robinson's model suggests that other VN neurons might have first-order lag characteristics with the time constant of the VOR. In Raphan's model, the direct path must occur through a VN neuron, since afferents don't project to motoneurons in mammals (Wilson and Melvill Jones 1979). Thus, Raphan's model suggests that other VN neurons might have the first-order lag characteristics of the canal afferents. Neither model really has anything to say about the gains of VN neurons relative to afferents, or about their spontaneous discharge rates (SRs). Also, both models are linear; Raphan's model is piecewise linear in that it treats right and left VOR differently (Raphan et al. 1979).

Many real VN neurons do have time constants that are close in value to that of the VOR (Fuchs and Kimm 1975; Buettner et al. 1978), in accordance with both control theory models. Other VN neurons have time constants closer to that of the canals, lending support to Raphan's parallel path model. However, individual VN neurons can vary considerably in their time constants, and many appear to have dynamics intermediate between those of the canals and the overall VOR (ibid.). This diversity is not explained by the lumped-parameter control theory models (Figure 2).

Relative to canal afferents, VN neurons tend to have higher gains and lower SRs (Fuchs and Kimm 1975; Buettner et al. 1978). These properties cause VN neuron response asymmetry, with inhibitory responses being forced into cut-off (rectification). Another nonlinear feature of VN neurons is skew in response to sinusoidal head rotations, with the rising phase steeper than the falling phase (ibid.). None of these properties are reproduced by the linear control theory models of the VOR (Figure 2).

Dynamic Neural Network Models of the VOR

To gain further insight into the nature of VN neurons, dynamic processing in the horizontal VOR was modeled using recurrent neural networks (Anastasio 1991). The architecture of the networks (Figure 4) is consistent with the known neuroanatomy of the mammalian VOR (Wilson and Melvill Jones 1979). The networks are arranged to operate reciprocally, as does the real, bilaterally symmetric VOR. The networks have two each of input and output units, and either two, four (as shown in Figure 4), or eight hidden units. Input units represent afferents from the left and right horizontal canals (**lhc**, **rhc**). Output units represent motoneurons of the lateral and medial rectus muscles of the left eye (**lr**, **mr**). Hidden (middle layer) units correspond to VN neurons on the left and right sides of the brainstem (**lvn**, **rvn**). All of the units are nonlinear, computing the weighted sum of their inputs and passing it through the sigmoidal squashing function (Rumelhart et al. 1986). As such, unit responses are sigmoidally bounded from zero to one.

To represent the VOR relay, forward connections connect input to hidden units and hidden to output units. Inhibitory vestibular commissural connections, which link the real VN bilaterally (Wilson and Melvill Jones 1979), are modeled as recurrent connections that occur between hidden units on opposite sides. All units make both excitatory and inhibitory connections. The latter are assumed to take place via inhibitory interneurons that are not represented explicitly in the networks. Because it is generally believed that plastic changes occur only at the VN, hidden-to-output connections are fixed in a reciprocal pattern.

The networks are trained on continuous input/desired-output patterns using a recurrent backpropagation learning algorithm (Williams and Zipser 1989). This algo-

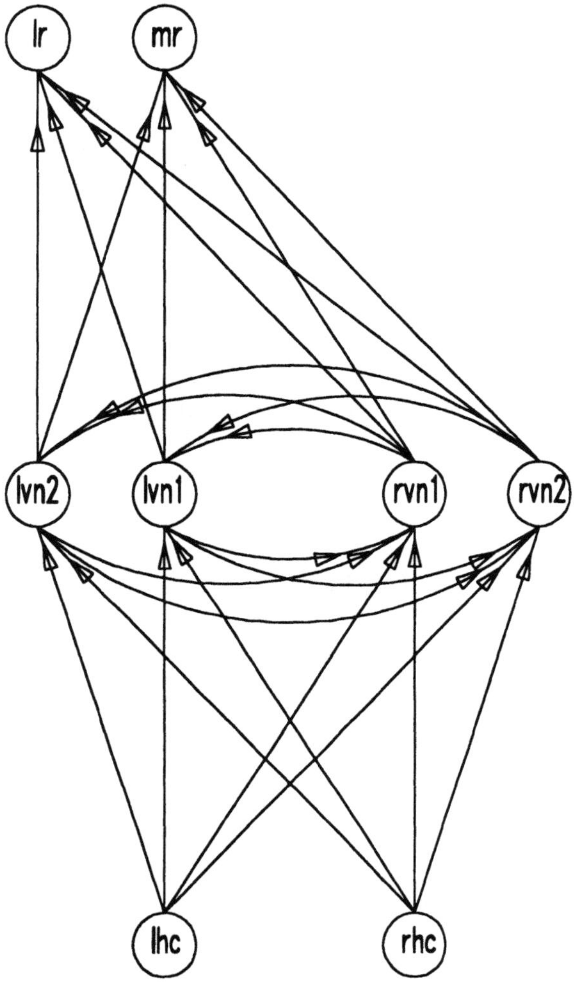

rithm is a dynamic version of backpropagation (Rumelhart et al. 1986), in which the gradient of network error in weight space is computed as a function of time, and modifiable weights are updated continually. The training set corresponds to the canal afferent head velocity signals (Fernandez and Goldberg 1971; Büttner and Waespe 1981) and motoneuron eye velocity commands (Skavenski and Robinson 1973) that mediate the horizontal VOR response to two impulse head accelerations, one to the left and the other to the right. These are the same input and output signals that are shown for the control systems models (Figure 3A). Input (solid) and desired-output (dashed) signals for the networks are shown in Figure 5A and B (5A, **lhc** and **lr**; 5B, **rhc** and **mr**).

Figure 4. Architecture of dynamic (recurrent) neural network models of the VOR (redrawn from Anastasio 1991). Different networks can have two, four (as shown here), or eight hidden units. **lhc**, **rhc**, left and right horizontal canal afferents; **lvn**, **rvn**, left and right vestibular nuclei neurons; **lr**, **mr**, lateral and medial rectus motoneurons of the left eye.

A VOR Network With Two Hidden Units

The first network has only two hidden units, one on each side. Since the networks are constrained so that recurrent connections can occur only between hidden units on opposite sides, these units cannot make self-connections. They can only form feedback loops by interconnecting. They both also receive a connection from each of the input units. Before training, all of the connection weights at the hidden units were randomized. The incorrect motoneuron responses from the randomized network are shown in Figure 5A and B (dotted). The actual-output (solid)

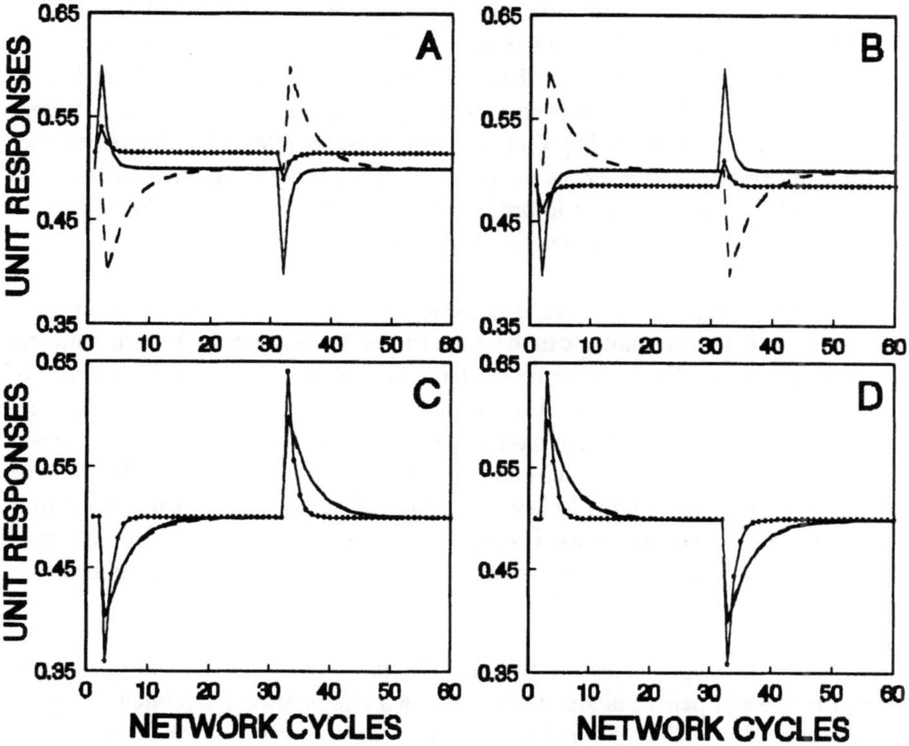

Figure 5. Training a recurrent VOR neural network model to produce the horizontal VOR with velocity storage. Unit responses to a left (network cycles 1 -- 30) followed by a right (network cycles 31 -- 60) impulse head rotational acceleration are shown. Subsequent plots will have the same format unless otherwise specified. A and B, pretraining; inputs, solid; desired-outputs, dashed; actual-outputs of initially randomized network, dotted; **lhc** and **lr** in A; **rhc** and **mr** in B. C and D, post-training; desired-outputs, dashed; actual-outputs, solid; actual-outputs of commissurotomized network, dotted; **lr** in C; **mr** in D.

matched the desired-output (dashed) after about 10,000 training set presentations (Figure 5C and D). The network develops reciprocal input-to-hidden connections, with **lhc** exciting and inhibiting **lvn** and **rvn**, respectively; **rhc** has the opposite hidden unit projection pattern. This crossed excitatory/inhibitory projection pattern reflects actual VOR organization (Wilson and Melvill Jones 1979) and its development is expected from the nature of the required VOR transformation. The recurrent connections form a lateral inhibitory loop between the hidden units. Cutting the recurrent connections eliminates velocity storage in the network, reducing output unit time constant to that of the inputs (Figure 5C and D, dotted). Paradoxically, removal of recurrent connections also increases output unit gain. These effects are observed for the real VOR following vestibular commissurotomy (Blair and Gavin 1981).

It is interesting that a network with only two hidden units (one per side), can learn the VOR transformation. By developing mutual inhibitory connections, the hidden units effectively exert positive feedback on themselves, which allows them to produce a leaky integration of the push-pull canal signals, but not their SRs. A similar mechanism was suggested by Cannon and coworkers (1983) for the velocity-to-position integrator. However, as pointed out above, leaky integration of the canal signal is not sufficient by itself to produce the VOR transformation. Such a signal would lack the rapid onset that is crucial for effective VOR function.

The mechanism that the network discovers, to produce the VOR transformation with only two hidden units, can be understood by examining hidden unit responses. These are shown in Figure 6A (**lvn**, dashed; **rvn**, solid); the responses of one input unit (**rhc**, dotted) are also shown for comparison. It is immediately apparent that the hidden units both have higher gains and lower SRs than the inputs, just as for real VN neurons (Fuchs and Kimm 1975; Buettner et al. 1978). These factors cause the hidden units to have asymmetrical responses, as real VN neurons have also (ibid.). This nonlinear behavior is what allows the network to both pass the sharp peak of the canal signal and integrate its tail with only one hidden unit per side.

Because of the reciprocal nature of the VOR network, the commissural loop is formed between a pair of hidden units, of which one is always inhibited when the other is excited. As the inhibited hidden unit is driven down into the low gain region of the squashing function, the gain of the loop, which mediates velocity storage, is also reduced. In this way, velocity storage in the network is rendered magnitude dependent -- as input magnitude and hidden unit rectification increase, loop gain and velocity storage integration decrease. Thus at peak, when input response magnitude is highest, velocity storage is reduced and this allows the sharp, initial portion of the canal signal to pass. After the peak, when input response magnitude decreases, velocity storage increases again and integrates the rapidly decaying tail of the canal response. These modeling results suggest that VN neurons have high gains, low SRs, and consequent rectified responses because this enhances signal processing in the VOR.

This nonlinear behavior can explain the magnitude dependency of the real VOR. The time constant of the VOR (τ_V) decreases as head rotation magnitude increases (Fetter and Zee 1988). Also, a plateau in the initial response of the VOR appears at lower stimulus intensities, indicating that the low-pass characteristics of the response increase as stimulus intensity decreases (Raphan et al. 1979). These are two different ways of stating the same fact: that velocity storage decreases as input magnitude increases.

Magnitude dependency in the two-hidden-unit network is shown in Figure 6 for output (6B) and hidden (6D) units, when the input magnitude is at the training

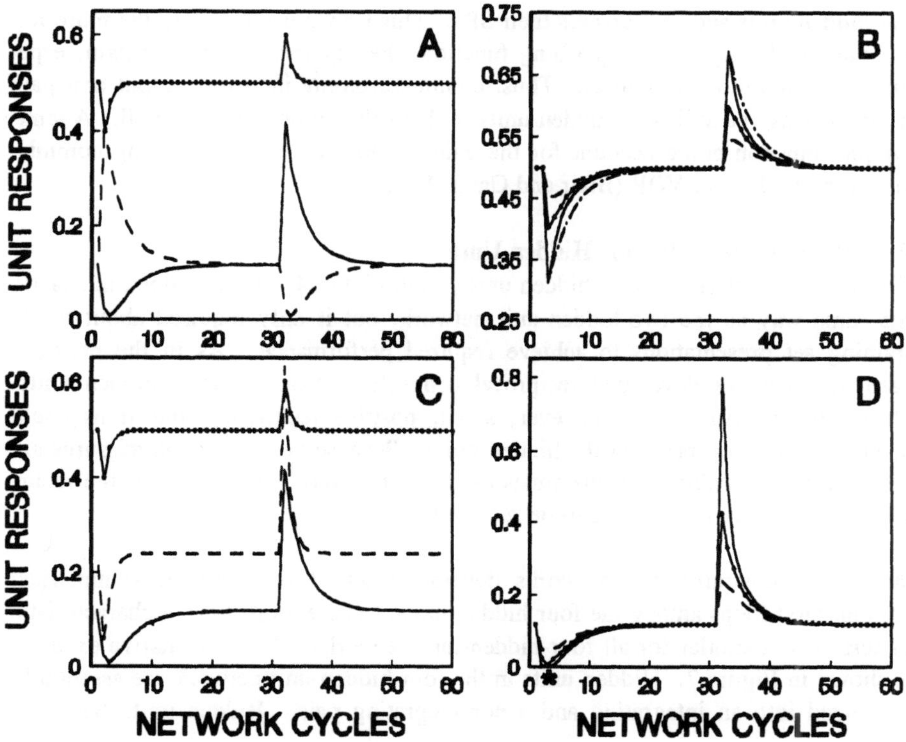

Figure 6. Impulse responses of hidden and output units in a recurrent VOR network having only two hidden units. A, responses of hidden units in the intact network (**rvn**, solid; **lvn**, dashed). C, responses of **rvn** before (solid) and after (dashed) commissurotomy; responses of **rhc** shown in A and C for reference. B and D, responses of output unit **lr** (B) and hidden unit **rvn** (D) to inputs at the training magnitude (dotted) or at twice (solid) or half (dashed) this level. Expected **lr** response at high magnitude shown in B (dot-dash).

level (dotted), or at twice (solid) or half (dashed) this value. When input magnitude is doubled, output response magnitude also approximately doubles (Figure 6B, solid), but decay rate is increased relative to what would be expected if the time constant had remained at τ_V (Figure 6B, dot-dash). The reason for this time constant decrease is again due to hidden unit rectification, which is more pronounced at the higher input intensity (Figure 6D, solid, $*$). When input magnitude is halved, the initial responses of hidden and output units become more rounded, indicating an increase in velocity storage (Figure 6B and D, dashed).

As for the outputs, cutting the recurrent connections (corresponding to inhibitory vestibular commissures), eliminates velocity storage and so reduces hidden unit time constant. This is shown for **rvn** in Figure 6C, before (solid) and after (dashed) commissurotomy; the responses of **rhc** are also shown for reference (dotted). In addition, commissurotomy also removes a source of inhibition from the hidden units and so increases their SRs. This brings them up into the more linear part of the sigmoidal squashing function, thereby increasing their gain, especially for inhibitory responses. Thus, commissurotomy increases output unit gain by increasing the ability of hidden units to drive the outputs in push-pull. A similar phenomenon could account for the gain increase observed following commissurotomy in the real VOR (Blair and Gavin 1981).

A VOR Network with Four Hidden Units

The second network has four hidden units, as in Figure 4. It was trained in exactly the same way as the two-hidden-unit network, but it only required about 5,000 training set presentations to achieve required performance. As in the previous case, the network developed reciprocal connections from input to hidden units. Unlike the previous case, however, small, positive recurrent connections could sometimes develop between the hidden units. Because vestibular commissures are believed to be inhibitory in mammals (Wilson and Melvill Jones 1979), the recurrent weights were constrained to be zero or less.

With this constraint, the networks develop a set of overlapping, symmetrical commissural loops among the four hidden units. These loops have a characteristic pattern that is similar for all four-hidden-unit networks. A representative example is shown in Figure 7. Hidden units in the four-hidden-unit network are essentially organized into an integrating and a nonintegrating pair. Hidden units **lvn1** and

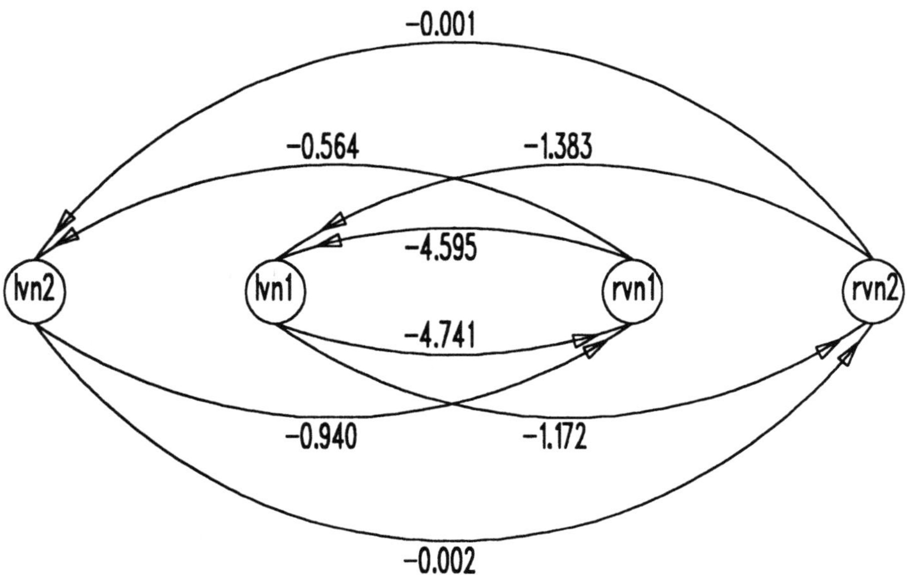

Figure 7. Recurrent (commissural) organization in a VOR neural network having four hidden units. Numbers indicate the values of the recurrent connection weights. **lvn1**, **lvn2**, left side hidden units; **rvn1**, **rvn2**, right side hidden units.

rvn1 strongly inhibit each other. This allows them to exert net positive feedback on themselves and produce a leaky integration of the canal inputs, as in the two-hidden-unit network. In contrast, hidden units **lvn2** and **rvn2** have almost no mutual inhibition. They do not integrate the canal signal by themselves and tend to pass it unaltered. Thus, it appears that the four-hidden-unit network develops parallel direct and leaky integrating paths to the outputs. This parallel arrangement was originally suggested by Raphan and coworkers (1979) (Figure 2A). However, hidden units **lvn2** and **rvn2** are coupled to units **rvn1** and **lvn1**, respectively, through moderately strong connections. This coupling endows units **lvn2** and **rvn2** with longer decay times than they would have by themselves. This mechanism is similar to that of feedback through a leaky integrator, as first proposed by Robinson (1981) (Figure 2B). Thus, the four-hidden-unit network combines both of the mechanisms that had been identified in control systems models, and appears to do so in a way that optimizes the utilization of the available elements.

The responses of the hidden units in the four-hidden-unit network are shown in Figure 8, along with those of input unit **rhc** (dotted) for comparison. The responses of hidden units **rvn1** (solid) and **lvn1** (dashed) appear as low-pass filtered versions of the canal signals, with rounded initial responses and long decay times (Figure 8A). This is due to their strong mutual inhibition (Figure 7). The responses of hidden units **rvn2** (solid) and **lvn2** (dashed) exhibit the peaked initial segment of the input response (Figure 8B), but have prolonged decays, due to their coupling to **lvn1** and **rvn1** (Figure 7). Although the waveforms for the two types of hidden units differ somewhat, they both have longer time constants than the inputs. They also have lower SRs, higher gains, and more asymmetrical responses than the inputs. This means that they also utilize the mechanism of nonlinear

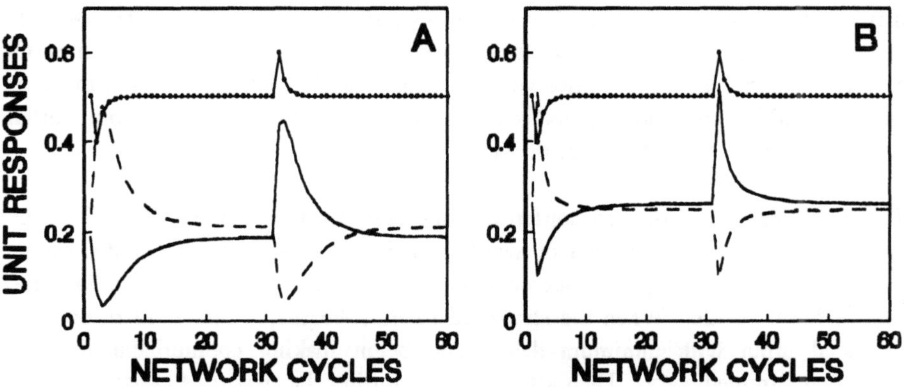

Figure 8. Impulse responses of units in a four-hidden-unit neural network model of the VOR (redrawn from Anastasio 1991). A, responses of **rvn1** (solid) and **lvn1** (dashed); B, responses of **rvn2** (solid) and **lvn2** (dashed); responses of **rhc** (dotted) shown in both A and B for comparison.

enhancement to realize the VOR transformation, as in the two-hidden-unit network.

Magnitude dependency in response to head rotational acceleration impulses is similar in four- as in two-hidden-unit networks (Figure 6B and D). The reason is that in the four-hidden-unit network, the integrating hidden units (**lvn1** and **rvn1**), which provide velocity storage in the network, also utilize the mechanism of nonlinear enhancement. They are therefore subject to rectification, and so velocity storage in the larger network is also magnitude dependent (Anastasio 1991). Nonlinear behavior can be further explored by subjecting the networks to types of stimuli for which they had not been trained (Anastasio, in press). The responses of hidden and output units in the four-hidden-unit network to step accelerations and sinusoidal rotations are presented in Figure 9.

Because the stimulus to the VOR is mechanical rotation of the head, it is unlikely that the real VOR ever experiences anything like an acceleration impulse. Instead, most real head movements probably consist of a short step of rotational acceleration followed by a period of constant velocity rotation. The response of output unit **lr** in the four-hidden-unit network to a rotational acceleration step is shown in Figure 9A. Also shown is the expected response of **lr** at the VOR time constant, τ_V (dotted), and a response at the canal time constant, τ_C (dashed). The actual response of **lr** does not follow the expected response. Instead, the rise time constant of the actual response is close to that of the canals, while the fall constant is closer to that expected of the VOR.

The reason for this nonlinear behavior is, again, due to events at the hidden unit level. The step responses of the integrating pair of hidden units (**rvn1**, solid; **lvn1** dashed) are shown in Figure 9C. At stimulus onset, hidden unit **lvn1** is quickly driven into rectification. This disrupts velocity storage and decreases the rise time constants of hidden and output units network wide. At step offset, **lvn1** comes out of rectification. This reactivates velocity storage and prolongs the fall time constants of hidden and output units as the network relaxes from the push-pull initial condition in which it is left after the step.

This nonlinear behavior may confer an important advantage upon the VOR. The nonlinear enhancement mechanism would switch off velocity storage at step onset, and switch it back on again at step offset. This would allow the VOR to have the best of both worlds. When the step head rotation begins, VOR could respond to its canal input with minimum delay, and begin making counterbalancing eye movements immediately. Then after the step, during the period of constant velocity, the VOR response could be prolonged even as the canal response decays. This nonlinear step response, with a short rise time constant and long fall time constant, has actually been observed for VN neurons (Waespe and Henn 1979). However,

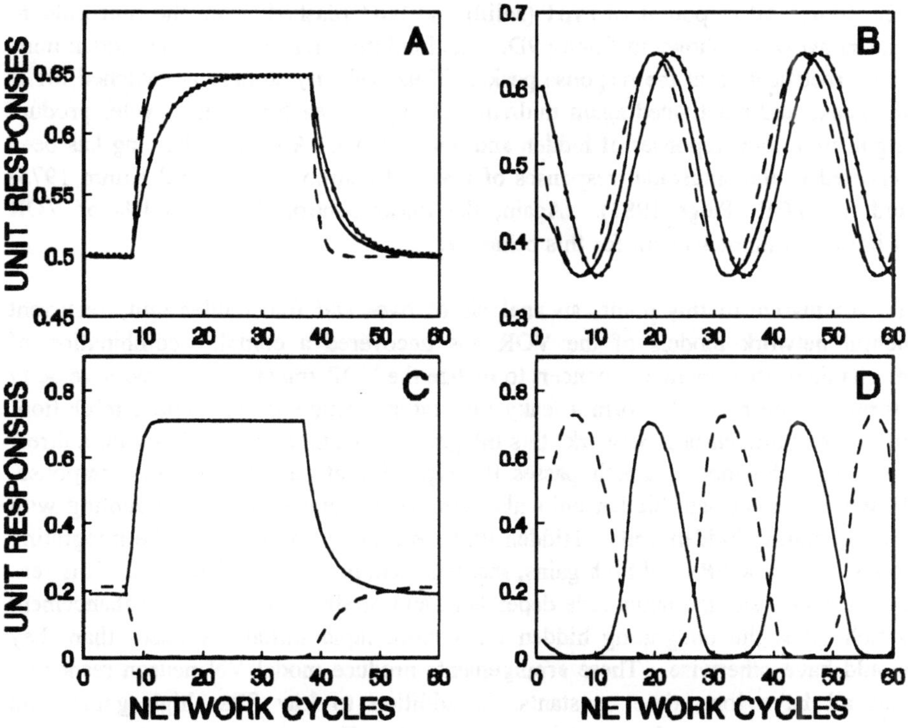

Figure 9. Nonlinear responses of output and hidden units in the four-hidden-unit network (redrawn from Anastasio, in press). A and C, acceleration step responses; B and D, sinusoidal responses; A and B, actual response of output unit **lr** (solid) along with responses at the VOR time constant, τ_V (dotted), or the canal time constant, τ_C (dashed); C and D, responses of integrating hidden units **rvn1** (solid) and **lvn1** (dashed).

the linear control systems models of VOR (Figure 2), which specify a linear acceleration step response (equation 7), cannot account for this behavior.

The four-hidden-unit model likewise produces nonlinear responses to sinusoidal inputs. The control theory models specify that the VOR response to sinusoidal input should also be sinusoidal and have a phase lag commensurate with its time constant, τ_V (equation 5) (Milsum 1966). The actual sinusoidal response of **lr** (solid) in the four-hidden-unit network is shown in Figure 9B, along with the expected, linear response of **lr** at the VOR time constant, τ_V (dotted), and a response at the canal time constant, τ_C (dashed). Rather than following the response expected from the linear control theory models, the actual sinusoidal response of **lr** is skewed, alternating between the responses at the two time constants. This nonlinear behavior is again due to hidden unit rectification.

The sinusoidal responses of **rvn1** (solid) and **lvn1** (dashed) from the four-hidden-unit network are shown in Figure 9D. The inhibitory responses of the hidden units show rectification at the response peaks. Thus, velocity storage is switched off at the peaks and reactivated again midrange throughout each response cycle, producing skew in the responses of hidden and output units. A similar skewing has been reported in the sinusoidal responses of real VN neurons (Fuchs and Kimm 1975) and the VOR (Paige 1983). Again, the linear control theory models of VOR (Figure 2) cannot account for this behavior.

To summarize to this point, an analysis of two- and four-hidden-unit, recurrent neural network models of the VOR has uncovered a complex combination of mechanisms that operate in concert to realize the VOR transformation with velocity storage. The networks form a leaky integrator through strong mutual inhibition. In the four-hidden-unit network, this integrator is arranged in parallel with a direct path from the inputs, which passes the rapid initial part of the canal response. However, direct path hidden units also achieve prolonged decay by coupling with the integrating hidden units. Hidden units, especially those within the integrating loops, have low SRs and high gains, making them prone to rectification. This renders velocity storage magnitude dependent, and produces a nonlinear enhancement which gives the integrating hidden units more rapid initial responses than they would have otherwise. These arrangements produce model VN neuron responses that all have long time constants, in addition to low SRs, high gains, and asymmetries, compared to model canal afferents. These features are also observed for real VN neurons (Fuchs and Kimm 1975; Buettner et al. 1978).

A VOR Network with Eight Hidden Units

The four-hidden-unit model suggests a novel, complex neural organization for VOR, and it can account for many of the properties of real VOR neurons. One drawback of this network, however, is that it produces hidden units with only two types of dynamic responses (Figure 8A and B). Real VN neurons, on the other hand, are known to vary widely in their dynamics (Fuchs and Kimm 1975; Buettner et al. 1978). To better account for VN neuron dynamic variability, the recurrent VOR neural network was expanded to eight hidden units. Again, the network was trained as in the previous examples, requiring about 2,500 training set presentations to achieve required performance. The learned weights are generally similar to the previous cases, with reciprocal input-to-hidden connections and symmetrical recurrent connections. As in the four-hidden-unit network, commissural interactions include strong lateral inhibition, producing leaky integration, and coupling to these low-pass filters. However, the loops overlap and interact to a much greater extent in the larger network. This results in a set of hidden units that vary in their dynamic behavior (Figure 10).

As in the previous networks, hidden units in the eight-hidden-unit network have longer time constants, lower SRs, higher gains, and asymmetrical responses as

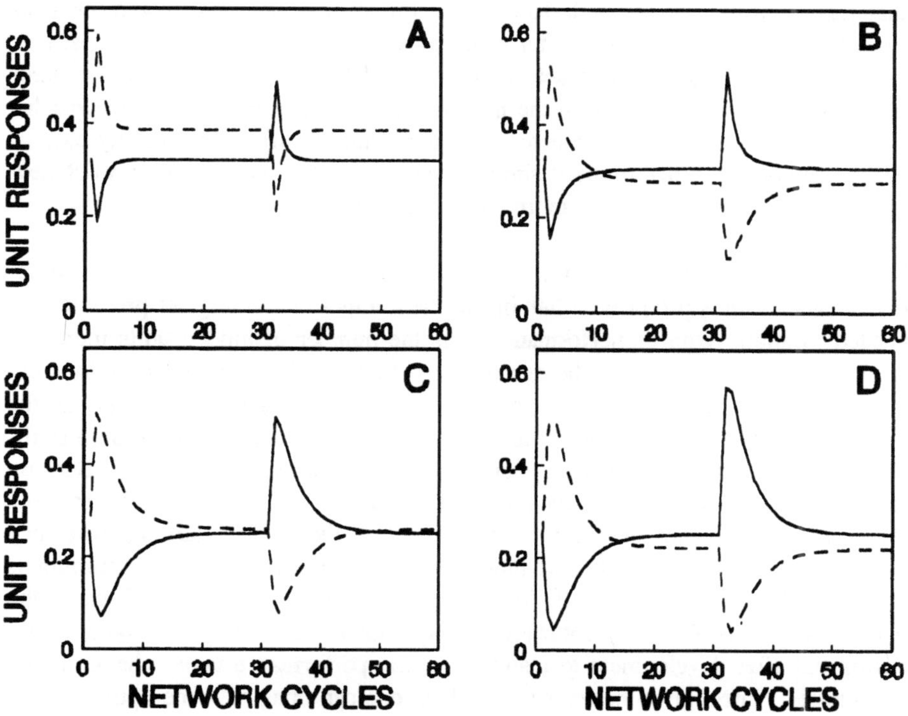

Figure 10. Impulse responses of the eight hidden units in an expanded, recurrent neural network model of the VOR. Right hidden units, solid; left hidden units, dashed.

compared to the inputs (Figure 10; right hidden units, solid; left hidden units, dashed). However, hidden units are more diverse in their properties in the larger network. This is due primarily to the more complex commissural interactions in the larger network. Hidden unit impulse responses in the eight-hidden-unit example can range from those having peaked initial responses and relatively short decay times (Figure 10A), to those having more rounded initial responses and long decay times (Figure 10D). Intermediate forms also occur (Figure 10B and C). It appears as though the impulse response of each unit is a combination of the canal response and its leaky integral, with the relative strengths of each varying in a continuous but nonuniform manner over the population. This result suggests that the diversity in the dynamics of the real VN population may be a consequence of distributed processing among a large number of highly interconnected neurons. This neural behavior could not have been reproduced in the lumped-parameter control systems models (Figure 2).

Discussion

The VOR neural network models presented above were programmed with a recurrent backpropagation learning algorithm (Williams and Zipser 1989). It is unlikely that this algorithm corresponds to a real neurophysiological learning mechanism (ibid.). However, the models offer many plausible mechanisms and explanations. Also, they adhere closely in structure to the actual VOR. For this reason, testable predictions can be derived from the models (Anastasio, in review), so the mechanisms they propose can be subjected to experimental verification. In any case, it will be important to determine if similar mechanisms also emerge from VOR neural networks programmed with more realistic learning strategies.

Using recurrent backpropagation, it became clear that every doubling of the number of hidden units produced a halving in the number of training set presentations needed to attain required performance. As the number of hidden units increased from two, to four, to eight, the number of required training set presentations decreased from about 10,000, to 5,000, to 2,500. This happened because as the number of hidden units increased, the network had more degrees of freedom available to find the solution, and the result was greater learning efficiency. Because the algorithm is computationally intensive (Williams and Zipser 1989), training larger networks is not practical. However, it is likely that learning efficiency would increase as the number of hidden units increases, until some optimum is reached, after which learning efficiency might start to decrease again. Because the nervous system is obviously not static, some adaptive capability is crucial not only during but after development to maintain system performance under changing circumstances. For a given neural system, then, one can imagine some optimal number of elements, specific for its particular function, that ensures efficient adaptive capability in an ever changing environment.

As clearly shown above, recurrent neural networks are descendant from a long line of vestibular and VOR models. Another area that has experienced a development similar to the one described here is the modeling of the velocity-to-position integrator, which produces an eye position command from the afferent head velocity signal (Skavenski and Robinson 1973). Models of this mechanism have moved from control systems theory (Kamath and Keller 1976; Robinson 1981; Galiana and Outerbridge 1984), to linear lateral inhibitory networks (Cannon et al. 1983), and on to learning networks (Arnold and Robinson 1991). Ironically, although Robinson proposed feedback through an integrator for velocity storage, he proposed a parallel direct/integrated path for velocity-to-position integration (Robinson 1981), similar to the arrangement Raphan originally proposed for velocity storage (Raphan et al. 1979). Galiana and Outerbridge (1984) took the position that velocity-to-position integration could be accomplished by feedback through a leaky integrator. As for the recurrent VOR models presented here, learning network models discovered that the neural organization of velocity-to-position integration may involve a melding of both the parallel path and feedback mechanisms (Robinson, personal communication). Happily, it is possible that everyone may be right!

One important difference between velocity storage and velocity-to-position integration concerns the involvement of the cerebellum. Cerebellectomy reduces the time constant of the velocity-to-position integrator by an order of magnitude, but has little effect on the time constant of velocity storage (Robinson 1974; Zee et al. 1981). Indeed it has been suggested that, if velocity-to-position integration is mediated by feedback loops, than at least some of these may pass through the cerebellum (Galiana and Outerbridge 1984).

In contrast, the cerebellum may actively suppress velocity storage. In a process known as habituation, the VOR time constant is actively suppressed with repeated presentations of certain stimuli (Blair and Gavin 1979). This suppression is reversed if parts of cerebellum, such as the vermis (ibid.) or nodulus/uvula (Waespe et al. 1985) are removed. If velocity storage had been previously habituated, then cerebellar ablation would actually increase velocity storage.

In the neural network models, all inhibitory connections are mediated by inhibitory interneurons which are not represented explicitly (see above). In the actual VOR, inhibitory commissural connections are also mediated by inhibitory interneurons: so called type II VN neurons (Wilson and Melvill Jones 1979). It is possible that the cerebellum, which sends inhibitory Purkinje cell projections to the VN (ibid.), could suppress (habituate) velocity storage by inhibiting type II VN neurons. This could open the loops and reduce velocity storage, but leave intact the VN relay from canal afferents to motoneurons. Removing the cerebellum would then remove type II VN neuron suppression, reclose the loops, reactivate velocity storage, and produce an increase in VOR time constant. This hypotheses could also be tested experimentally.

Acknowledgments
Special thanks to Dr. M. Nelson and Dr. B. Wheeler for reviewing the manuscript.

References

Anastasio TJ (1991) Neural network models of velocity storage in the horizontal vestibulo-ocular reflex. Biol Cybern 64:187-196

Anastasio TJ (in press) Implications of vestibular nucleus neuron rectification for signal processing in the horizontal vestibulo-ocular reflex. Ann N Y Acad Sci

Anastasio TJ (in review) Testable predictions from recurrent neural network models of the vestibulo-ocular reflex.

Arnold DB, Robinson DA (1991) A learning network model of the integrator of the oculomotor system. Biol Cybern 64:447-454

Blair SM, Gavin M (1979) Modification of the macaque's vestibulo-ocular reflex after ablation of the cerebellar vermis. Acta Otolaryngol 88:235-243

Blair SM, Gavin M (1981) Brainstem commissures and control of time constant of vestibular nystagmus. Acta Otolaryngol 91:1-8

Buettner UW, Büttner U, Henn V (1978) Transfer characteristics of neurons in vestibular nuclei of the alert monkey. J Neurophysiol 41:1614-1628

Buettner UW, Henn V, Young LR (1981) Frequency response of the vestibulo-ocular reflex (VOR) in the monkey. Aviat Space Environ Med 52:73-77

Büttner U, Waespe W (1981) Vestibular nerve activity in the alert monkey during vestibular and optokinetic nystagmus. Exp Brain Res 41:310-315

Cannon SC, Robinson DA, Shamma S (1983) A proposed neural network for the integrator of the oculomotor system. Biol Cybern 49:127-136

Fernandez C, Goldberg JM (1971) Physiology of peripheral neurons innervating semicircular canals of the squirrel monkey. II. response to sinusoidal stimulation and dynamics of peripheral vestibular system. J Neurophysiol 34:661-675

Fetter M, Zee DA (1988) Recovery from unilateral labyrinthectomy in rhesus monkey. J Neurophysiol 59:370-393

Fuchs AF, Kimm J (1975) Unit activity in vestibular nucleus of the alert monkey during horizontal angular acceleration and eye movement. J Neurophysiol 38:1140-1161

Galiana HL, Outerbridge JS (1984) A bilateral model for central neural pathways in vestibuloocular reflex. J Neurophysiol 51:210-241

Kamath BY, Keller EL (1976) A neurological integrator for the oculomotor system. Math Biosci 30:341-352

Mayne R (1950) The dynamic characteristics of the semicircular canals. J Comp Physiol Psychol 43:304-319

Milsum JH (1966) Biological control systems analysis. McGraw-Hill, New York

Paige GD (1983) Vestibuloocular reflex and its interaction with visual following mechanisms in the squirrel monkey. I. response characteristics in normal animals. J Neurophysiol 49:134-151

Raphan Th, Matsuo V, Cohen B (1979) Velocity storage in the vestibulo-ocular reflex arc (VOR). Exp Brain Res 35:229-248

Robinson DA (1975) Oculomotor control signals. In: Lennerstrand G, Bach-y-Rita P (eds) Basic mechanisms of ocular motility and their clinical implications. Pergammon Press, Oxford, pp 337-374

Robinson DA (1974) The effect of cerebellectomy on the cat's vestibulo-ocular integrator. Brain Res 71:195-207

Robinson DA (1981) The use of control systems analysis in the neurophysiology of eye movements. Ann Rev Neurosci 4:463-503

Rumelhart DE, Hinton GE, Williams RJ (1986) Learning internal representations by error propagation. In: Rumelhart DE, McClelland JL, PDP Research Group (eds) Parallel distributed processing: explorations in the microstructure of cognition, vol 1: foundations. MIT Press, Cambridge, pp 318-362

Skavenski AA, Robinson DA (1973) Role of abducens neurons in vestibuloocular reflex. J Neurophysiol 36:724-738

Steinhausen W (1933) Über die beobachtung der cupula in den bogengängsampullen des labyrinths des lebenden hechts. Pflügers Arch Ges Physiol 232:500-512

Waespe W, Cohen B, Raphan Th (1985) Dynamic modification of the vestibulo-ocular reflex by the nodulus and uvula. Science 228:199-202

Waespe W, Henn V (1979) The velocity response of vestibular nucleus neurons during vestibular, visual, and combined angular acceleration. Exp Brain Res 37:337-347

Williams RJ, Zipser D (1989) A learning algorithm for continually running fully recurrent neural networks. Neural Comp 1:270-280

Wilson VJ, Melvill Jones G (1979) Mammalian vestibular physiology. Plenum Press, New York.

Zee DS, Atsumi Y, Butler PH, Gucer G (1981) Effects of ablation of flocculus and paraflocculus on eye movements in primate. J Neurophysiol 46:878-899

Movement Primitives in the Frog Spinal Cord

S. Giszter, F.A. Mussa-Ivaldi and E. Bizzi
Department of Brain and Cognitive Sciences
Massachusetts Institute of Technology
Cambridge, MA 01239-4703

Abstract

We present evidence for the generation of stable convergent force field patterns and muscle synergies in the spinal cord of the frog. These synergies may form the bases of postural and trajectory adjustments. We present recent analyses which show that (1) similarly structured force fields underlie natural behaviors, (2) the active fields underlying both microstimulation and natural behavioral force fields are structurally invariant, but may be modulated in overall force amplitude and stiffness, (3) these fields are drawn from a limited set of such fields, (4) these invariant fields can be used to predict the termination position of limb endpoint trajectories in the unrestrained limb, (5) in those instances tested, multiple stimulations resulted in fields proportional to the vectorial summation of the fields resulting from single stimulations. This recent body of work suggests that we may view the spinal cord as possessing a small number of movement primitives: circuits that may specify invariant force fields, which may be combined in a more or less flexible manner to produce adjustable behaviors.

Introduction

A fundamental question in motor control is the nature of the basic units of motor behavior and their control (see Windhorst, 1991). A large body of work supports the notion that a variety of complex mechanisms for motor output reside in the spinal cord. Both locomotor pattern generation and phasic limb positioning mechanisms have large proportions of their circuitry in the spinal cord (Grillner and Wallen 1985; Stein 1983; Alstermark and Sasaki 1986). The frog's spinal cord is known to support a variety of movements and complex kinematic adjustments even in isolation (Fukson et al. 1980; Giszter et al. 1989; Berkinblitt et al. 1989; Schotland, et al. 1989; Ostry et al. 1991). The frog's spinal cord operating in isolation from the brain is thus an ideal preparation in which to study spinal mechanisms of motor control. Using this preparation we have obtained evidence that the generation of stable convergent force field (CFF) patterns and muscle synergies underlie motor acts which are elicited in the spinal cord of the frog by microstimulation (Mussa-Ivaldi et al. 1991; Giszter et al. 1991 a,b,c; Bizzi et al. 1991). These synergies may form the bases of both posture and trajectory adjustments. This set of observations support the equilibrium point framework for examining movement and posture (Feldman, 1979; Bizzi et al. 1984). In this article we will concentrate on recent analyses and extensions of this work which show that (1) similarly structured force fields underlie natural behaviors, (2) the active fields underlying both microstimulation and natural behavioral force fields are structurally invariant, but may be modulated in amplitude and stiffness, (3) these fields are drawn from a limited set of such fields, (4) these invariant fields can be used to predict the termination position of endpoint trajectories in the unrestrained limb, (5) in those instances, tested multiple stimulations resulted in fields proportional to the vectorial summation of the fields resulting from single stimulations.

This recent body of work suggests that we may view the spinal cord as possessing a small number of movement primitives: circuits that may specify invariant force fields, and that these primitives may be combined in a more or less flexible manner to produce adjustable behaviors.

Methods

We transected frogs at the caudal termination of the fourth ventricle and removed several lumbar vertebrae to allow access to the lumbar elargement of the spinal cord. Eleven muscles were implanted with pairs of recording electrodes. The spinal column of the frog was immobilized in a clamping system to allow stable microelectrode placements. Besides the effects of these surgeries, the frog's limbs, and the motor and afferent systems of the spinal cord were intact. Spinal behaviors could be elicited in all preparations used.

We elicited motor responses by microstimulating the lumbar spinal cord. Our electrode placements in the work described here were always in the grey matter. They were usually at depths and lateral positions that placed the tip in the premotor and proprioceptive regions in the grey matter of the spinal cord. For each spinal stimulation, we collected the EMG signals from 11 implanted muscles and the mechanical force at the ankle.

To measure the mechanical responses, we attached the right ankle of the frog to a six-axes force transducer, as shown in Fig. 1. The force transducer was mounted on a two-axes cartesian manipulator. We determined the location of the frog's ankle (with a resolution of 1 mm) by independently setting the x and y coordinates of this manipulator. We measured the mechanical response at different ankle locations to the same spinal stimulation. Typically, we recorded the force vectors in a set of from 9 to 16 locations forming a grid (3 x 3 or 4 x 4) in the limb's workspace. We divided this recording area into triangles. Within each triangle, we applied a linear interpolation to the force vectors measured at the corners. We applied this interpolation procedure to each collection of data vectors measured at a given latency from the onset of the stimulus. Using data sets obtained at different latencies, we thus derived estimates of the temporal development of the force field generated by a single spinal stimulation. This procedure is described in more detail in Mussa-Ivaldi, Giszter and Bizzi (1991) and Giszter, Bizzi and Mussa-Ivaldi (1991). The method is shown in Fig. 1.

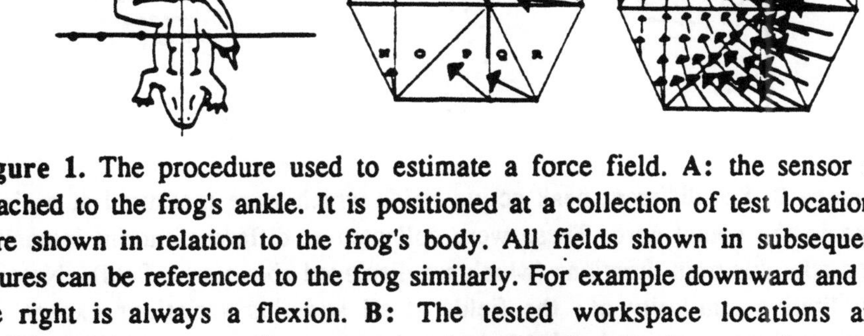

Figure 1. The procedure used to estimate a force field. **A:** the sensor is attached to the frog's ankle. It is positioned at a collection of test locations here shown in relation to the frog's body. All fields shown in subsequent figures can be referenced to the frog similarly. For example downward and to the right is always a flexion. **B:** The tested workspace locations are partitioned into a set of non overlapping triangles. Each vertex is a test location. Measured force vectors are displayed as solid arrows. The dashed line is a single interpolated vector. **C:** the field samples in B are interpolated.

Results

The principal result of these experiments is that microstimulation of the spinal grey of the frog elicits repeatable and well-organized patterns of muscle activation. These generate patterns of forces in the limb which can be summarized as force fields. These fields are always simply organized and are most often found to be convergent.

A number of controls described elsewhere were used to determine (1) the origins and spread of activity underlying convergent force fields and (2) the circuit elements that were both necessary and sufficient to support the convergent force fields (Mussa-Ivaldi et al., 1991; Giszter et al., 1991a,b; Giszter et al., manuscript in preparation). Briefly, the results of these control procedures can be summarized as follows: (1) Random activation of motoneurons does not account for the results. (2) The known motoneuron topography does not account for the results. (3) Afference is not necessary to generate a field structure, although it has an influence on the fine structure of

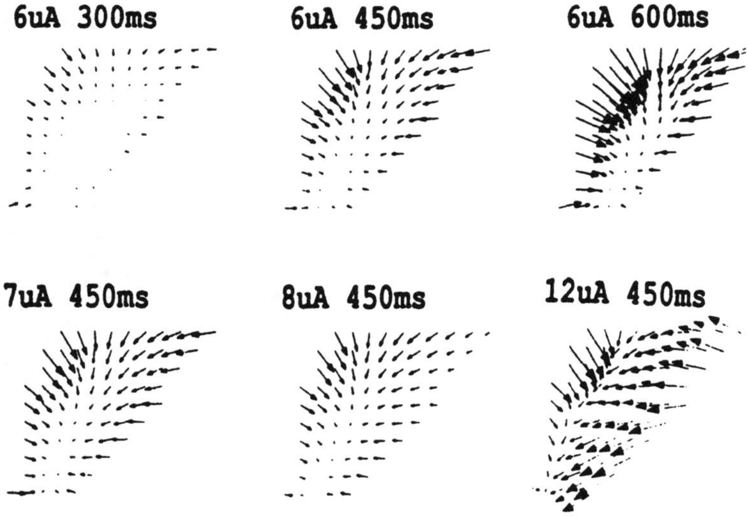

Figure 2. Six different peak active fields which were measured at a single site in the spinal cord. These were collected at different current intensities and stimulus train durations (which are displayed above each field). Note the qualitative similarity of the fields at all stimulus variations except that collected at a strength of 12mA. At this stimulus strength we believe additional cord structures or tracts were activated.

the force field. However afferent effects can be more than modest, and may cause variations of EMG magnitude as large as 80% of the peak EMG magnitude. (4) Activation of motoneurons occurs over a far broader area than can be accounted for by local current spread; activation occurs in elongated patches, even extending to motoneuron groups in the brachial cord.

CFFs and the decomposition into active passive and force fields.

We found that the collection of force field data was most simply comprehensible after the following procedure was applied. We divided the net field which resulted from stimulation into the pre-stimulus quiescent field and a post-stimulus active field. We performed this separation by a point-by-point vector subtraction of forces in the pre-stimulus postural field from the field which resulted from stimulation. The operation is thus equivalent to removing the baseline forces. This analysis of our data has a strong basis in the literature on frog spinal cord physiology, which has usually divided the frog motor system into a tonic or graded postural system (which would support the postural field) and a phasic twitch system (Simpson 1976). Our stimulation is likely to be activating the twitch system. In our experiments, we believe this twitch system which is activated by our stimulation is expressed in parallel with and overlaid on the postural field. While equilibria of the total recorded field were spread across the workspace we found that following the division into active and quiescent fields, our active field data fell into a few classes or force field types as described below. This considerable simplification of interpretation gives us confidence in this procedure. An example is shown in Fig. 2.

The pre-stimulus postural field of our preparations remained stable over several days of recording, so long as the preparations were reflexive and undamaged. We found that the extracted active force fields following stimulation were also usually convergent. Surprisingly, as we varied the duration, and amplitude of stimulation we found the peak active field structure remained constant (i.e., the orientation and relative proportions of force vectors and the equilibrium location were stable), while the maximum force amplitudes varied as a function of the stimulation parameters. Thus, we were able to modulate the maximum amplitude and stiffness of the field without any alteration of its qualitative spatial structure.

Following the decomposition of the field at a spinal cord locus into active and quiescent postural fields we also found that the rise and fall of the active field showed the stability in force pattern seen in the peak fields. The structure of the field remained almost constant as it evolved in time. The amplitude of forces and, where applicable, the stiffness of the field, as indicated by vector magnitude, rose and fell, but to a good approximation the peak field was a scaled version of the field at lower force magnitudes. The trajectory of the equilibrium point of the net force field was thus determined by the relative balance between the active field and the postural field at a point in time. This invariance of the active field's structure, the equilibrium trajectory and total field's variation is shown in Fig. 3 and Fig. 4.

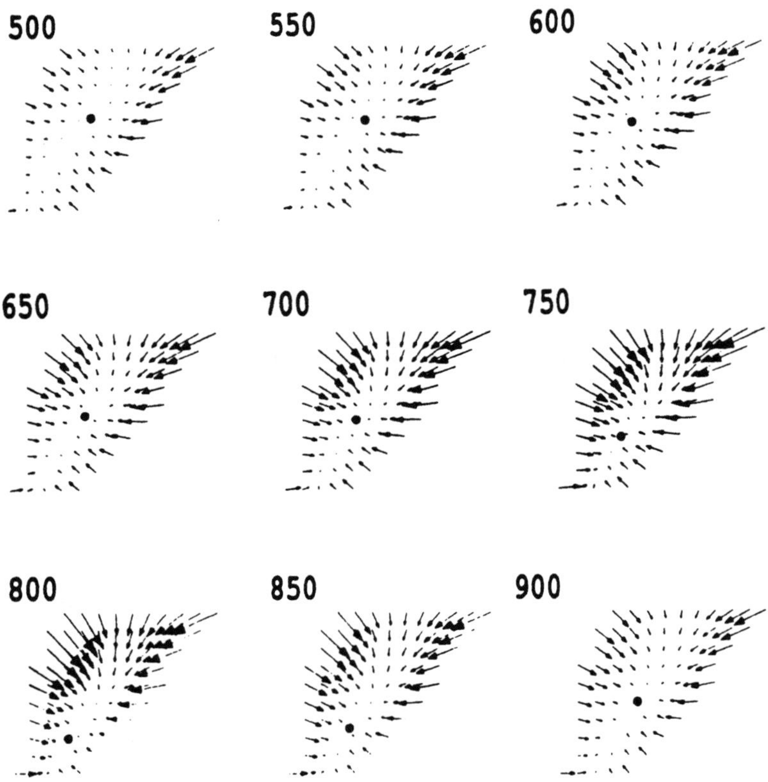

Figure 3 The temporal evolution of the total field structure and the virtual trajectory of the total field's equilibrium is shown. The time from recording onset (500 to 900 ms) at which the field was measured is shown to the upper left of each field. Stimulus delay was 300 ms. The first record shown is thus 200 ms after stimulus onset.

A small collection of force fields are found in the cord.

We found that the types of active fields fell into a few classes. These were very similar within a class or type. Some of these types are presented here as a sort of force field "manifest" for the spinal cord in Fig. 5. They were each obtained at different rostro-caudal and medio-lateral loci and slightly different depths in the cord. However detailed mapping of their distributions in relation to one another is yet to be performed. These fields may represent a collection of "movement primitives," or indivisible elements

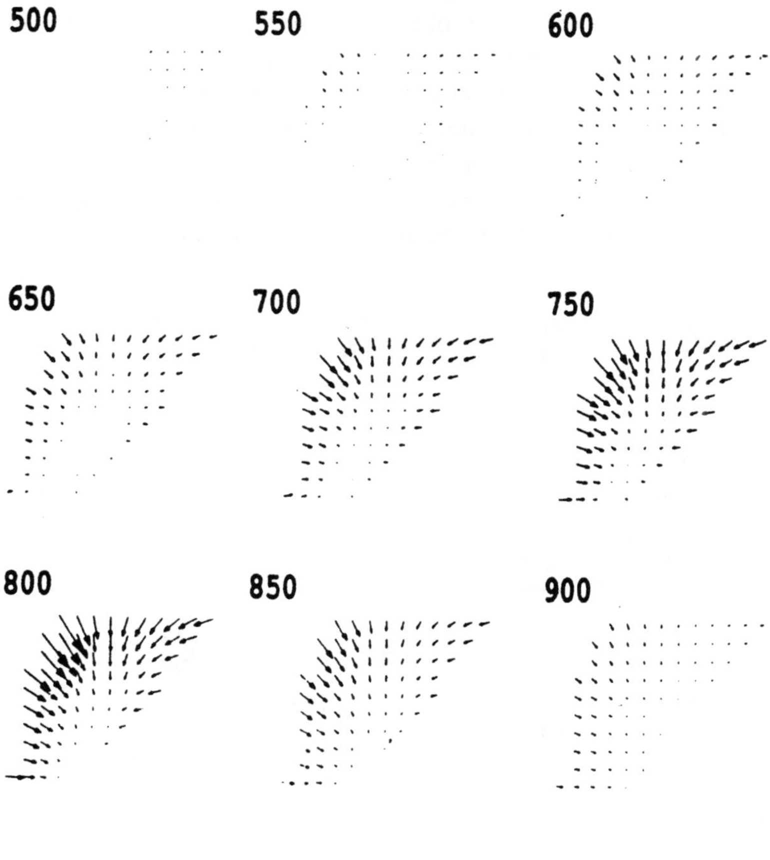

Figure 4 The temporal evolution of the active field which underlay the total field and equilibrium trajectory which is shown in Fig. 3. Note the structural similarity of the field throughout its increase and decrease. Forces retain their orientation and balance throughout.

from which motor patterns such as wiping and the more adjusted behaviors of the intact frog are built up. Fig. 5 also shows a very preliminary 2-dimensional map of the four most common field types found in the lumbar enlargement. These fields can organize movement to four different directions. It can be seen that these fields can be elicited in several discrete zones. Within a zone the fields which are elicited are similar. Switching between types at the zone boundaries is rapid and discontinuous (that is to say, less than the expected resolution of our stimulation technique: i.e., switching occurs within 200 microns). The repertoire of force field types observed following microstimulation of the spinal grey consisted of force field types varying from simply convergent to fields with broad parallel regions. Divergent fields were never observed however. The collection included invariant force fields which would induce movements to extension, to flexion, to rostral positions, and to caudal positions in several different ways. Force fields induced by microstimulation could be expressed either unilaterally or bilaterally. In bilateral responses, the ipsilateral response could be coupled either to the same or to a different movement in the other limbs. Each type of field was invoked relatively discretely depending on electrode site.

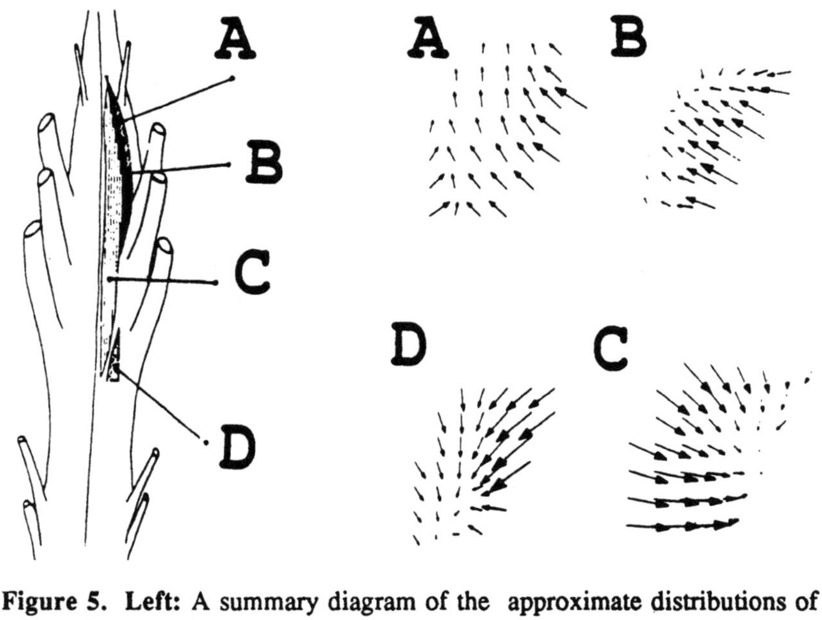

Figure 5. Left: A summary diagram of the approximate distributions of four types of fields. The sites shown are those producing the active fields shown to the right. Depths of the electrode tip ranged 500 to 800 microns. This is based on 40 penetrations in 3 frogs. **Right:** Four types of active field associated with each of the regions . These four fields would direct the limb in four different directions.

Naturally elicited fields.

We have examined behaviors elicited by cutaneous stimulation to determine whether similar convergent and patterned force fields underlie more natural behaviors. Flexion withdrawal, hindlimb-hindlimb wiping and forelimb withdrawal, forelimb wiping, back wiping and either turning or ambulatory motions can all be activated by cutaneous stimulation in the spinal frog. Those of these which we have examined could be described as convergent or regularly patterned force fields in the spinal frogs we tested. Here we will present a 2-dimensional description of wiping. Examples of these types of fields are shown in Fig. 6. However, we have also recorded such cutaneously elicited fields in 3-dimensions (Giszter, Mussa-Ivaldi and Bizzi, manuscript in preparation).

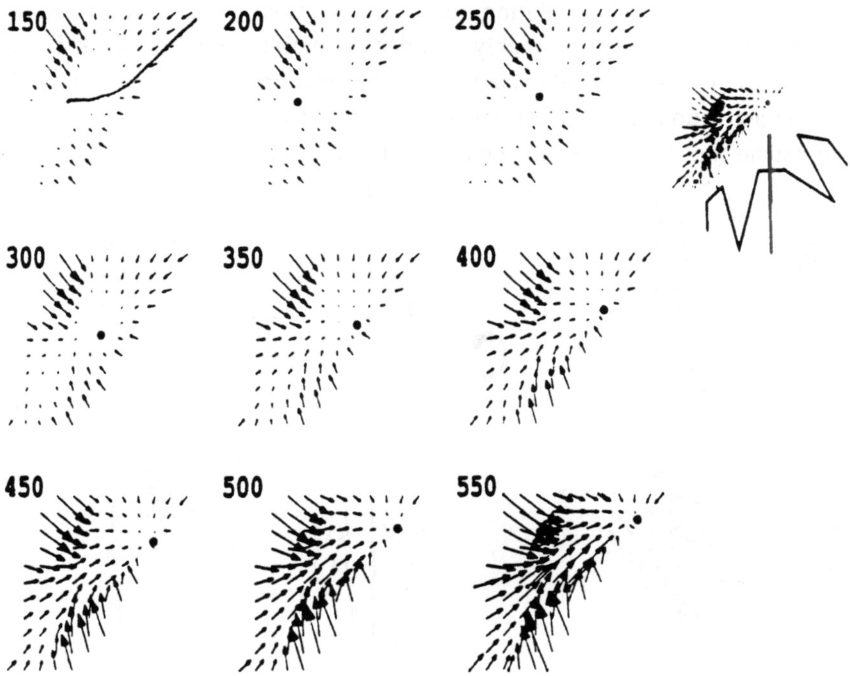

Figure 6. Wiping force field. This force field was measured in the restrained leg following a stimulation which would elicit wiping in the unrestrained frog both before and after field collection. The virtual trajectory is shown and this matches previously published wiping kinematics very well.. The stick figure inset shows the frogs hindlimbs and body axis in relation to the displayed field.

Differences and similarities of natural and microstimulated fields.

It can be seen from a comparison of records in Fig. 7 A and B that wiping fields induced by stimulating the skin and particular CFFs induced via microstimulation were very similar. It should be emphasized that this similarity is between the microstimulation based **active** field and the behavior's field. However, a preliminary comparison of motoneuron activation patterns in the spinal cord using rhodamine as a marker indicated occasional differences. For example the marking of motoneurons in the microstimulation induced flexion activity was sometimes unilateral. In contrast, in the cutaneously induced flexion behavior, we observed that the activity in motoneuron pools was bilateral, with a contralateral activation of what are likely to be extensors (diagrammed in Fig. 7 C and D). This suggests that either (1) microstimulation activates a subset of the ensemble of synergies induced by cutaneous stimulation, or (2) activation by microstimulation generated a more labile or weakly coupled ensemble. Further work is in progress to improve our understanding of the relationships between natural cutaneous activation and microstimulation, to examine the underlying circuits and to extend the rhodamine motoneuron marking results.

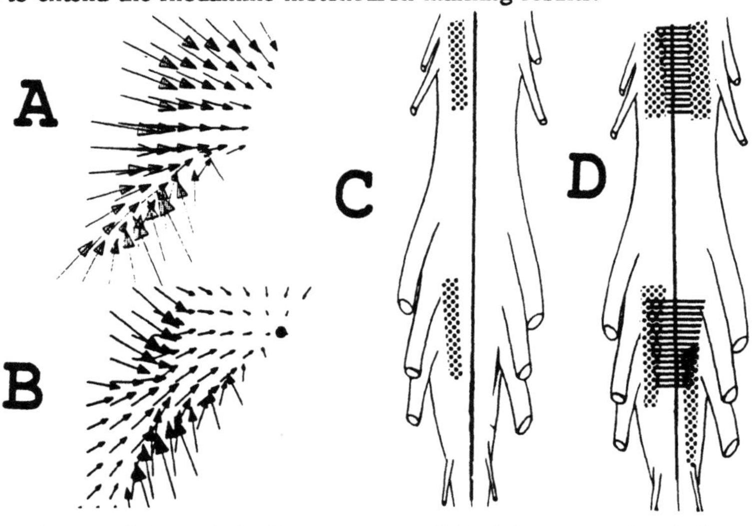

Figure 7 A,B fields of similar structure elicited by microstimulation (A) and by skin stimulation for wiping (B). Note that data points in field A from microstimulation cover a smaller total area than B and hence A does not include the equilibrium. **C, D:** Diagrams which summarize the distributions of motoneurons underlying flexion directed active fields, as revealed by rhodamine uptake in two labelling experiments. The flexion pattern was activated in C by microstimulation and in **D** by cutaneous stimulation. The hatched regions are highly labelled fibers. The stippled areas are highly labelled motoneurons.

Force fields and free limb kinematics.

The force fields and synergies induced under isometric conditions in our experiments need not have a simple relationship to the synergies and force regimes that underlie unrestrained movement. To examine this issue, we have now recorded both movement kinematics and underlying force fields resulting from cutaneous stimulation in eight frogs. The match of the CFF for hindlimb wiping and the stereotyped kinematics of the movements is very good. The movements are both similar to those recorded in the same trial and similar to the data previously reported in Giszter, McIntyre and Bizzi (1989).

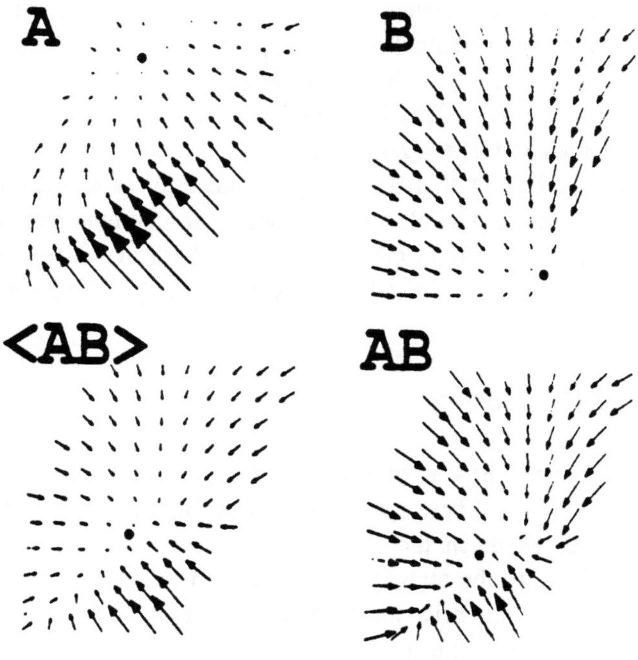

Figure 8 Summation of total force fields elicited by dual stimulation. **A** an extension field and **B:** a flexor field. These fields resulted from individual stimulations at two separate sites. **<AB>:** the prediction of a vector sum of fields A and B. **AB:** the actual measured result of a simultaneous dual stimulation

Vectorial summation or "blending" in microstimulation based fields.

The work described here suggests that there may a very limited number of force field types available in the spinal cord. Thus, to obtain a repertoire of adjusted movements, these primitive elements would need to be combined or enhanced in various ways. We envisage several mechanisms which might enlarge the the repertoire of outputs. Individual force fields could be modulated directly by descending pathways or indirectly by modifying afferent effects. Force fields could be altered by overlaying additional novel muscle activity. Alternatively, several force field types could be combined together. We have made preliminary tests of this latter mechanism. We found that multiple stimulations in two different zones produced well-organized fields in two dimensions. As shown in Fig. 8, the stimulation of two sites simultaneously resulted in a field proportional to the vector sum of the two fields elicited singly. Thus a "blending" of the two fields elicited in different zones can occur. This is reminiscent of the kinematic blending that has been observed in turtle scratch (Mortin et al., 1985, Robertson et al. 1985) and frog (Berkinblitt et al. 1989). It still remains for us to show (a) that this vectorial summation is possible for all zones in all combinations, and (b) that the summation rule holds in three dimensions, both of which would be needed in order to support a completely general positioning mechanism.

Discussion and Conclusions

The work described here suggests that the frog spinal cord contains a number of fixed synergies. These synergies are coordinative groupings which generate stable force fields in the limb. We hypothesize that these are movement primitives. By the term movement primitive, we seek to indicate indivisible elements of motor behavior which form a basis for the combination of sensory and descending control, and that generate a fixed, regulated and stable mechanical response. Presumably, an organization of convergent force fields or movement primitives supported by specific muscle synergies may also allow the central nervous system to utilize simpler feedback schemes, possibly of an invariant pattern. This might considerably simplify the problem of sensor fusion and the central control of afferent effects.

These results should be viewed in the context of other work on spinal cord mechanisms and the evolutionary position of the frog. The frog spinal cord differs from the mammalian cord in several respects. The finely differentiated laminar structure of the mammalian cord is missing. In contrast, only two areas of cutaneous and proprioceptive afferent arbors can be easily distinguished in the areas of frog spinal cord in which we usually stimulate (Jhaveri and Frank, 1983). The gamma fusimotor system is completely absent in the frog. The frog motoneuron arbors are more distributed in the cord than in the mammal (Simpson 1976). Lastly, the behavioral repertoire of the frog is simpler. In what ways should we relate this work on frog spinal cord to mammalian cord physiology? The frog spinal cord may give us insight into both the capabilities and limitations of a solely beta fusimotor organization. Thus the frog spinal cord probably represents a good system in which to examine the primitive tetrapod cord on which the more sophisticated control mechanisms of other tetrapods and ultimately the mammalian organization was layered, to "subsume" the amphibian mechanisms. The amphibian cord in turn is likely to represent a more sophisticated development of the controls found in pre-terrestrial spinal cord as seen, for example, in a primitive chordate such as the lamprey (see Cohen 1987, McClellan and Sigvardt 1988).

We have shown that structured convergent force fields (CFFs) almost identical to those induced by microstimulation underlie cutaneously elicited behaviors in the spinal frog. A natural question, therefore, is the relationship of the circuits activated in cutaneously induced behaviors and the circuits underlying the CFFs generated by microstimulation. We imagine two possible organizations of these CFF circuits in the spinal cord: (1) the microstimulation might simply induce a (partial or complete) behavior just as cutaneous stimulation would. The circuits activated by microstimulation might only be a subset of the elements of the complete behavior, but this subset might nonetheless be unavailable for use in other behaviors. (2) In contrast to this arrangement the circuits activated by microstimulation might be shareable and might be used in several natural behaviors either in isolation or in combination with other such circuits. These two contrasting possibilities may be clearly distinguished by a series of experiments which examine the full set of rules of combination and interaction of force fields. If we were to discover the same sets of blends, exclusions, switching and transitions among CFFs activated by microstimulation as have been observed among the natural behaviors (see e.g., Berkinblitt 1989) then microstimulation based CFFs would probably simply represent the behaviors activated by a different route. In contrast, if the exclusions do not operate and couplings of microstimulation based CFFs are types of vectorial summation then CFFs induced by microstimulation could represent a far less restrictive collection of mechanisms. While such recombinable mechanisms could still be utilized in stereotyped movements, they could also form a general positioning mechanism. Such a mechanism could be used to transform the various sensory and intermediate representations found in the higher nervous system structures (e.g. Georgopoulos et al. 1988a,b, Grobstein 1990, Masino and Grobstein 1989, Masino and Knudsen 1990) into muscle activations and movement patterns. The observation that, at least in some instances, the CFFs induced by multiple stimulation in different zones often do produce fields proportional to the vector sum suggests these may be a general positioning mechanism. The vectorial summation mechanism provides a simple means for the central nervous system to use CFF movement primitives in a very flexible manner (Bizzi et al. 1991).

To conclude, this work suggests that the spinal cord possesses a small number of movement primitives: circuits that may specify invariant force fields. Some of these force field types clearly also underlie phases of natural behaviors. Preliminary results suggest that these movement primitives' force fields may be combined by vectorial summation. It is therefore likely that the CFFs or movement primitives can often be coupled intra-limb, can be coupled between limbs or can be chained sequentially together. This may occur in a more or less flexible manner in pattern generators, or in response to descending control systems, to produce adjustable behaviors.

Acknowledgements: We thank Eric Loeb and Paul Borghesani for sharing occasional frogs, and Margo Cantor, Sylvester Sczepankowsi and Pat Harlan for essential technical assistance. Grants: Office of Naval Research (N00014/88/K/0372) and NIH (NS09343 and AR26710).

References

Alstermark B and Sasaki S (1986) Integration in descending motor pathways controlling the forelimb in the cat 15. Comparison of the projection from excitatory C3-C4 propriospinal neurones to different species of forelimb motoneurones. Exp. Brain Res. 63: 543-566.

Berkinblitt MB, Feldman AG, and Fukson OI (1989) Wiping reflex in the frog: Movement patterns,receptive fields and blends. pp. 615-630 in Visuomotor Coordination: amphibians, comparisons, models and robots. (J.-P. Ewert and M.A. Arbib Eds). Plenum Press.

Bizzi E, Mussa-Ivaldi FA, and Giszter SF (1991) Computations underlying the execution of movement: a novel biological perspective. Science 253:287-291.

Bizzi E, Accornero N, Chapple W and Hogan N (1984) Posture control and trajectory formation during arm movement. J. Neurosci. 4:2738-2744.

Cohen AH (1987) The structure and function of the intersegmental coordinating system of the lamprey central pattern generation for locomotion. J. Comp. Physiol. 160: 181-193.
Feldman AG (1979) Central and reflex mechanisms of motor control. Nauka, Moscow
Fukson OI, Berkinblitt MB and Feldman AG (1980) The spinal frog takes into account the scheme of its body during the wiping reflex. Science 209: 1261-1263.
Georgopoulos AP, Kettner RE and Schwartz AB (1988a) Primate motor cortex and free arm movements to visual targets in three-dimensional space. I. Relations between single cell discharge and direction of movement. J. Neurosci. 8: 2913-2927.
Georgopoulos AP, Kettner RE and Schwartz AB (1988b) Primate motor cortex and free arm movements to visual targets in three-dimensional space. II. Coding of the direction of movement by a neuronal population. J. Neurosci. 8: 2928-2937.
Giszter SF, McIntyre J and Bizzi E (1989) Kinematic strategies and sensorimotor transformations in the wiping movements of frogs. J. Neurophysiol. 62: 750-767.
Giszter SF, Mussa-Ivaldi FA, and Bizzi E (1991a) Equilibrium point mechanisms in the spinal frog. In: M. Arbib and J.P. Ewert (eds.), Visual structures and integrated functions. Plenum Press, New York
Giszter SF, Mussa-Ivaldi FA, and Bizzi E (1991b) The organization of motor space in the frog spinal cord. Exp. Brain Res (In press).
Giszter SF, Bizzi E, and Mussa-Ivaldi FA (1991) Motor organization in the frog spinal cord. In: F.H. Eeckman and C.D. Deno (eds.), Analysis and Modelling of neural systems. Kluwer Press, Moffett Field, CA, 1991
Grillner S and Wallen P (1985) Central pattern generators for locomotion, with special reference to vertebrates. Annual Rev. Neurosci. 8: 233-261.
Grobstein,P (1990) Strategies for analyzing complex organization in the nervous system. II. A case study: Directed movement and spatial representation in the frog. pp. 242-255 in Computational Neuroscience. (E. L. Schwartz Ed.) M.I.T. Press, Cambridge, MA.
Jhaveri S and Frank E (1983) Central projections of the brachial nerve in bullfrogs: muscle and cutaneous afferents project to different regions of the spinal cord. J. Comp. Neurol. 221: 304-312.
Lichtman JW, Wilkinson RS and Rich MM (1985) Multiple innervation of tonic endplates revealed by activity-dependent uptake of florescent probes. Nature 314: 357-359.
Masino T and Grobstein P (1989a) The organization of descending tectofugal pathways underlying orienting in the frog, *Rana pipiens*. I. Lateralization, pracellation, and an intermediate spatial representation. Exp. Brain Res. 75:

227-244.
Masino T and Knudsen EI (1990) Horizontal and vertical components of head movement are controlled by distinct neural circuits in the barn owl. Nature 345: 434-437.
McClellan A and Sigvardt K (1988) Features of entrainment of spinal pattern generators for locomotor activity in the lamprey spinal cord. J. Neurosci. 8: 133-145.
Mortin LI, Keifer J and Stein PSG (1985) Three forms of the scratch reflex in the spinal turtle: Movement Analyses. J. Neurophysiol. 53: 1501-1516.
Mussa-Ivaldi FA, Giszter SF and Bizzi E (1991) Motor-space coding in the central nervous system. In: Cold Spring Harbor symposium on quantitative biology, vol. LV Cold Spring Harbor Laboratory Press 53.
D.J. Ostry, A.G. Feldman, J.R. Flanagan, J. Neurophysiol. 65, 547 (1991).
Preparata FP and Shamos MI (1988) Computational Geometry. Addison Wesley, N.Y.
Robertson GA, Mortin LI, Keifer J and Stern PSG (1985) Three forms of the scratch reflex in the spinal turtle: central generation of motor patterns. J. Neurophysiol. 53: 1517-1534.
Schotland JL, Lee WA and Rymer WZ (1989) Wiping and flexion withdrawal reflexes display different EMG patterns prior to movement onset in the spinalized frog. Exp. Brain Res. 78: 649-653.
Simpson JI (1976) Functional synaptology of the spinal cord pp. 728-749 in Frog Neurobiology LLinas R and Precht W, Eds, Springer-Verlag, Berlin.
Stein PSG (1983) The vertebrate scratch reflex. Symp. Soc. Exp. Biol. 37: 398-403.
Stein PSG, Mortin LI and Robertson GA (1986) The forms ot a task and their blends. In: S. Grillner, P. Stein, DG Stuart, H Forssberg and RM Herman (Eds), Neurobiology of Vertebrate Locomotion. Macmillian, London, 201-216.
Windhorst UR (1991) Group report: What are the units of motor behavior and how are they controlled? with Burke RE, Dieringer N, Evinger C, Feldman AG, Hasan Z, Hultborn H, Illert M, Lundberg AP, Macpherson JM, Massion J, Nichols TR, Schwartz M, Vilis T. pp101-120 in Motor Control: Concepts and Issues. Humphrey DR and Freund H-J Eds. Wiley, New York.

Model and Simulation of a Simplified Cerebellar Neural Network for Classical Conditioning of the Rabbit Eye-blink Response

E.K. Blum, **P.M. Khademi, and** **R. F. Thompson**
Dept. of Mathematics, Dept. of Computer Science, Dept. of Psychology
University of Southern California, Los Angeles, CA 90089

Abstract

Experimental results from several laboratories over the past twenty years have suggested that the cerebellum is necessary for the learning and memory of the classically conditioned rabbit nictitating membrane response (NMR). In particular, results from R.F. Thompson's laboratory strongly suggest that the conditional stimulus (CS) pathway involves the mossy fibres (mf), the unconditional stimulus (US) involves the climbing fibres (cf), and associated neurons in the cerebellar cortex and cerebellar nuclei form a network for the main neuronal substrates of this basic associative learning/memory paradigm. Thompson has proposed a schematic circuit involving these cerebellar neurons and associated brain stem neurons. Blum and Thompson have refined this schematic circuit into a simplified neural network confined to the cerebellum, since experiments by Thompson et al have demonstrated that the tone CS can be replaced by direct mf stimulation in the pontine nucleus (pn) and the air-puff US can be replaced by direct cf stimulation in the inferior olive (IO). In this paper, we present a simplified cerebellar network based on the Blum-Thompson network and give two mathematical models of this simplified network. We present results of computer simulations of these two models. We compare the simulation results with experimental test data from Thompson's laboratory, and find reasonable agreement.

Thompson's Cerebellar Model for NMR Conditioning

Over the last ten years Thompson and a number of collaborators have worked on identifying the neuronal substrates involved in classical conditioning of the nictitating membrane response (NMR) and localizing the 'memory trace', the processes of neuronal plasticity that serve to code memories in the brain. The results of their lesion studies and microstimulation experiments -- which will be briefly described here -- provide evidence that an essential part of the memory trace for the classical conditioning of the NMR in particular and simple associative learning of discrete adaptive responses in general, resides in the cerebellum.

The basic experiment consists of training an animal (rabbit) with repeated paired stimuli. An unconditioned stimulus (US) -- in this case a corneal airpuff -- evokes an unconditioned response (UR) -- nictitating membrane extension and eyelid closure. A conditioned stimulus (CS) -- a 350 msec tone -- is associated with the US, by applying the US during the last 100 msec of the CS. During early trials, the only response is the UR. However, after repeated trials (50-200) the onset of the NMR occurs during the 250 msec interstimulus interval (ISI) before the onset of the US, i.e., the CS evokes a conditioned response (CR) [Tho83].

It should be noted that other brain systems, for example the hippocampus -- by reason of increased pyramidal cell activity correlating to the acquisition of eyeblink response -- and brain stem nuclei are also implicated as putative plasticity sites for the classical conditioning of the NMR [BT78], [TON81], [DM82], [HGC85], [KZB90]. However, decerebrate animals retain the ability to acquire CR[Ens76], and animals decerebrated subsequent to training are able to retain CR [MT87], implying that a primary memory trace circuit exists below the level of the thalamus, i.e., in the cerebellum and its major afferent/efferent systems in the brain stem. On the other hand cerebellar lesions ipsilateral to the trained eye -- including electrolytic lesions of the interpositus nucleus (IP) -- permanently abolish the learned CR, and prevent subsequent learning by the ipsilateral eye, but have no effect on the UR. Furthermore, learning-related unit activity increases appear in certain regions in cerebellar cortex and nuclei as well as the pontine nuclei and the red nucleus, respectively parts of the afferent and efferent systems of the cerebellum. The increased activity in the IP of the trained animal exhibits the requisite memory trace properties since it precedes the NMR by as much as 60 msec, which is consistent with the minimum latency of the CR in well-trained animals (about 100 msec). These results indicate that the cerebellum is part of the learned response circuit for classical conditioning of the NMR, and strongly implicate the cerebellum as the locus of the memory trace.

The NMR conditioning circuit proposed by Thompson is shown in Figure 1, which is adapted from Thompson's original sketch [Tho86]. It includes parts of the cerebellar cortex and nuclei and their afferent/efferent systems, consistent with well-established cerebellar anatomy [PCP74] and physiology [Ito84]. Paired-microstimulation of the DAO and PN regions was found to be sufficient for the CR acquisition [SLT89]. This result allows for the introduction of external stimuli to the network in the form of current or frequency pulses, without the requirement of a network-interpretation of peripheral input stimuli. The US and CS pathways converge coincidentally on the Purkinje cell layer and the IP. The implication here is that collaterals of the mossy fibres relaying the tone CS to a particular Purkinje cell, say cell A, go to the same general area in the interpositus as do the collaterals of the climbing fibre ascending to cell A, and further that cell A inhibits that same region of the interpositus. The essential efferent CR pathway is believed to consist of excitatory interpositus projections to the red nucleus and relayed to motor neurons [Tho88]. It is hypothesized that inhibitory projections from the interpositus to the inferior olive are involved in the conditioning circuit by causing the phasic response in the inferior olive to be attenuated as conditioning takes place [SS91].

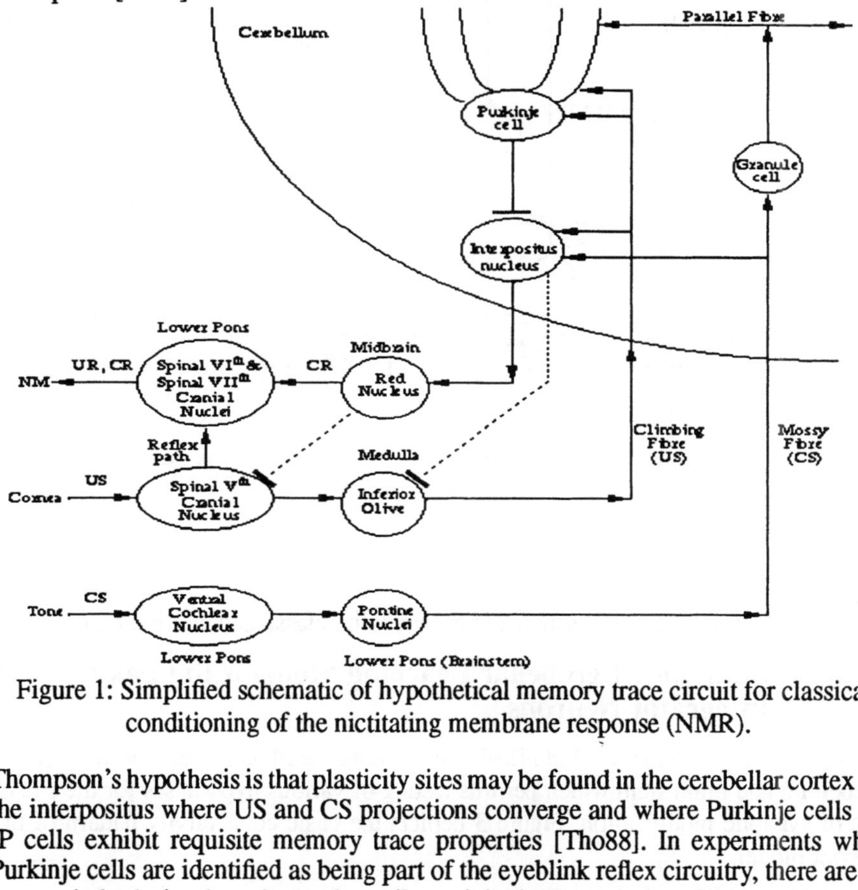

Figure 1: Simplified schematic of hypothetical memory trace circuit for classical conditioning of the nictitating membrane response (NMR).

Thompson's hypothesis is that plasticity sites may be found in the cerebellar cortex and the interpositus where US and CS projections converge and where Purkinje cells and IP cells exhibit requisite memory trace properties [Tho88]. In experiments where Purkinje cells are identified as being part of the eyeblink reflex circuitry, there are decreases in both simple and complex spike activity in the majority of these cells as conditioning takes place. In these same experiments, IP cells are observed to have increased firings as a result of conditioning. Blum and Thompson [BT90] have refined the schematic circuit of Figure 1 into a simplified neural network confined to the cerebellum, since as we have mentioned CS stimulation in the form of direct mossy fibre stimulation in the pontine nucleus and US stimulation in the form of direct stimulation in the inferior olive have proved sufficient for reproducing the conditioning phenomenon. Figure 2 is a sketch of the Blum-Thompson network, showing the more detailed connections in the cerebellar cortex and the hypothesized roles of two adjacent Purkinje cells working in tandem. In this paper, we present a simplified cerebellar network based on a modified Blum-Thompson network and give two mathematical models of this simplified network. Results of the computer simulations are presented as are comparisons of the simulation results with experimental test data from Thompson's laboratory.

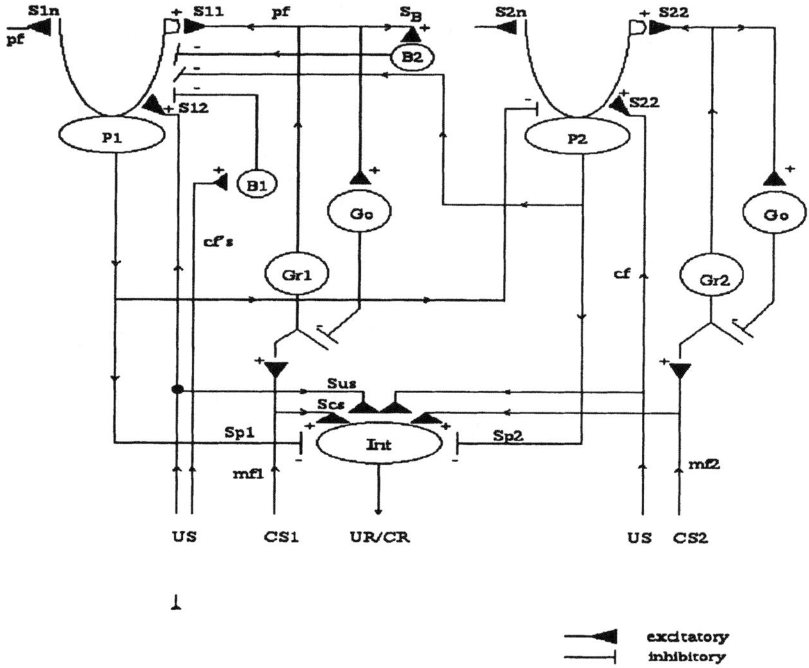

Figure 2: Blum-Thompson schematic of cerebellar classical conditioning network

Model 1: Simplified Cerebellar Learning Model with Leaky Integrator Neurons

We propose a simplified model of NMR conditioning based on a neural network in the cerebellum and brain stem which permits testing of various hypotheses for the site or sites of synaptic modification. Figure 3 depicts the basic connectivity pattern of the network model.

Each node can be interpreted as representing either a cell or group of cells. In this first model, we ignore dendritic structures as well as various connections between the areas of the brain represented by single nodes.

The input to the network consists of the US which is afferent to the inferior olive (IO), the CS afferent to the PN, and background activity (BA) which is hypothesized to cause spontaneous firing in the Purkinje cell layer (PC). Both the IO and PN nodes have excitatory pathways to PC -- the former directly -- and send collaterals to the network output node (IP), which is also inhibited by PC. The basket cell (BC) receives excitation from IO (via a collateral of the climbing fibre) and inhibits PC. G1 and G2 represent (groups of) granule cells which relay mossy fibre signals from PN to PC by the parallel fibre pathways.

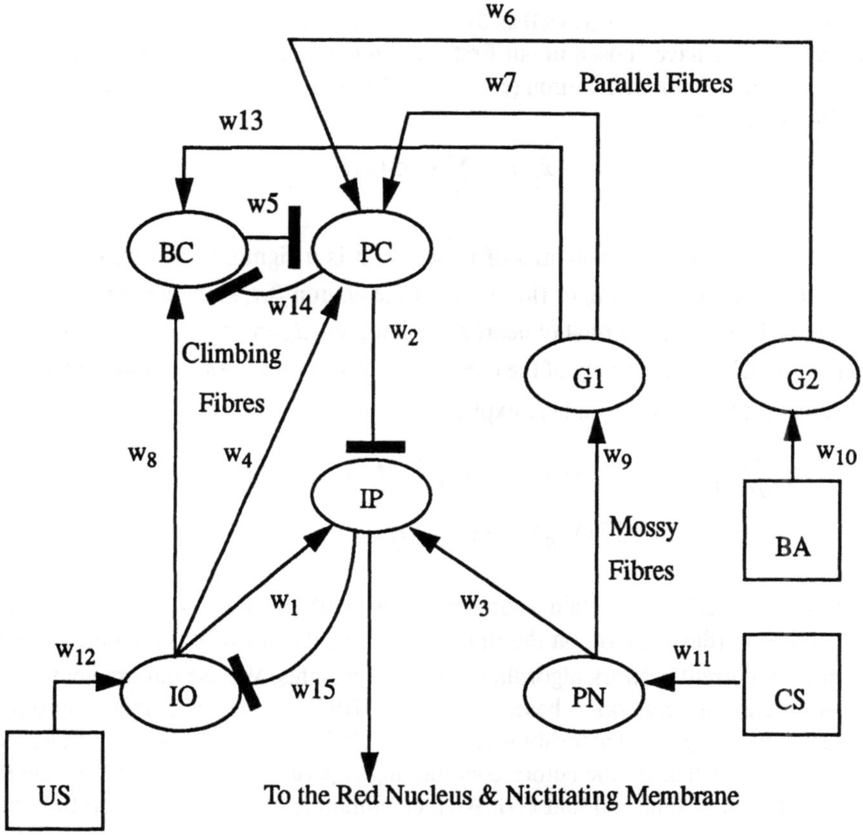

Figure 3: Schematic Network of Connectivity of Related Cell Types. w_i: weights; PC: Purkinje Cell Layer; IP: Interpositus Nucleus; IO: Inferior Olive; PN: Pontine Nucleus; G1 and G2: Granule Cells; BC: Basket Cell; CS, US and BA: Conditioned, Unconditioned and Background Stimuli.

Experimental data on the frequency of firing of the IP and Purkinje cells before and after conditioning [Tho88] implicates both regions as participants in the conditioning-type learning. However, this data does not single out particular synaptic connections as sites of plasticity. Posing this problem as one of system identification, we use this data as target output and the stimuli as the input to train the network first to find the weights that exist before conditioning, and then the learned weights after conditioning. The comparison of these two sets of weights allows us to hypothesize on the site or sites of plasticity. Thus with this approach we can identify synapses whose modification from before-conditioning values to after-conditioning ones can account for the learning of the conditioned NMR. We make no attempt, as yet, to model possible biological learning mechanisms.

A System Identification Algorithm for Recurrent Networks

The weight-determination algorithm is of the non-linear, least-squares, gradient-descent variety. Topologically as well as dynamically, the network in Figure 3 is a recurrent network. We have chosen in our first version to model each node of the network as a type of leaky integrator neuron (e.g., [Cwn68]) with constant delays, described by the differential equation:

$$\tau_i \frac{d}{dt} x_i(t) = -x_i(t) + \sum_j w_{ji} \sigma(x_j(t - d_{ji})) + h_i,$$

where x_i is the membrane potential of neuron i, σ is a sigmoid function (thus $\sigma(x_j)$ yields the frequency of firing or firing rate of the neuron j),τ_i is membrane time constant and h_i the resting potential of neuron i, and w_{ji} and d_{ji} are the synaptic weight and propagation delay respectively of the connection from x_i to x_j. As an example the network equation for IP is shown here explicitly:

$$\tau_{ip} \frac{d}{dt} x_{ip}(t) = -x_{ip}(t) + w_1 \sigma(x_{io}(t - d_1)) + w_2 \sigma(x_{pc}(t - d_2)) + w_3 \sigma(x_{pn}(t - d_3)) + h_{ip} \qquad (1)$$

By using this model we maintain a correspondence with the biological data. By virtue of the self-loop (the - x_i term on the right-hand side), the network is a recurrent one, requiring a recurrent network algorithm to find the weights. We use our gradient algorithm for recurrent networks -- based on ideas in [Blu72] -- which differs from other recurrent network gradient algorithms [Pin87], [WZ89], [Prl89]. It will be sufficient to describe the algorithm for the before-conditioning experiment. To solve the system of ode's, we discretize time and find $x_i(t)$. We then minimize the following discretized error function:

$$E(W) = \sum_m \left[IP^T(t_m) - \sigma(x_{ip}(t_m))\right]^2 + \left[PC^T(t_m) - \sigma(x_{pc}(t_m))\right]^2$$

where IP^T and PC^T are the before-conditioning target (experimentally determined) firing-rate outputs for IP and PC; $x_{ip}(t_m)$ and $x_{pc}(t_m)$ are the discrete-time values for IP and PC membrane potentials as computed by the model, and m is the time index over an entire conditioning trial (typically $t_m = 0,...,750$ in steps of 0.1 msec).

It should be noted that this is a particular form of the error function in which only the IP and Purkinje nodes are "tracked". In fact, the simulation results in the next section are of an experiment in which the IO is also tracked. In the more general case, any or all of the network nodes for which target data are available may be included in the network error function, and the algorithm will attempt to find the set of weights that would best reproduce the overall targeted dynamical behavior.

The discretization of E follows our numerical method for solving the differential equations. The gradient descent is formulated as follows:

$$\nabla w_l E = \left[\frac{\partial E}{\partial w_l}\right]_{l=1\ldots k}$$

$$w_l^{r+1} = w_l^r - \eta(\nabla w_l E) \tag{2}$$

where r is the iteration index, and η is the gradient step constant and,

$$\frac{\partial E}{\partial w_l} = 2 \cdot \sum_m \Bigg(\left[IP^T(t_m) - \sigma(x_{ip}(t_m))\right] \cdot \frac{\partial}{\partial w_l}\sigma(x_{ip}(t_m)) + \left[PC^T(t_m) - \sigma(x_{pc}(t_m))\right] \cdot \frac{d}{dw_l}\sigma(x_{pc}(t_m)) \Bigg)$$

$$\frac{\partial}{\partial w_l}\sigma(x_{ip}(t_m)) = \frac{\partial \sigma}{\partial x_{ip}(t_m)} \cdot \frac{\partial}{\partial w_l} x_{ip}(t_m),$$

$$\frac{\partial}{\partial w_l}\sigma(x_{pc}(t_m)) = \frac{\partial \sigma}{\partial x_{pc}(t_m)} \cdot \frac{\partial}{\partial w_l} x_{pc}(t_m)$$

We let $\sigma(x(t)) = (1 + e^{-a(x(t)-b)})^{-1}$, be the sigmoid function, where a/4 is the maximum slope and b is the point at which the maximum slope occurs (the point of inflection). Differentiating with respect to $x(t)$, we get: $\partial\sigma/\partial x(t) = a\,\sigma(x(t))\,(1 - \sigma(x(t)))$. We shall use Euler's method to solve the ode's. We can obtain the general solution for $\partial x_{ip}(t_m)/\partial w_l$, $\partial x_{pc}(t_m)/\partial w_l$ and other partials by differentiating both sides of the Euler difference equations with respect to w_l. We will show this for $\partial x_{ip}(t_m)/\partial w_1$:

$$x_{ip}(t_m) = x_{ip}(t_{m-1}) + \Delta t \cdot F_{ip}(X(t_{m-1}), W) \tag{3}$$

where $X(t_{m-1}) = (x_{ip}(t_{m-1}),\ldots,x_{io}(t_{m-1}))$ is the state vector at time t_{m-1}, $W = (w_1,\ldots,w_{13})$ is the weight vector, and $F_{ip}(X(t_{m-1}), W)$ is obtained by multiplying both sides of equation (1), by $(1/\tau_{ip})$:

$$F_{ip}(X(t_{m-1}), W) = \frac{d}{dt} x_{ip}(t_{m-1})$$

$$F_{ip}(X(t_{m-1}), W) = \left(\frac{1}{\tau_{ip}}\right)[-x_{ip}(t_{m-1}) + w_1\sigma(x_{io}(t_{m-1} - d_1)) + w_2\sigma(x_{pc}(t_{m-1} - d_2)) + w_3\sigma(x_{pn}(t_{m-1} - d_3)) + h_{ip}]$$

Then, differentiating both sides of equation (3) with respect to w_l, we get,

$$\frac{\partial}{\partial w_1} x_{ip}(t_m) = \frac{\partial}{\partial w_1} x_{ip}(t_{m-1}) + \Delta t \cdot \frac{\partial}{\partial w_1} F_{ip}(X(t_{m-1}), W) \tag{4}$$

Finally,

$$\frac{\partial}{\partial w_1} F_{ip}(X(t_{m-1}), W) = \left(\frac{1}{\tau_{ip}}\right) \cdot \Big(-1 \cdot \frac{\partial}{\partial w_1} x_{ip}(t_{m-1}) +$$
$$w_1 \cdot \frac{\partial}{\partial x_{io}(t_{m-1} - d_1)} \sigma \cdot \frac{\partial}{\partial w_1} x_{io}(t_{m-1} - d_1) +$$
$$w_2 \cdot \frac{\partial}{\partial x_{pc}(t_{m-1} - d_2)} \sigma \cdot \frac{\partial}{\partial w_1} x_{pc}(t_{m-1} - d_2) +$$
$$w_3 \cdot \frac{\partial}{\partial x_{pn}(t_{m-1} - d_3)} \sigma \cdot \frac{\partial}{\partial w_1} x_{pn}(t_{m-1} - d_3) +$$
$$\sigma(x_{io}(t_{m-1} - d_1))\Big)$$

where the $\partial x_i / \partial w_l$ are obtained by computing at successive t_m's using equations illustrated in (4).

Thus during each trial run, we compute $x_i(t_m)$, $\partial x_i(t_m)/\partial w_l$, and $\nabla w_l E(t_m)$, for all i, l, and m. We use the gradient iteration -- equation (2) --to determine the W that minimizes $E(W)$. Thus after each gradient step, the network will better approximate the observed dynamical behavior as given by the input functions US(t), CS(t) and BA(t), and the output functions $IP^T(t)$ and $PC^T(t)$ before and after conditioning. Through successive iterations, the gradient algorithm will find a local minima. The problem of escaping local minima is not as critical here, since we are not as interested in finding a global minimum as we are in getting an acceptable approximation to the target data before and after conditioning and comparing of the weight vectors. We have discussed this problem further in the next section in the context of the network simulation results.

Simulation Results

We have written a FORTRAN program based on the above algorithm and simulated the network given in Figure 3, using input and output target data from Thompson's laboratory. An aspect of the simulation which is not discussed above is the introduction of constraints on synaptic weight modifications. Quite clearly, the biology imposes certain restrictions, for example a synapse which is excitatory will not as a result of modification become inhibitory, or as this translates to our network, a weight may not change its sign. Theoretically, this will restrict the algorithm from descending along certain gradients in the solution space, however it does not introduce any errors in the algorithm.

Another aspect of the simulation which bares on the issue of local minima is the value of η (eta), the gradient step constant. "Large" η's cause the algorithm to take a large gradient step enabling it to escape local minima, whereas a "small" η enables it to "slowly" descend down a particular local minimum. In watching our simulations and "manually" changing the value of η during runs, we have developed what might best be described as an expert-system set of rules for changing η which we have not as yet implemented in the program.

We will discuss the results of an experiment the input and target data of which are shown in Figure 4. The CS is turned on at 250 ms, the US is at 500 ms and both are turned off at 600 ms. The target data for the Purkinje, IP and IO cells are computer-generated approximations of histograms of multi-cell extracellular recordings over many trials gathered in Thompson's [MT84], (unpublished results) and other laboratories [SM91], and for the purposes of this model are interpreted as frequency of firings. Notice that after conditioning the IP responds to the CS onset and is at a peak around the onset of the US. Background activity is a somewhat random computer-generated function. The model parameters were set as follows: $d_1 = 25$; $d_2 = 10$; $d_3 = 20$; $d_4 = 40$; $d_5 = 1$; $d_6 = 3$; $d_7 = 3$; $d_8 = 40$; $d_9 = 20$; $d_{10} = 0$; $d_{11} = 0$; $d_{12} = 0$; $d_{13} = 3$; $d_{14} = 1$; $d_{15} = 25$; $\tau_i = 5$ for all i, except for $\tau_{ip} = 15$; $h_{ip} = -.98$; $h_{pc} = 5.9$; $h_{bc} = 5.9$; $h_{g1} = -3.8$; $h_{g2} = 10$; $h_{pn} = -3.8$; $h_{io} = -3.8$; a = 0.1; b = 10;

Figure 5 shows the comparison of the network outputs (dotted lines) after training to the set of before-conditioning target data, and similarly Figure 6 for the after-conditioning case. In both cases the weight-determination algorithm was able to find a set of weights to reproduce the target output quite accurately. The weight vector profile for the before and after-conditioning cases (respectively the first and second columns), as well as the percent changes in weights were as follows (for an inhibitory weight, a positive change signifies stronger inhibition):

w_1	27.8	22.4	-19%
w_2	-19.3	-31.8	65%
w_3	5.0	33.3	566%
w_4	5.0	5.77	14%
w_5	-23.3	-24.7	6%
w_6	15.1	7.9	-48%
w_7	20.7	5.0	-76%
w_8	35.1	10.0	-72%
w_9	19.6	14.7	-22%
w_{10}	8.5	35.4	319%
w_{11}	15.8	24.3	54%
w_{12}	22.1	7.6	-66%
w_{13}	5.1	45.9	809%
w_{14}	-20.5	-40.2	96%
w_{15}	-19.3	-6.6	-66%

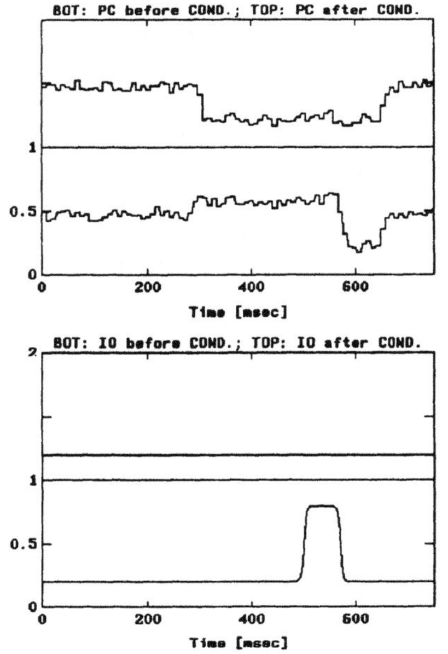

Figure 4: Input and Target data

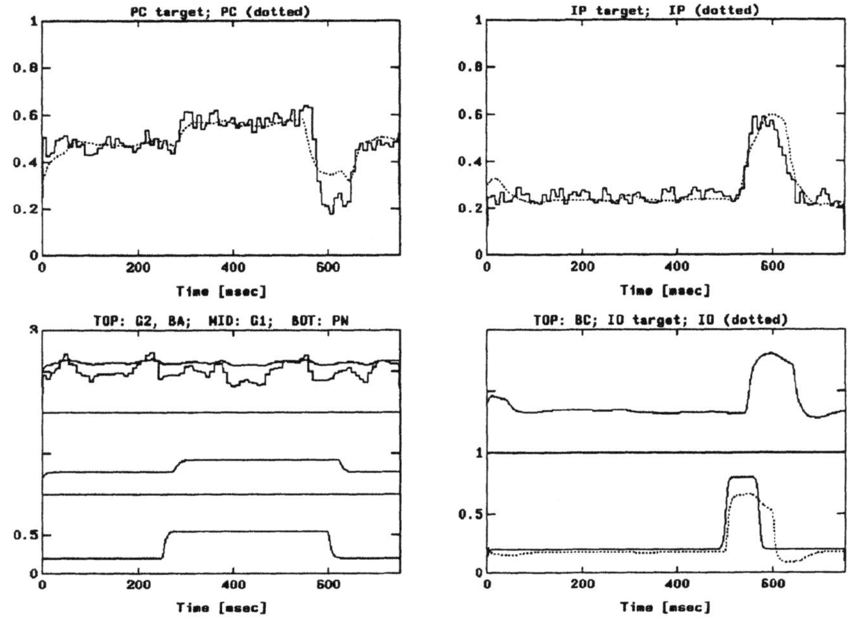

Figure 5: Results of fitting network to before-conditioning target data

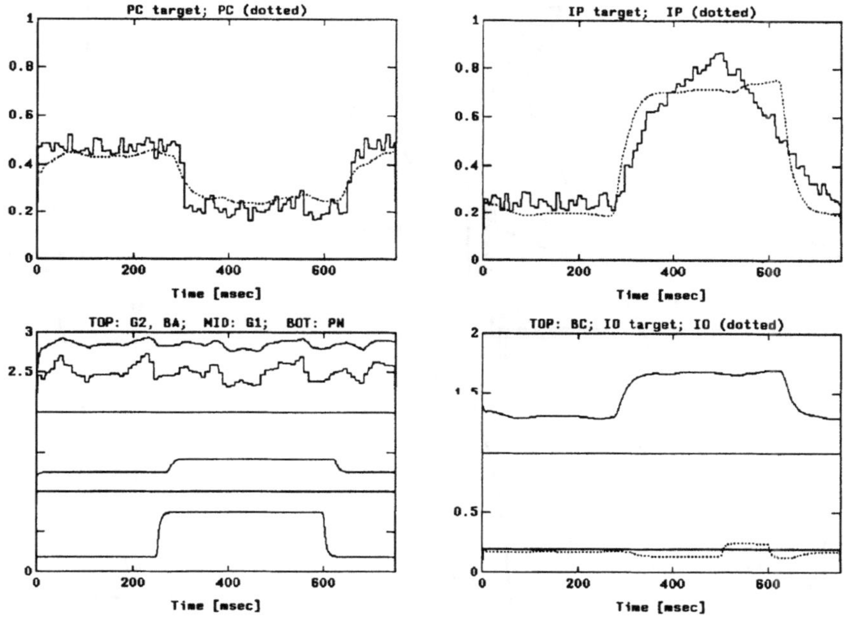

Figure 6: Results of fitting network to after-conditioning target data

Model 2: Simplified Cerebellar Network Modeled with CAJAL

Given the dense arborization of the Purkinje cell and the structural regularity of the cerebellar cortex, we intend to expand the first model to include higher-level representations of the neuron matter and a population of each of the neuron types. Model 2 is a first attempt at this expansion of the basic model, where the Purkinje cell dendrite is represented by a compartmentalized tree and each neuron's activity is represented by a Hodgkin-Huxley axon model. We have used CAJAL, a neurobiological network simulator [BK90], to simulate the model. Figure 10 depicts the connectivity of the network.

We have not as yet introduced synaptic plasticity into this model, rather we can use CAJAL to do a parametric study of the effect of different synaptic parameter values. We are currently engaged in such a study. Figure 8 is the output of a CAJAL run of the model in Figure 7. The inputs are not shown; the CS is on from 50 to 350 msec, and the US from 150 to 300 msec. The parameters have been set to reproduce a before-conditioning scenario: The Purkinje cell fires from the beginning due to Background activity. At 50 msec, the granule cell starts firing in response to the CS, and at 150 msec the basket cell starts firing in response to the US. During the interval 50-150 msec, the interpositus is quiescent by reason of the inhibition from the Purkinje cell outweighing the excitation from the CS, but the firing of the basket cell at 150 msec inhibits the Purkinje cell, disinhibiting the interpositus.

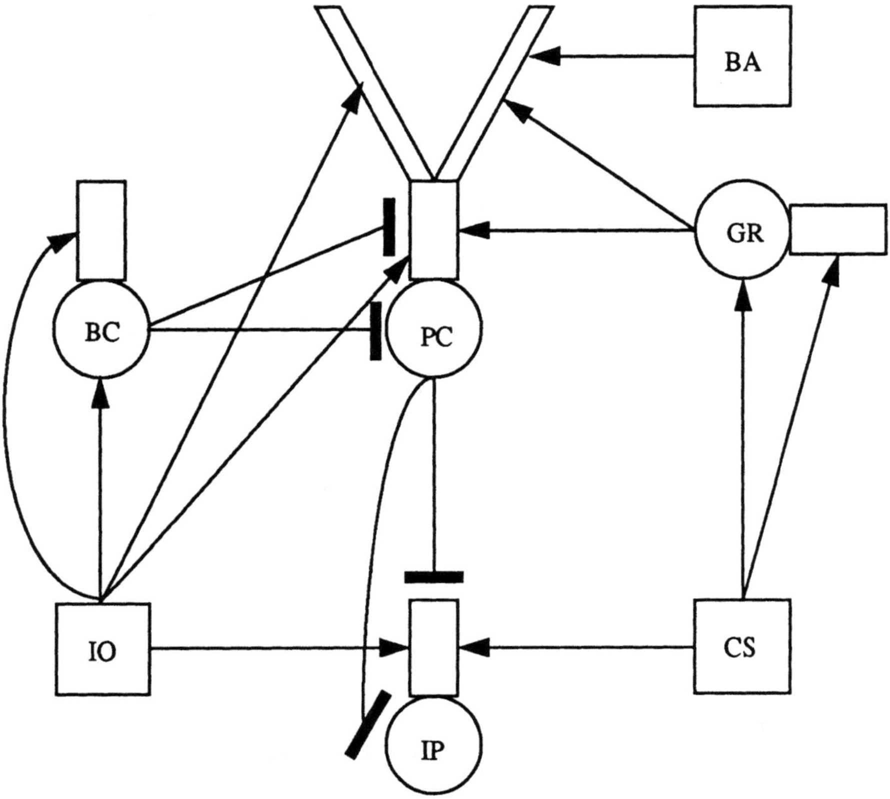

Figure 7: Schematic connectivity of a cerebellar network. The Cell types are the same as in Figure 3. Each neuron is represented by a circular soma and rectangular dendritic compartments. Input neurons are represented by squares.

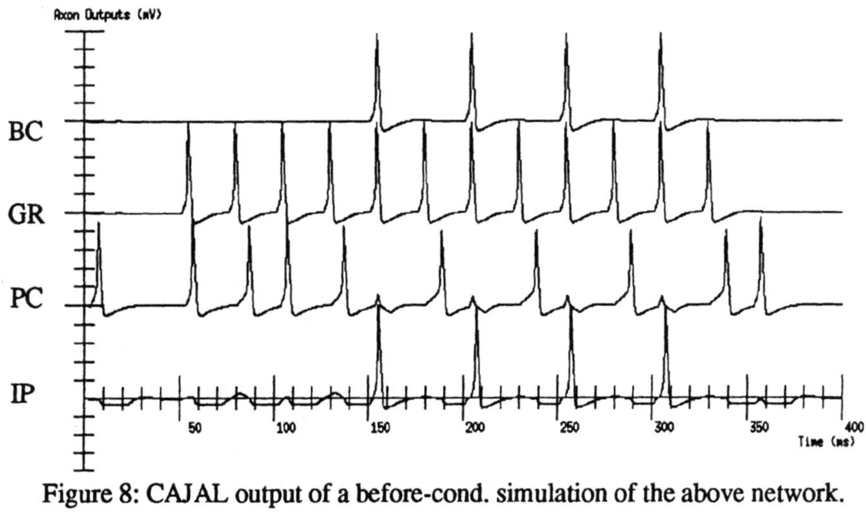

Figure 8: CAJAL output of a before-cond. simulation of the above network.

Discussion

The Model 1 results show, primarily, that it is possible with a model under reasonable biological constraints to reproduce experimental results with good accuracy. The comparison of the set of weights before and after conditioning is somewhat revealing. The greatest increases in synaptic strength are seen in w_{13} (G1 to BC), w_3 (PN to IP) and w_{10} (BA to G2), and the greatest decreases in w_7 (G1 to PC), w_8(IO to BC), w_{12} (US to IO) and w_{15} (IP to IO). The changes in w_7 and w_{13} account for how before conditioning the Purkinje cell responds to the CS input by increasing its activity and thus inhibiting the IP, and after conditioning there is a decrease in PC activity following CS input, thus disinhibiting the IP; at the same time the increase in w_3 causes the IP to respond to the CS input. Also in the absence of a target, BC learns to respond to the US and not the CS in the before-conditioning case (Figure 5), but after conditioning (Figure 6) it responds mainly to the CS and only slightly to the US, which explains the decrease in w_8 and further explains the increase in w_{13}. By decreasing w_{12} after conditioning the model has reduced the effectivity of the US input, thus attenuating the activity in the IO, however a strange consequence of this is that we no longer need inhibition from IP to cause the phasic response of IO, thus w_{15} is also decreased. The increase in w_{10} serves to shape G2 to greater reflect the profile of BA, however with the decrease in w_6, there is little consequence to the circuit dynamics.

These results rather than leading to any conclusive statements about plasticity, point out ways to refine our modeling and analysis. As we have mentioned, for this first attempt at modeling the circuit, we have chosen a single node to represent the average behavior of a cell type. In future models we will use multiple nodes to represent each cell type, while maintaining anatomical plausibility of circuit connections, and employing random-noise-added representations of target data for each targeted node. In particular, the modeling of the basket cells as one node seems insufficient given their off-beam axonal extensions. We will also conduct sensitivity studies by selectively restricting some of the weights obtained from the before-conditioning experiment from changing in order to isolate key weights whose plasticity is sufficient for conditioning. Further, we will use Monte Carlo simulations to derive mean characterizations of network weight fluctuations in order not to be misled by local minima phenomena.

Regardless of what we hope to learn from Model 1, the size and density of dendritic trees in the cerebellar cortex, and the spatial distribution of the different afferents synapsing onto the Purkinje cell impel us to further develop the compartmental modeling we have started with Model 2. In this model we need to first develop small sized models to arrive at reasonable parameters for the circuit elements, a stage for which we have shown some initial results, then assemble large working networks and finally introduce different plasticity hypotheses.

We view our models of cerebellar NMR conditioning to be at an intermediate stage of research and as such it would be premature to either base strong claims on our current results, or to dismiss them as inconclusive. On the contrary, we are encouraged by the proximity of our model results to experimental data and we intend to pursue this style of modeling and report on future results.

References

[BT78] Berger, T.W. and Thompson, R. F. "Neuronal plasticity in the limbic system during classical conditioning of the rabbit nictitating membrane response: I. The Hippocampus." *Brain Research*, 145, 323-346, 1986.

[Blu72] Blum, E. K. *Numerical Analysis and Computation: Theory and Practice.* Addison-Wesley, 1972.

[BK90] Blum, E. K. and Khademi, P. M. "CAJAL: A biological neural network simulator." To appear in *Proc. Berkeley Workshop on Analysis and Modeling of Neural Systems*, 1990.

[BT90] Blum, E. K. and Thompson, R. F. "Neurobiological models of classical conditioning." *Proposal to NIMH*, 1990.

[Cwn68] Cowan, J.D. "Statistical mechanics of nervous nets." In *Neural Networks*, ed. Carianiello, E.R., Springer-Verlag, Berlin, 181-188, 1968.

[DM82] Desmond, J.E., Moore, J.W. "A brain stem region essential for classical conditioning but not unconditioned nictitating membrane response." *Physiology & Behavior*, 28, 1029-1033, 1982.

[Ens76] Enser, D. "A study of classical nictitating membrane conditioning in neodecorticate, hemidecorticate and thalamic rabbits." Ph. D. thesis, University of Iowa, 1976.

[HGC85] Harvey, J.A., Gormezano, I. and Cool-Hauser, V. A. "Relationship between heterosynaptic reflex facilitation and acquisition of the NMR in control and scopolamine-injected rabbits." *Journal of Neuroscience*, 5, 596-602, 1985.

[Ito84] Ito, M. *The Cerebellum and Neural Control.* Raven Press, New York, 1984.

[KZB90] Kelly, T.M., Zuo, C. and Bloedel, J.R. "Classical conditioning of the eyeblink reflex in decerebrated decerebellate rabbit." *Behavioral Brain Research* 38, 7-18, 1990.

[MST86] Mauk, M.M., Steinmetz., J. E. and Thompson, R. F. "Classical conditioning using stimulation of the olive as the unconditional stimulus." *Proceedings of N.A.S.*, 83, 5349-5353, 1986.

[MT87] Mauk, M.M. and Thompson, R. F. "Retention of classical conditioned eyelid response following acute decerebration." *Brain Research*, 493, 89-95, 1987.

[MST85] McCormick, D. A., Steinmetz., J. E. and Thompson, R. F. "Lesions of the inferior olivary complex cause extinction of the classically conditioned eyeblink response." *Brain Res.*, 359, 120-130, 1985.

[MT84] McCormick, D. A. and Thompson, R. F. "Neuronal responses of the rabbit cerebellum during acquisition and performance of a classically conditioned nictitating membrane-eyelid response." *Journal of Neuroscience*, 4, 2811-2822, 1984.

[PCP74] Palay, S.L. and Chan-Palay, V. *Cerebellar Cortex.* Springer-Verlag, New York, Heidelberg, Berlin, 1974.

[Prl89] Pearlmutter, B.A. "Learning state space trajectories in recurrent networks." *Neural Computation*, Vol. 1, 263-269, 1989.

[Pin87] Pineda, F.J. "Generalization of backpropagation to recurrent neural networks." *Physical Review Letters*, 18, 2229-2232, 1987.

[SS91] Sears, L.L. and Steinmetz., J. E. "Dorsal accessory inferior olive activity diminishes during acquisition of the rabbit classically conditioned eyelid response." *Brain Research*, 545, 114-122, 1991.

[SLR87] Steinmetz, J. E., Logan, C. G., Rosen, D. J. and Thompson, J. K. "Initial localization of the acoustic conditioned stimulus projection system to the cerebellum essential for the classical eyelid conditioning." *Proc.of the Nat. Academy of Sciences*, 84, 3531-3535, 1987.

[SLT89] Steinmetz, J. E., Lavond, D. G.and Thompson, R. F. "Classical conditioning in rabbits using pontine nucleus stimulation as a conditioned stimulus and inferior olive stimulation as an unconditioned stimulus." *Synapse*, 3, 225-233, 1989.

[SRC86] Steinmetz, J. E., Rosen, D.J., Chapman, P.F., Lavond, D. G.and Thompson, R. F. "Classical Conditioning of the Rabbit Eyelid Response with a Mossy-Fiber Stimulation CS: I Pontine Nuclei and Middle Cerebellar Peduncle Stimulation." *Behavioral Neuroscience*, Vol. 100, No. 6, 878-887, 1986.

[Tho83] Thompson, R. F. "Neuronal substrates of simple associative learning: classical conditioning." *TINS*, 270-275, 1983.

[Tho86] Thompson, R. F. "The neurobiology of learning and memory." *Science*, 233, 941-947, 1986.

[Tho88] Thompson, R. F. "The neural basis of basic associative learning of discrete behavioral responses." *TINS*, Vol. 3, No. 4, 152-155, 1988.

[TON81] Tsukahara N., Oda, Y., Notsu, T. "Classical conditioning mediated by red nucleus in the cat." *Journal of Neuroscience*, 1, 72-79, 1981.

[WZ89] Williams, R.J. and Zipser, D. "A learning algorithm for continually running fully recurrent neural networks." *Neural Computation*, Vol. 1, 270-280, 1989.

INDEX

MIX
Papier aus verantwortungsvollen Quellen
Paper from responsible sources
FSC® C105338

If you have any concerns about our products,
you can contact us on
ProductSafety@springernature.com

In case Publisher is established outside the EU,
the EU authorized representative is:
Springer Nature Customer Service Center GmbH
Europaplatz 3, 69115 Heidelberg, Germany

Printed by Libri Plureos GmbH
in Hamburg, Germany